ST. MARY'S COLLEGE OF MARYLAND
ST. MARY'S CITY, MARYLAND 20686

Carbon-13 NMR Spectroscopy of Biological Systems

Carbon-13 NMR Spectroscopy of Biological Systems

EDITED BY

Nicolau Beckmann

Biophysics Unit
Preclinical Research
Sandoz Pharma
Basel, Switzerland

ACADEMIC PRESS
San Diego   New York   Boston
London   Sydney   Tokyo   Toronto

This book is printed on acid-free paper.

Copyright © 1995 by ACADEMIC PRESS, INC.

All Rights Reserved.
No part of this publication may be reproduced or transmitted in any form or by any means, electronic or mechanical, including photocopy, recording, or any information storage and retrieval system, without permission in writing from the publisher.

Academic Press, Inc.
A Division of Harcourt Brace & Company
525 B Street, Suite 1900, San Diego, California 92101-4495

United Kingdom Edition published by
Academic Press Limited
24-28 Oval Road, London NW1 7DX

Library of Congress Cataloging-in-Publication Data

Carbon-13 NMR spectroscopy of biological systems / edited by Nicolau Beckmann.
    p.   cm.
   Includes bibliographical references and index.
   ISBN 0-12-084370-6
    1. Nuclear magnetic resonance spectroscopy.   2. Carbon--Isotopes--Spectra.   I. Beckmann, Nicolau.
   QP519.9N83C37   1994
   574'.072--dc20   94-31849
                                                                        CIP

PRINTED IN THE UNITED STATES OF AMERICA
95 96 97 98 99 00 QW 9 8 7 6 5 4 3 2 1

The little girl gave a cry of amazement and looked about her, her eyes growing bigger and bigger at the wonderful sights she saw . . .

L. Frank Baum
The Wizard of Oz, 1899

(Dorothy upon emerging from her
home when she landed in Oz)

Contents

Contributors   xv

Foreword   xvii
JOACHIM SEELIG

Preface   xix
NICOLAU BECKMANN

1. **Introduction**   1
   NICOLAU BECKMANN
   1.1   The Beginning   2
   1.2   Essential Developments for ^{13}C NMR   2
   1.3 ^{13}C NMR in the Solid State: Cross-Polarization and Magic-Angle Spinning   3
   1.4   Heteronuclear Polarization Transfer   4
   References   4

2. **Methodology and Applications of Heteronuclear and Multidimensional ^{13}C NMR to the Elucidation of Molecular Structure and Dynamics in the Liquid State**   7
   GERD GEMMECKER AND HORST KESSLER
   2.1   Introduction   7
   2.2   General Considerations   8
         2.2.1   Inverse or Direct Detection?   10

- 2.3 Heteronuclear Coherence Transfer   13
  - 2.3.1 INEPT   13
  - 2.3.2 DEPT   17
  - 2.3.3 Heteronuclear Cross-Polarization   18
- 2.4 Multidimensional Correlation Experiments   19
  - 2.4.1 Natural ^{13}C Abundance   19
    - *2.4.1.1 Heteronuclear Single-Quantum and Multiple-Quantum Coherence   19*
    - *2.4.1.2 BIRD-Filtered HSQC and HMQC   22*
    - *2.4.1.3 Sensitivity-Enhanced HSQC   25*
    - *2.4.1.4 Heteronuclear Multibond Correlation   26*
    - *2.4.1.5 Gradient Selection   31*
    - *2.4.1.6 Inverse Correlation Experiments with Additional Transfer Steps   36*
    - *2.4.1.7 Heteronuclear 3D NMR in Natural Abundance   38*
  - 2.4.2 ^{13}C-Enriched Samples   41
    - *2.4.2.1 Isotope Filters   42*
    - *2.4.2.2 2D ^{13}C-TOCSY   46*
    - *2.4.2.3 3D and 4D Experiments   47*
- 2.5 Determination of ^{13}C Relaxation Times   51
- 2.6 Heteronuclear Coupling Constants   53
  - 2.6.1 $^1J_{H,C}$ Coupling Constants   54
  - 2.6.2 Heteronuclear Long-Range Coupling Constants   54
    - *2.6.2.1 Lineshape Analysis   55*
    - *2.6.2.2 Simplification of Multiplet Patterns (E.COSY)   56*
    - *2.6.2.3 Long-Range Couplings from Cross-Peak Intensities   58*
- 2.7 Concluding Remarks   61
- References   61

3. *Measurement of Internuclear Distances in Biological Solids by Magic-Angle Spinning ^{13}C NMR*   65
JOEL R. GARBOW AND TERRY GULLION
- 3.1 Introduction   65
- 3.2 Measuring Distances between Isolated Heteronuclear Spin Pairs by MAS NMR   69

- 3.2.1 A Simple Analogy   70
- 3.2.2 Rotational-Echo, Double-Resonance NMR   72
- 3.2.3 A General REDOR Strategy for the Elucidation of Protein Backbone Conformations: ^{13}C–^{15}N REDOR of Melanostatin   79
  - 3.2.3.1 Conformational Analysis   83
  - 3.2.3.2 General Labeling Strategy   84
  - 3.2.3.3 Natural-Abundance Corrections   86
- 3.2.4 ^{31}P–^{13}C REDOR of the EPSPS/S3P/Glyphosate Complex   87
- 3.2.5 Transfer-Echo, Double-Resonance NMR   90
- 3.2.6 Rotary-Resonance Recoupling   93
- 3.3 ^{13}C–^{13}C Distances by MAS NMR   95
  - 3.3.1 Dipolar Recovery at the Magic Angle   96
  - 3.3.2 Dipolar Recovery by Synchronously Applied π Pulse Trains   99
  - 3.3.3 Rotational-Resonance NMR   101
    - 3.3.3.1 A Simple Description of R^2 101
    - 3.3.3.2 The Spin-Exchange Experiment   103
    - 3.3.3.3 The $n=1$ Experiment   106
    - 3.3.3.4 ^{13}C–^{13}C Distances in Bacteriorhodopsin by Rotational-Resonance NMR   108
- 3.4 Concluding Remarks   112
- References   113

4. *^{13}C NMR Studies of the Interactions of Fatty Acids with Phospholipid Bilayers, Plasma Lipoproteins, and Proteins*   117

    JAMES A. HAMILTON

- 4.1 Introduction   117
- 4.2 Properties of Fatty Acids   118
- 4.3 Interactions of Fatty Acids with Serum Albumin   119
  - 4.3.1 Mole Ratio Studies   120
  - 4.3.2 pH Studies   123
  - 4.3.3 Locations of High-Affinity Binding Sites   125
  - 4.3.4 Molecular Motions and Environments of Different Fatty Acid Carbons   128
  - 4.3.5 Exchange of Fatty Acids between Binding Sites   130

- 4.4 Binding of Medium-Chain Fatty Acids to Human Serum Albumin   132
- 4.5 Binding of Fatty Acids to Human Serum Lipoproteins   134
- 4.6 Binding of Fatty Acids to Cytosolic Fatty Acid Binding Proteins   138
- 4.7 Binding of Fatty Acids to Model Membranes   139
- 4.8 Exchange of Fatty Acids between Albumin and Model Membranes   143
- 4.9 Exchange of Fatty Acids among Proteins, Model Membranes, and Lipoproteins   151
- 4.10 Concluding Remarks   154
- References   155

5. *Application of ^{13}C NMR Spectroscopy to Metabolic Studies on Animals*   159
   BASIL KÜNNECKE
   - 5.1 Introduction   159
   - 5.2 ^{13}C MRS and the Concepts of ^{13}C Labeling   161
     - 5.2.1 Natural Abundance Spectroscopy   161
     - 5.2.2 Specific ^{13}C Labeling of a One-Carbon Site   162
     - 5.2.3 Specific ^{13}C Labeling of Several-Carbon Sites   163
     - 5.2.4 Isotopomer Analysis   166
     - 5.2.5 Higher-Order Effects of ^{13}C Labeling   167
   - 5.3 ^{13}C MRS, ^{13}C Labeling, and Models of Intermediary Metabolism   168
     - 5.3.1 Metabolic Modeling and Labeling   168
     - 5.3.2 Metabolic Modeling and Isotopomer Analysis   169
     - 5.3.3 Metabolic Modeling and the Tricarboxylic Acid Cycle   170
     - 5.3.4 Isotopomer Analysis and Modeling with Differential Equations   171
     - 5.3.5 Isotopomer Analysis and Modeling with Infinite Convergent Series   172
     - 5.3.6 Isotopomer Analysis and Modeling with Input–Output Equations   173
     - 5.3.7 Simplified Models of Isotopomer Analysis   174
     - 5.3.8 Limitations of Models Based on Isotopomer Analysis   175

- 5.4 Experimental Considerations   176
  - 5.4.1 Experimental Systems and Sample Preparation   176
  - 5.4.2 Data Acquisition, Spatial Localization, and Spectral Discrimination of Tissues   179
  - 5.4.3 Decoupling and Nuclear Overhauser Effect   182
  - 5.4.4 Quantification   183
- 5.5 Metabolism in the Liver   186
  - 5.5.1 Lipogenesis and Fatty Acid Stores   187
  - 5.5.2 Glycogen, the Major Carbohydrate Store   190
  - 5.5.3 Regulation of Glycogen Metabolism   193
  - 5.5.4 Metabolic Pathways for Glycogen Synthesis and the "Glucose Paradox"   195
  - 5.5.5 Gluconeogenesis   201
  - 5.5.6 Substrate Cycling/Futile Cycling   203
  - 5.5.7 The Pentose Phosphate Pathway   204
  - 5.5.8 Malate–Fumarate Equilibrium   206
  - 5.5.9 The Tricarboxylic Acid Cycle   207
  - 5.5.10 β-Oxidation of Fatty Acids and Ketogenesis   210
  - 5.5.11 ω-Oxidation and Dicarboxylic Fatty Acids   213
- 5.6 Metabolism in the Heart   214
  - 5.6.1 Substrate Selection in the Myocardial Tricarboxylic Acid Cycle   215
  - 5.6.2 Anaplerosis   217
  - 5.6.3 Metabolic Consequences of Altered Workload   218
  - 5.6.4 Ischemia, Hypoxia, and the Tricarboxylic Acid Cycle   220
  - 5.6.5 Ischemia, Hypoxia, and Glycogen Synthesis   222
  - 5.6.6 Postischemic Recovery, Cardioprotectants, and Cardiotoxins   224
  - 5.6.7 Metabolic Compartmentation   226
- 5.7 Metabolism in the Brain   226
  - 5.7.1 Uptake and Metabolism of Cerebral Fuels *in Situ*   228
  - 5.7.2 The Tricarboxylic Acid Cycle   231
  - 5.7.3 Substrate Selection in Neuronal and Glial Tricarboxylic Acid Cycles   233
  - 5.7.4 Pyruvate Recycling   234

                5.7.5   4-Aminobutyrate Synthesis   237
                5.7.6   Anaplerosis   238
                5.7.7   N-Acetylaspartate Metabolism   240
                5.7.8   Choline Metabolism   240
        5.8   Metabolism in the Ocular Lens   241
                5.8.1   The Polyol Pathway   241
                5.8.2   The Hexose Monophosphate Shunt   243
        5.9   Metabolism in the Kidneys   244
                5.9.1   Gluconeogenesis   244
                5.9.2   The Tricarboxylic Acid Cycle   246
                5.9.3   Renal Osmolytes   249
        5.10 Metabolism in the Skeletal Muscles   250
                5.10.1 Metabolite Profile   250
                5.10.2 Glycogen   250
        5.11 Metabolism in Parasites   251
                5.11.1 Glucose Metabolism in Cestodes   251
                5.11.2 Glucose Metabolism in Trematodes   254
                5.11.3 Glucose Metabolism in Nematodes   256
        5.12 Metabolism in Tumors   257
                5.12.1 Metabolite Profile   258
                5.12.2 Glucose Metabolism   258
        5.13 Concluding Remarks   260
        References   261

6. *^{13}C Magnetic Resonance Spectroscopy as a Noninvasive Tool for Metabolic Studies on Humans*   269
    NICOLAU BECKMANN
        6.1   Introduction   269
        6.2   Coils   271
        6.3   Decoupling   272
        6.4   Schemes for the Acquisition and Enhancement of ^{13}C Signals and for ^{13}C Editing   273
                6.4.1   Pulse-Acquire Scheme   273
                6.4.2   Frequency-Selective Pulses   274
                6.4.3   Adiabatic Pulses   275
                6.4.4   Polarization Transfer and Spectral Editing   275
                6.4.5   Nuclear Overhauser Effect   276

6.5 Spatially Localized ^{13}C Spectroscopy   276
    6.5.1 Suppression of Surface Signals   277
    6.5.2 ^{13}C Chemical-Shift Imaging   278
    6.5.3 Localization and Signal Enhancement Based on Polarization Transfer, Heteronuclear Editing, and Gradient-Enhanced Heteronuclear Multiple-Quantum Coherence   281
        6.5.3.1 *Direct Polarization Transfer from ^1H to ^{13}C*   281
        6.5.3.2 *Proton-Observed ^{13}C-Edited Spectroscopy*   283
        6.5.3.3 *Gradient-Enhanced Heteronuclear Multiple-Quantum Coherence*   283
        6.5.3.4 *Polarization Transfer in the Rotating Frame*   286
    6.5.4 Frequency-Selective Pulses in Combination with B_0 Gradients Applied Directly to ^{13}C Nuclei   287
        6.5.4.1 *Depth-Resolved Surface Coil Spectroscopy (DRESS)*   288
        6.5.4.2 *Image-Selected in Vivo Spectroscopy (ISIS)*   288
6.6 Absolute Quantification of Metabolites   288
6.7 *In Vivo* Metabolic Studies with ^{13}C MRS   289
    6.7.1 Human Brain Function   290
        6.7.1.1 *Glucose Transport and Metabolism*   291
    6.7.2 Carbohydrate Metabolism in the Liver   296
        6.7.2.1 *Glucose Metabolism*   296
        6.7.2.2 *Galactose Metabolism*   300
    6.7.3 Glucose Metabolism in the Muscle   303
6.8 Noninvasive Determination of the Degree of Unsaturation of Fatty Acids from Adipose Tissue   304
6.9 Body Fluids and Isolated Tissues   305
    6.9.1 Mechanisms of Hepatic Glycogen Repletion   306
6.10 Diseased States   307
    6.10.1 Genetic Diseases   307
        6.10.1.1 *Non-Insulin-Dependent Diabetes Mellitus*   307
        6.10.1.2 *Glycogen Storage Disease*   308

    6.10.1.3 *Fructose Intolerance* 311
    6.10.1.4 *Methylmalonic Acidemia* 313
    6.10.1.5 *Cystic Fibrosis* 313
  6.10.2 Tumors 314
    6.10.2.1 *Differentiation of Human Tumors from Nonmalignant Tissue by Natural Abundance ^{13}C NMR* 314
  6.10.3 Steatosis 315
  6.10.4 Rheumatic Diseases 316
 6.11 Concluding Remarks 316
References 318

Index 323

Contributors

Numbers in parentheses indicate the pages on which the authors' contributions begin.

Nicolau Beckmann (1, 269), Biophysics Unit, Preclinical Research, Sandoz Pharma, CH-4002 Basel, Switzerland.

Joel R. Garbow (65), Monsanto Corporate Research, Monsanto Company, St. Louis, Missouri 63198.

Gerd Gemmecker (7), Institute for Organic Chemistry and Biochemistry, Technical University of Munich, D-85747 Garching, Germany.

Terry Gullion (65), Department of Chemistry, Florida State University, Tallahassee, Florida 32306.

James A. Hamilton (117), Department of Biophysics, Boston University School of Medicine, Boston, Massachusetts 02215.

Horst Kessler (7), Institute for Organic Chemistry and Biochemistry, Technical University of Munich, D-85747 Garching, Germany.

Basil Künnecke (159), Biocenter of the University of Basel, CH-4056 Basel, Switzerland.

Foreword

Nuclear magnetic resonance spectroscopy has become a standard analytical tool in biological sciences. A large variety of different isotopes has been used, each nucleus having its own advantages and disadvantages. The most conspicuous features of ^{13}C NMR are (i) the ubiquity of carbon in organic matter, (ii) the low natural abundance of the ^{13}C isotope which allows the specific labeling of organic molecules with ^{13}C, and (iii) the large chemical shift range of ^{13}C, which simplifies the structural assignments.

This book is intended for those wishing to obtain an in-depth understanding of ^{13}C NMR as a tool in biological research. After a short introduction into the basic principles of ^{13}C NMR spectroscopy, the succeeding chapters provide authoritative reviews on all important methodological aspects of ^{13}C NMR as well as its applications, written by acknowledged experts in their fields.

^{13}C NMR has provided unique information concerning complex biological systems, from molecules to whole organisms. The availability of isotopically enriched biomolecules, stimulated by advances in molecular biology and biotechnology, has recently opened an avenue for structural studies of molecules of increasing complexity using multidimensional heteronuclear techniques, both in the liquid and solid states. The investigation of interactions in complex systems such as model membranes has also been an important target of ^{13}C NMR. In addition, ^{13}C NMR has become the method of choice for the elucidation of metabolic mechanisms *in vivo* and *in vitro*. Upon metabolization of the ^{13}C-enriched substrates, the ^{13}C label is transferred to several intermediates, and hence metabolic pathways and fluxes may be analyzed. Finally, ^{13}C NMR has found its way into clinical studies on humans. Of particular interest have been investigations of liver and muscle glycogen in healthy and pathological states.

All these different aspects of ^{13}C NMR, regarded as a tool in biological research, are covered in this book. In summary, it is this unique mix of physical methods and biological applications that makes it a convenient starting point for those interested in research in this interdisciplinary area of physics, biology, and medicine.

Joachim Seelig

Preface

In the past two decades, nuclear magnetic resonance (NMR) has gone through a revolution, becoming a key technique for answering configurational, conformational, and many other questions about molecular structure on the one hand, and for obtaining anatomical, functional, and metabolic information from living beings on the other. Although the introduction of 2D proton NMR has without a doubt played the most important role in this revolution, significant contributions have also been made by ^{13}C methods. For instance, the drive to obtain structural information from biomolecules of increasing complexity has quite recently led to the development of multidimensional heteronuclear NMR techniques involving ^{13}C nuclei. The ubiquitous presence of carbon in biomolecules, the possibility of labeling these molecules with ^{13}C and other isotopes (e.g., ^{15}N), and the favorable spectral resolution make ^{13}C NMR a relevant approach in investigating complex biological systems.

This book introduces the reader to a wide range of tools for research in biological and biomedical fields with ^{13}C NMR techniques. NMR fits well into the interdisciplinary character of modern biology; it is not a push-button technique, and its sound application in biological research requires a strong interaction among individuals with different backgrounds. The contents of the present volume were therefore chosen and prepared in such a way as to be of potential interest to biochemists, biophysicists, chemists, medical doctors, molecular biologists, and physicists. Both the NMR methodology itself and the motivation for the use of ^{13}C NMR to address certain biological problems are the primary issues. Each chapter is accompanied by an extensive bibliography covering both aspects.

Chapter 1 presents from an historical point of view essential concepts that constituted the basis for the more recent developments on ^{13}C NMR. The succeeding chapters are presented in such a way as to familiarize the reader with ^{13}C NMR approaches available for research on peptides, proteins, and nucleic acids (Chapters 2 and 3); on the interactions of fatty acids with bilayers, lipoproteins, and proteins (Chapter 4); and for noninvasive metabolic studies in animals (Chapter 5) and humans (Chapter 6). In Chapters 2 and 3, which deal with the elucidation of molecular structure in the liquid and solid states, respectively, emphasis is placed on recent multidimensional, heteronuclear techniques. Rather than pretending to explain these methods thoroughly, a more pragmatic approach is taken in an effort to provide concise information about their scope, necessity, and range of applicability. The central role played by the physiological and metabolic questions in determining the rationale for ^{13}C NMR experiments becomes clearer in the remaining chapters. For this reason significant parts of Chapters 4 to 6 are devoted to biological and biomedical issues in addition to describing ^{13}C NMR methodology.

Since NMR constitutes a field of very active research, no book dealing with this technique and its applications can achieve a state of completeness. Therefore, my primary intention was to delineate the framework of the experiments. It is my hope that this book will stimulate dialogue between researchers with different backgrounds and consequently improve the mutual understanding of the possibilities and limitations in the individual areas.

I want to thank all authors for their pleasant collaboration. Due to their cooperation, many difficulties encountered during the preparation of the book were circumvented despite the geographical distances. The kind permission of the researchers who allowed the reproduction of figures from their work is gratefully acknowledged. I am also indebted to Marvin Yelles of Academic Press for assistance and editorial supervision. And last but not least, I want to express my profound gratitude to Professor Joachim Seelig from the Biocenter of the University of Basel, whose support enabled me to discover the beauty and importance of *in vivo* NMR.

Nicolau Beckmann

CHAPTER 1

Introduction

Nicolau Beckmann
Biophysics Unit, Preclinical Research, Sandoz Pharma, CH-4002 Basel, Switzerland

The large number of publications and journals devoted to the technical improvements and applications of nuclear magnetic resonance (NMR) demonstrates the vitality of this technique. Since its discovery by Bloch and Purcell in 1946, NMR has been widely used by chemists and biochemists to investigate molecular structures. Novel approaches significantly broadened the aims of NMR, which has become a powerful spectroscopic and imaging tool applied in different areas, from materials science to chemistry, molecular biology, and medicine.

It is not an easy task to keep track of all recent developments, even if attention is focused on a particular nucleus, as is done for ^{13}C in this book. The principles for these developments, which are compiled and well explained in the marvelous book by Ernst, Bodenhausen, and Wokaun (1987), have been known for a long time. However, some of the new approaches had to wait for developments in hardware and computer technology to become feasible in practice.

An example is provided by the magnetic field gradients, which are the heart of imaging and spectroscopy *in vivo*. The use of actively shielded gradients resulted in significant improvements in the quality of *in vivo* spectra and of images acquired in a subsecond scale. The recent introduction of these gradients into high-resolution NMR is helping to extend the application of NMR to molecules of increasing complexity and to reduce significantly the acquisition time (see Sections 2.4.1.5 and 6.5.3.3).

Because a technique can be better appreciated by following its progress, aspects that constitute the base of the present state of the art of ^{13}C NMR are going to be presented from a historical point of view.

1.1 The Beginning

The first publications on ^{13}C NMR spectroscopy were presented in 1957 by Holmes and Lauterbur[1] in the same volume of the *Journal of Chemical Physics*. These pioneering papers contained ^{13}C chemical shifts and J couplings to protons in some compounds. Slow-passage (or continuous-wave, cw) natural abundance ^{13}C spectra were acquired at magnetic fields of 0.8 and 1.0 T generated by electromagnets. With sweep rates between 50 and 100 mG/s, signal-to-noise ratios ranging from 20 to 50 were reported.

During the 1960s NMR spectroscopists became aware of the proportional relationship between sensitivity and the square root of the measurement time. Field-frequency control was implemented, and signal averaging became possible. Slow-passage experiments were then substituted by averaging of rapid scans, thereby increasing the sensitivity of the measurements.

Nevertheless, the early applications of ^{13}C NMR were hindered by the low sensitivity of the ^{13}C signal compared to that of protons, the poor spectral resolution, and the necessity of working with highly soluble, low-molecular-weight materials. Therefore, ^{13}C spectroscopy was initially restricted to a few groups, including those of D. M. Grant, P. C. Lauterbur, J. D. Roberts, and J. B. Stothers. The interested reader is referred to the books by Breitmaier and Voelter (1974), Levy and Nelson (1972), and Stothers (1972) for detailed accounts of the early applications of ^{13}C NMR.

1.2 Essential Developments for ^{13}C NMR

Two major developments contributed significantly to the establishment of ^{13}C NMR as a routine technique: the Fourier transform (FT) approach and broadband proton decoupling. With the introduction of FT NMR as described and first implemented by Ernst and Anderson (1966), ^{13}C spectra of much better quality could be obtained in a shorter period of time. On the other hand, the usefulness of double resonance experiments had been realized quite early, as is testified to by, for example, the fact that broadband proton decoupling was already available in 1965. The following facts were already known:

- The splittings of resonance lines caused by spin-spin coupling to a group of nuclei could be removed if a sufficiently strong transverse

---

1. With the first experimental demonstration of the feasibility of macroscopic imaging by NMR, Lauterbur (1972, 1973) pioneered another field that has developed extremely rapidly in the last years: that of magnetic resonance imaging.

radiofrequency (rf) magnetic field were applied near the resonance frequency of these nuclei (Anderson and Freeman, 1962; Bloom and Shoolery, 1955). This early work revealed enhancements of ^{13}C resonances upon proton decoupling that exceeded those expected from a simple collapse of the multiplets due to ^{13}C-^{1}H spin coupling. The phenomenon appeared to be similar to that discovered by Overhauser (1953) in electron spin resonance spectra and to that suggested, but not observed, by Bloom and Shoolery (1955) in their pioneering paper on heteronuclear double resonance.

- The information contained in the spin-spin couplings could be used for structural analysis. In this application, rather weak perturbations were necessary to systematically perturb the splittings but not to remove them. These so-called "spin-tickling" experiments allowed one to trace the network of coupled transitions (Freeman and Anderson, 1962).

Double resonance with an incoherent rf magnetic field, originally proposed by Ernst and Primas (1963), was primarily intended for the decoupling of strongly coupled systems. Ernst (1966) later showed that double resonance with random noise was of particular advantage if the couplings were relatively weak, even when the resonance frequencies of the nuclei causing the splittings covered a wide spectral range. Also off-resonance decoupling proved to be a very useful approach for spectral assignment in heteronuclear spin systems by scaling the spin-spin interactions (Reich *et al.*, 1969).

1.3 ^{13}C NMR in the Solid State: Cross-Polarization and Magic-Angle Spinning

Acquisition of high-resolution spectra of low-sensitivity nuclei in solids was pioneered by the group of J. S. Waugh. By means of a method called *proton-enhanced NMR* conceived by Pines *et al.* (1973), signals of dilute nuclei were enhanced by repeatedly transferring polarization from a more abundant species to which they were coupled, for example, by Hartmann-Hahn cross-polarization (Hartmann and Hahn, 1962). High resolution was attained through decoupling of the abundant spins, usually protons.

Although with this approach spectra were dominated by chemical shifts, thereby permitting a degree of separation of different chemical groups, normally the anisotropies of the chemical shifts still led to relatively broad powder lineshapes, which tended to overlap in all but the simplest polymers. The supplementation of the proton-enhanced method with magic-angle spinning (MAS) (Andrew and Eades, 1953) removed the anisotropy broadening and resulted in a spectra of bulk polymers that resembled the highly resolved NMR spectra of liquids (Schaefer et al., 1975).

1.4 Heteronuclear Polarization Transfer

Cross-polarization as a means of enhancing the signals of low-sensitivity nuclei in solids was soon extended to measurements in the liquid state. Heteronuclear polarization transfer techniques have played a central role in ^{13}C NMR. Basically three approaches have been explored:

1. enhancement of the initial ^{13}C polarization
2. indirect detection of the ^{13}C resonances by looking at proton signals
3. editing of spectra by selecting resonances belonging to specific subunits in a spin system, such as CH, CH_2, and CH_3 groups.

In solution, heteronuclear polarization transfer has been achieved efficiently by Overhauser polarization (Noggle and Schirmer, 1971), adiabatic J cross-polarization in the rotating frame (Chingas et al., 1980), and much more commonly by rf pulses. Sequences such as INEPT (Morris and Freeman, 1979) and DEPT (Doddrell et al., 1982) were developed based on the demonstration that coherence transfer could be attained by applying a 90° pulse to each of the two spin species coupled (Bodenhausen and Freeman, 1977; Maudsley et al., 1977).

This brief introduction should remind the reader about the unity of NMR despite the broadness of its applications. We owe this unity to the introduction of pulsed NMR, which allows the use of similar techniques for high-resolution NMR in solution, for solid-state NMR, and for *in vivo* NMR. The following chapters document only a small fraction of the possibilities provided by this fascinating technique.

References

Anderson, W. A., and Freeman, R. (1962). *J. Chem. Phys.* **37**, 85.
Andrew, E. R., and Eades, R. G. (1953). *Proc. R. Soc. London Ser. A*, **216**, 398.
Bloom, A. L., and Shoolery, J. N. (1955). *Phys. Rev.* **97**, 1261.
Bodenhausen, G., and Freeman, R. (1977). *J. Magn. Reson.* **28**, 471.
Breitmaier, E., and Voelter, W. (1974). "^{13}C NMR Spectroscopy." Verlag Chemie, Weinheim.
Chingas, G. C., Garroway, A. N., Moniz, W. B., and Bertrand, R. D. (1980). *J. Am. Chem. Soc.* **102**, 2526.
Doddrell, D. M., Pegg, D. T., and Bendall, M. R. (1982). *J. Magn. Reson.* **48**, 323.
Ernst, R. R. (1966). *J. Chem. Phys.* **45**, 3845.
Ernst, R. R., and Anderson, W. A. (1966). *Rev. Sci. Instrum.* **37**, 93.

Ernst, R. R., and Primas, H. (1963). *Helv. Phys. Acta* **36**, 583.
Ernst, R. R., Bodenhausen, G., and Wokaun, A. (1987). "Principles of Nuclear Magnetic Resonance in One and Two Dimensions." Clarendon Press, Oxford.
Freeman, R., and Anderson, W. A. (1962). *J. Chem. Phys.* **37**, 2053.
Hartmann, S. R., and Hahn, E. L. (1962). *Phys. Rev.* **128**, 2042.
Holm, C. H. (1957). *J. Chem. Phys.* **26**, 707.
Lauterbur, P. C. (1957). *J. Chem. Phys.* **26**, 217.
Lauterbur, P. C. (1972). *Bull. Am. Phys. Soc.* [2] **18**, 86.
Lauterbur, P. C. (1973). *Nature (London)* **242**, 190.
Levy, G. C., and Nelson, G. L. (1972). "Carbon-13 NMR for Organic Chemists." Wiley (Interscience), New York.
Maudsley, A. A., Müller, L., and Ernst, R. R. (1977). *J. Magn. Reson.* **28**, 463.
Morris, G. A., and Freeman, R. (1979). *J. Am. Chem. Soc.* **101**, 760.
Noggle, J. H., and Schirmer, R. E. (1971). "The Nuclear Overhauser Effect: Chemical Applications." Academic Press, New York.
Overhauser, A. W. (1953). *Phys. Rev.* **92**, 411.
Pines, A., Gibby, M. G., and Waugh, J. S. (1973). *J. Chem. Phys.* **59**, 569.
Reich, H. J., Jautelat, M., Messe, M. T., Weigert, F. J., and Roberts, J. D. (1969). *J. Am. Chem. Soc.* **91**, 7445.
Schaefer, J., Stejskal, E. O., and Buchdahl, R. (1975). *Macromolecules* **8**, 291.
Stothers, J. B. (1972). Carbon-13 NMR spectroscopy. *In* "Organic Chemistry" (A. T. Blomquist and H. Wasserman, eds.), Vol. 24. Academic Press, New York.

CHAPTER 2

Methodology and Applications of Heteronuclear and Multidimensional ^{13}C NMR to the Elucidation of Molecular Structure and Dynamics in the Liquid State

Gerd Gemmecker and Horst Kessler

Institute for Organic Chemistry and Biochemistry,
Technical University of Munich,
D-85747 Garching, Germany

2.1 Introduction

This chapter deals with modern multidimensional NMR techniques involving ^{13}C nuclei and their significance for structure elucidation. In Section 2.2 we address some general aspects of ^{13}C NMR spectroscopy that are necessary for the understanding of the text that follows. Among these aspects are the matters of sensitivity and resolution, which are connected to the question of direct or inverse detection. The type of detection used will itself affect the requirements on the sample, i.e., whether a natural abundance substrate can be used or if isotopic enrichment is necessary instead. Another important issue is the determination of homo- and heteronuclear coupling constants.

Section 2.3 deals with the basic heteronuclear coherence transfer steps: INEPT (Morris and Freeman, 1979), DEPT (Bendall *et al.*, 1981), and heteronuclear Hartmann-Hahn (Zuiderweg, 1990). In Section 2.4 we give an overview of important multidimensional NMR techniques involving ^{13}C, either for unlabeled (Section 2.4.1) or ^{13}C-enriched molecules (Section 2.4.2).

Section 2.5 describes the meaning of heteronuclear scalar couplings for structural studies and some NMR techniques for their experimental determination. The measurement of ^{13}C relaxation times and their implications for the molecular structure are briefly explained in Section 2.6, and, finally, in Section 2.7 we sum up the scope and limitations of ^{13}C NMR for the elucidation of molecular structures.

2.2 General Considerations

First, we discuss some general questions about ^{13}C NMR, namely:

1. Why ^{13}C NMR at all? What are its special advantages? What kind of information can we gain from it in terms of structural and dynamic parameters?
2. Which special problems will we encounter when working with ^{13}C? How can we circumvent them?
3. When is the use of isotopically enriched samples appropriate?
4. Which is better: direct or inverse detection?

The fundamental question, of course, is "Why ^{13}C NMR?" Certainly at this point the reader will agree that there are interesting features in carbon spectroscopy. Here we sum up the most important ones.

First, a very general feature of ^{13}C spectra is their high spectral dispersion. Unlike, e.g., ^{1}H spectra, very few signal overlaps occur in ^{13}C spectra even for molecules with several dozens of carbons. In many cases, merely counting the number of carbon signals in a (proton-decoupled) 1D ^{13}C spectrum can provide valuable information about the molecule under study. The reason for this is the large chemical shift range comprising more than 200 ppm for most organic molecules, in combination with the narrow lineshapes down to less than 0.01 ppm (i.e., < 1 Hz) because of slow relaxation. Usually there are distinct chemical shift regions for certain types of carbon atoms (such as carbonyls, aromatic carbons, aliphatic methyl groups, etc.) that allow for a quick orientation and rough assignment by the experienced spectroscopist (Kalinowski *et al.*, 1984).

This ease of interpretation is also the reason for the preference of most databases (i.e., CSEARCH and SPECInfo) for ^{13}C spectra. The information content of a 1D carbon spectrum can be easily stored by just providing the ppm values (and possibly multiplicities, corresponding to the number of directly bound protons) of each signal. By contrast, a proton spectrum would have to be stored as a complex ensemble of overlapping lines with different

---

1. CSEARCH ^{13}C data bank; Biorad, Sadtler Division, 3316 Spring Garden Street, Philadelphia, PA 19104, USA. SPECInfo data bank with ^{13}C, ^{31}P, ^{15}N, ^{19}F, and ^{17}O NMR data; Chemical Concepts, Weinheim, Germany, 1989.

lineshapes and often unresolved multiplicities. It is evident that the ^{13}C NMR data are much more suited to fast searches in huge databases.

Furthermore, carbon is the only heteronucleus that is practically ubiquitous in organic compounds (comparable to ^{1}H). Other heteroatoms such as ^{15}N, ^{19}F, and ^{31}P occur only sporadically in organic structures, except for very special substance classes. Therefore, the only stable and NMR active carbon isotope, ^{13}C, is *the* favorite NMR probe for carbon skeletons.

Modern ^{13}C NMR spectroscopy can also provide a wealth of parameters that can be used for conformational and dynamic studies. Among these are the direct and long-range ^{1}H–^{13}C coupling constants ^{i}J$_{H,C}$ with $i = 1-3$ (Marshall, 1983) that contain information about dihedral angles (Section 2.5), as well as ^{13}C–T$_1$ relaxation rates for molecular mobility (Section 2.6).

In addition, ^{13}C chemical shifts by themselves can hold structural information for certain classes of molecules, e.g., the cis/trans configuration of Xxx–Pro peptide bonds [indicated by δ(Pro–^{13}C$^\beta$/^{13}C$^\gamma$)] (Kessler, 1982) or the α/β configuration of O-glycosides [from δ(^{13}C) of the anomeric carbon] (Pavia and Ferrari, 1983). For many classes of compounds, increment systems exist that allow the fairly precise calculation of ^{13}C chemical shifts (Kalinowski *et al.*, 1984). This often enables one to distinguish between different proposed structures (e.g., substitution patterns) by just looking at the 1D ^{13}C spectrum.

In addition to these arguments, which use the typical properties of the ^{13}C nucleus, heteronuclear techniques can indirectly make use of the special features of ^{13}C as discussed below.

A heteroatom can be used to introduce a new dimension into proton spectra, because overlapping proton signals may split when correlated with the chemical shift of a directly bound heteroatom. As an example, the β protons of tryptophan (Trp) resonate in the same spectral region as the β protons of other amino acids, such as phenylalanin, histidine, or cysteine, whereas the ^{13}C$^\beta$ of Trp is typically shifted to a higher field (Kessler and Bermel, 1986). Also proton spin systems in crowded spectra can be more easily identified in a heteronuclear dimension, either in a long-range experiment [COrrelation via Long-Range Coupling (COLOC) (Kessler *et al.*, 1984), or its inverse version, Heteronuclear MultiBond Correlation (HMBC) (Bax and Summers, 1986)] or a TOCSY-HMQC (Lerner and Bax, 1986) (see also Section 2.4). A 3D TOCSY-HMQC with multiplicity selection provides an efficient way for this purpose (Kessler and Schmieder, 1991).

Isolated spin systems can also be connected via long-range couplings to a heteroatom (quaternary carbon) (Kessler *et al.*, 1984), which yields important connectivity information for the molecular skeleton.

For very large ^{13}C-enriched molecules, heteronuclear coherence transfer pathways are more efficient than direct proton-to-proton correlations (Clore and Gronenborn, 1991; Ikura *et al.*, 1990; Wagner, 1990). Thus, a COSY (Aue *et al.*, 1976) or TOCSY (Braunschweiler and Ernst, 1983) spectrum may be substituted advantageously by a proton-detected ^{13}C-COSY (Bax *et al.*, 1990a) or ^{13}C-TOCSY (Bax *et al.*, 1990b) in labeled proteins (this process usually requires ^{13}C enrichment). In combination with other heteroatoms, a variety of different coherence transfers is possible, e.g., with a HN(CA)H experiment (^{1}H–^{15}N–^{13}C$^\alpha$–H$^\alpha$) in ^{13}C- and ^{15}N-labeled proteins (Clubb *et al.*, 1992; Seip *et al.*, 1992). Because the heteroatoms have longer relaxation times and larger ^{1}J coupling constants (compared to ^{2}J$_{HH}$ and ^{3}J$_{HH}$), such heteronuclear experiments can help to increase significantly the sensitivity in correlation spectra of biopolymers (Clore and Gronenborn, 1991). These recently introduced techniques allow us to analyze proteins up to about 30 kDa. However, the required high isotopic enrichment with the NMR active nuclei ^{13}C (and ^{15}N), achievable for most proteins, is much more difficult to attain for other classes of substances.

2.2.1 Inverse or Direct Detection?

Traditionally, the term *inverse* is used for heteronuclear techniques that detect the nucleus with the higher magnetogyric ratio γ (usually ^{1}H), whereas detection of the low-γ spin is referred to as *direct* (*normal*). However, today inverse detection can be called normal; because it offers such a significant advantage in sensitivity for most applications, it must be considered state-of-the-art.

To understand the different sensitivities of normal and inverse detection, we only need to look at the basic equation describing the sensitivity (i.e., the signal-to-noise ratio, SNR) of an NMR experiment :

$$\text{SNR} = n \, \gamma_{\text{exc}} \, \gamma_{\text{det}}^{3/2} \, NS^{1/2} \, T^{-1} \, T_2^{-1}, \tag{2.1}$$

where n is the number of molecules, NS is the number of scans, T is the absolute temperature, and T_2 is the spin-spin relaxation time). Other things being equal, the sensitivity is proportional to $(\gamma_{\text{exc}})^1$ and $(\gamma_{\text{det}})^{3/2}$. To achieve a good SNR, it is important to detect the high-γ spin. Also, the maximum sensitivity is reached for exciting *and* detecting at the highest possible frequency. The differences in the *measuring times* required for an equal SNR are even more prominent because of the SNR being proportional to $NS^{1/2}$. Table 1 gives a comparison of the relative sensitivities and measuring times for different constellations involving ^{1}H and ^{13}C.

In addition, when the experiment starts from ^{1}H magnetization, the recycle delay depends on proton T_1 relaxation, which is generally faster than carbon relaxation. Furthermore, if the excited and the detected nuclei are the same, the acquisition period can already serve as part of the relaxation pe-

TABLE 1
Comparison of the Sensitivity of Experiments Involving ^{1}H and ^{13}C

Experiment	Excited/detected spin	Relative sensitivity	Time demand
1D ^{13}C{^{1}H}, no NOE	^{13}C/^{13}C	1	1024
1D ^{13}C-DEPT	^{1}H/^{13}C	4	64
H,C-COSY	^{1}H/^{13}C	4	64
^{1}H,^{13}C-HMQC or HSQC	^{1}H/^{1}H	32	1

riod, since there will be no pulses on this isotope during acquisition of the FID (the other spins are usually decoupled). Therefore, an experiment starting and ending on ^{1}H has the shortest possible recycle delay, leading to a superior SNR per time unit.

However, inverse spectroscopy also has several drawbacks. The digital resolution in any dimension is given by the longest time interval reached by the corresponding evolution period, i.e., t_1^{max} and t_2^{max} (for a 2D spectrum) (Ernst et al., 1987). A price has to be paid (in terms of measuring time) for a high resolution in the indirect dimension t_1, because more increments are needed if t_1^{max} is increased. This is especially true for very concentrated samples, where sensitivity is not a problem but a basic phase cycle has to be maintained to suppress artifacts (Kessler et al., 1988). Therefore, the number of scans per increment cannot be lowered below a certain minimum to compensate for a large number of increments. The situation is completely different in the direct dimension t_2: Additional resolution is "free," as long as t_2^{max} is shorter than the delay required anyway for relaxation (usually 3 to 5 times T_1).

Now the typical linewidth of a carbon resonance is much smaller than that of a proton resonance because of the slow T_2 relaxation and the absence of ^{13}C homonuclear coupling in natural abundance samples (in contrast to ^{1}H–^{1}H homonuclear coupling). To make use of this inherently higher resolution in the carbon dimension, $t_i^{max}(^{13}C)$ should be longer than $t_i^{max}(^1H)$. In addition, the spectral dispersion for carbon is also higher than for proton ($\sim$250 ppm : 12 ppm, corresponding to a ratio of $\sim$5 : 1 in hertz), which means that the increment for carbon must be approximately five times smaller than for ^{1}H. Therefore, putting ^{13}C into the indirect dimension (as in all inverse experiments) requires a very high number of increments to make full use of the carbon-inherent good resolution, making such multidimensional inverse experiments less attractive than suggested by Table 1.

Nevertheless, with dilute samples of natural ^{13}C abundance, which require many more scans than the minimum phase cycle per increment, this

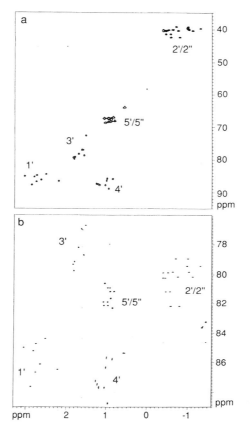

Figure 1. Two inverse 2D ^{13}C,^{1}H correlations of a natural abundance sample of the DNA duplex d(TTGGCCAA)$_2$, 6 mM single-strand concentration. (Reproduced with permission from Schmieder et al., 1992. Copyright 1992 Oxford University Press, Oxford.) (a) 600-MHz HMQC spectrum (Bax et al., 1983; Müller, 1979), acquired with a sweep width of 165.7 ppm in the ^{13}C dimension (1024 t_1 increments), measuring time 13 h. (b) 600-MHz HSQC spectrum (Bodenhausen and Ruben, 1980), acquired with a sweep width of 19.9 ppm in F$_1$ (1890 t_1 increments), measuring time 25 h. The 2'/2" and 5'/5" signals are folded in F$_1$, leading to a much higher resolution in the ^{13}C dimension.

aspect is irrelevant. Additionally, by folding in F$_1$ the spectral width can be significantly reduced (Figure 1) and hence so can the number of increments needed for a given digital resolution (Schmieder et al., 1991).

Another issue to be addressed is the problem of intense solvent peaks that can obscure signals from dilute samples. With highly deuterated organic solvents, inverse techniques are advantageous, because the almost neglectable ^{1}H content of the solvent allows the detection of even very low concentrated molecules. On the other hand, carbon-detected spectra of very dilute samples may suffer from the intense ^{13}C solvent signal, unless ^{13}C-depleted solvents are used. Undeuterated solvents, especially water, will cause serious problems with inverse techniques, but none at all in ^{13}C-detected experiments.

Many different procedures have been proposed over the years to suppress the water signal for measurement of biological samples (for a review, see Guéron et al., 1991), the most prominent ones being solvent presaturation (Kumar et al., 1980), jump-return (Plateau and Guéron, 1982) and other binomial excitation sequences (Sklenář and Bax, 1987), spin locks (Messerle et al., 1989), and most recently pulsed B_0-field gradients (Tyburn et al., 1992) and B_1-field gradient pulse trains (Canet et al., 1993). Although their detailed discussion is beyond the scope of this book, the reader should note here that these techniques, together with the superior performance of modern NMR spectrometers, have resulted in a dramatic improvement in this area in recent years, making indirect detection the more interesting.

Note that the theoretical sensitivity gains in Table 1 can only be obtained if certain requirements are met by the spectrometer hardware. Inverse experiments in particular detect *all* protons including the 99% bound to ^{12}C (at natural abundance) that have to be subtracted by the phase cycle in subsequent scans. This requires a high degree of spectrometer stability and reproducibility (e.g., in frequency synthesis, pulse phase and amplitude, receiver performance, temperature control, etc.) in order to reduce t_1 noise to a minimum. In addition, the spectrometer should be equipped with an inverse probe with the inner coil tuned to ^{1}H, and amplifiers and software for ^{13}C decoupling during the acquisition.

2.3 Heteronuclear Coherence Transfer

In this section we first describe the INEPT transfer (Morris and Freeman, 1979), which is by far the most common heteronuclear coherence transfer in multidimensional experiments. Sections 2.3.2 and 2.3.3 then briefly discuss the alternative techniques DEPT (Bendall et al., 1981) and heteronuclear Hartmann-Hahn cross-polarization (Zuiderweg, 1990), respectively.

2.3.1 INEPT

The INEPT technique (Figure 2a) is a very straightforward means of coherence transfer between *different nuclei* (Morris and Freeman, 1979). This definition contains all species to which rf pulses can be applied selectively.

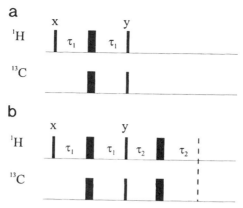

Figure 2. Pulse sequences (a) for the INEPT transfer and (b) the refocused INEPT transfer resulting in in-phase ^{13}C coherence (Morris and Freeman, 1979). The delays τ_1 have to be set to $(4\ ^1J_{H,C})^{-1}$, whereas the optimum value for τ_2 depends on the multiplicity of the carbon signal [compromise for CH, CH$_2$, and CH$_3$: $\tau_2 = (6\ ^1J_{H,C})^{-1}$].

This is, of course, the case with isotopes of different elements S(^{1}H/^{13}C or ^{15}N/^{13}C), but also for groups of signals of the same isotope that resonate in well-separated spectral regions (e.g., ^{13}C$^\alpha$ and ^{13}CO in proteins) and can be excited separately with frequency-selective pulses.

The majority of all NMR experiments are most conveniently described by Cartesian product operators, which we use throughout this chapter (Kessler et al., 1988; Sørensen et al., 1983). INEPT (Insensitive Nuclei Enhanced by Polarization Transfer) is based on the transfer of antiphase coherence via 90° pulses on each of the two spins involved:

$$2I_xS_z \xrightarrow{90°_y(I,S)} -2I_zS_x. \qquad (2.2)$$

Usually one starts with in-phase coherence $-I_y$, which first has to evolve into antiphase under the heteronuclear coupling J(I,S). The usually undesirable chemical shift evolution of I during the delay Δ can be suppressed by a pair of 180° pulses on both I and S in the center of the delay $\Delta = 2\tau_1$. For a weakly coupled IS spin system with coupling constant $J_{I,S}$ one gets:

$$I_y \xrightarrow{\Delta/2} \xrightarrow{180°_y(I,S)} \xrightarrow{\Delta/2} -2I_xS_z \sin\pi J_{I,S}\Delta + I_y \cos\pi J_{I,S}\Delta. \qquad (2.3)$$

This means that the optimum value is $\Delta = (2J_{I,S})^{-1}$, resulting in a complete conversion of I_y into the desired heteronuclear antiphase term $2I_xS_z$. However, in real systems there are additional effects that have to be taken into account. Generally, relaxation leads to a decrease of coherence amplitude; therefore, in fast relaxing systems (e.g., macromolecules) a shorter delay may be advantageous. If the spin system contains additional *passive* scalar couplings (that will not contribute to the desired coherence transfer), these will also lead to a loss of intensity because of the evolution of multiple antiphase terms. In the case of an $I^1(I^i)_n S$ system with nonvanishing couplings $J(I^1,I^i)$ ($i = 2 \ldots n$), the amplitude of the desired term $2I_x^1 S_z$ will be

$$\sin\pi J(I^1,S)\Delta \prod [\cos\pi J(I^1,I^i)\Delta] \qquad (2.4)$$

(e.g., for $CH_2=CHCl$ $n = 2$ for each proton, because it is coupled to two other protons). This means that, depending on the size and number of couplings $J(I^1,I^i)$, the optimum value for Δ will be shorter than $(2J_{I^1,S})^{-1}$. However, since the value of the direct proton-carbon coupling $^1J_{^1H,^{13}C}$ (~120 to 160 Hz) exceeds the normal $^1H-^1H$ couplings by more than an order of magnitude, the evolution of homonuclear proton couplings during Δ can be neglected when transferring from ^1H to ^{13}C.

Another issue that has to be dealt with is the uniformity of heteronuclear coupling constants $^1J_{H,C}$. For an optimum transfer, the delay Δ in the INEPT sequence should be tuned to $(2\ ^1J_{H,C})^{-1}$ for a CH group. However, the heteronuclear couplings can cover a certain range. Depending on the specific molecule under study, a compromise value has to be chosen for Δ. For the normal range of $^1J_{H,C}$ (from ~120 Hz for methyl groups to more than 160 Hz in aromatic systems) this corresponds to a sensitivity of ~95% for the extreme cases, according to an intensity modulation of $(\sin \pi\ ^1J_{H,C}\Delta)$ for the magnetization transferred onto ^{13}C (Kessler *et al.*, 1988).

The INEPT transfer described thus far starts with in-phase coherence on the I spin and then creates antiphase coherence on the S spin. If there is a need for *in-phase* S coherence, a second J-evolution period must follow to refocus the heteronuclear coupling (so-called "refocused INEPT," Figure 2b):

$$-2I_z S_x \xrightarrow{\Delta'/2} \xrightarrow{180°(I,S)} \xrightarrow{\Delta'/2} S_y \sin\pi J_{I,S}\Delta' \prod_i [\cos\pi J_{I^i,S}\Delta']. \qquad (2.5)$$

Again, the delay $\Delta' = 2\tau_2$ will have to be shorter than $(2\ ^1J_{I,S})^{-1}$ if there are passive I^i spins.

In the case of a reverse transfer from ^{13}C to ^1H (sometimes referred to as REVINEPT; Freeman *et al.*, 1981), the larger homonuclear $^{13}C-^{13}C$ couplings can be neglected in the case of natural abundance, but should be taken into account for ^{13}C-enriched molecules. But now there is another problem, because in CH_2 and CH_3 groups, the ^{13}C spin couples to *all* directly connected protons with a practically identical large $^1J_{I,S}$ coupling. The resulting amplitude factor for the creation of single antiphase terms is then for a CH_2 group

$$S_y \xrightarrow{\Delta/2} \xrightarrow{180°(I,S)} \xrightarrow{\Delta/2} -2I_zS_x \sin\pi J_{I,S}\Delta \cos\pi J_{I,S}\Delta \qquad (2.6)$$

and for methyl groups

$$S_y \xrightarrow{\Delta/2} \xrightarrow{180°(I,S)} \xrightarrow{\Delta/2} -2I_zS_x \sin\pi J_{I,S}\Delta \cos^2\pi J_{I,S}\Delta, \qquad (2.7)$$

which requires delays $\Delta < (2\ ^1J_{I,S})^{-1}$ for maximum transfer. Usually a value of $(3\ ^1J_{I,S})^{-1}$ is chosen as a compromise yielding almost equal efficiency for CH, CH_2, and CH_3 groups.

We have already mentioned the problem of suppressing signals from protons not bound to ^{13}C nuclei in inverse experiments. An efficient way of accomplishing this suppression in a refocused INEPT is the use of a spin-lock pulse (typically 2 to 3 ms in duration with high rf power) after the delay Δ immediately before the 90° pulses for the coherence transfer, as shown in Figure 3a (Messerle et al., 1989).

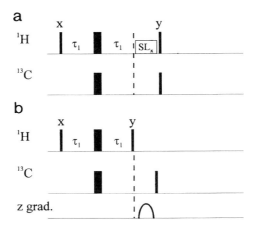

Figure 3. INEPT pulse sequences with additional $^{12}C-^1H$ and water suppression; $\tau_1 = (2\ ^1J_{H,C})^{-1}$. (a) INEPT with insertion of a spin-lock period (Messerle et al., 1989). The strong inhomogeneous B_1 field of the spin-lock pulse (SL, duration ~2 to 3 ms) dephases the y components created by the first $90°_x(^1H)$ pulse. Only antiphase terms $2I_xS_z$ caused by $^1J_{^1H,^{13}C}$ are not affected and are subsequently transferred to carbon magnetization (see text). (b) INEPT sequence with water suppression by a homospoil pulse (Bax and Pochapski, 1992). After the second 90° ^1H pulse with y phase, all heteronuclear antiphase terms $2I_xS_z$ have been converted to $-2I_zS_z$, while protons *not bound* to ^{13}C remain as I_y magnetization. The homospoil pulse destroys all coherences, i.e., I_y, but *not* $-2I_zS_z$. The substitution of the standard homospoil pulse by a z gradient pulse with actively shielded gradient coils leads to a clearly superior performance (see Section 2.4.1.5).

At this point, magnetization of protons directly bound to ^{13}C is oriented along the x axis ($2I_xS_z$), whereas all other protons assume an orthogonal orientation (I_y). A spin-lock pulse on the I spins with phase x does not affect $2I_xS_z$, but all y components—including the water resonance!—are rapidly defocused due to the inhomogeneity of the field.

In a similar way, a homospoil pulse or z gradient (see Section 2.4.1.5) can be employed instead (Figure 3b). It has to be placed *after* the second 90° pulse on ^{1}H, but *before* the first 90° pulse on ^{13}C (Bax and Pochapski, 1992). At this point the antiphase magnetization that evolved during the previous delay has been converted to $2I_zS_z$, a term insensitive to field inhomogeneity. However, the water resonance, still oriented in the x,y plane (I_y), will be quickly dephased by the homospoil/gradient pulse.

2.3.2 DEPT

The DEPT (Distortionless Enhancement by Polarization Transfer) sequence (Bendall *et al.*, 1981) represents an alternative to the widely used INEPT transfer. Like the INEPT sequence it starts with a delay $\Delta = (2\,^1J_{H,C})^{-1}$ for the evolution of heteronuclear antiphase magnetization I_xS_z, which is then converted into heteronuclear multiquantum coherence I_xS_y (Figure 4):

$$-I_y \xrightarrow{\Delta} 2I_xS_z \xrightarrow{90^\circ_x(S)} -2I_xS_y. \qquad (2.8)$$

In the following period Δ *no coupling evolution* occurs between I_x and S_y (but other protons couple to I or S). The chemical shift evolution of ^{1}H and ^{13}C is refocused by the two 180° pulses. After the second 90° pulse on ^{1}H ($\theta = 90°$), the coherence transfer from proton onto carbon is complete and the heteronuclear coupling refocuses during the last Δ delay:

$$-2I_xS_y \xrightarrow{90^\circ_y(I)} 2I_zS_y \xrightarrow{\Delta} -S_x. \qquad (2.9)$$

Although the DEPT sequence with a duration of $1.5(^1J_{H,C})^{-1}$ is slightly longer than the refocused INEPT with $(^1J_{H,C})^{-1}$, it contains only two 180°

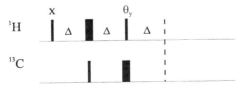

Figure 4. Pulse sequence for the DEPT transfer (Bendall *et al.*, 1981). The delays Δ have to be set to $(2\,^1J_{H,C})^{-1}$; multiplicity editing can be performed by adjusting the pulse angle θ. For optimum transfer of CH, CH$_2$, and CH$_3$ a value of $\theta = 60°$ is chosen.

pulses (INEPT: four) and five pulses overall (refocused INEPT: seven). Therefore, it is less sensitive to artifacts from improper pulse lengths and requires a shorter phase cycle.

The DEPT sequence, although only rarely found in multidimensional experiments, resembles the heteronuclear multiquantum coherence (HMQC) technique described in Section 2.4.

2.3.3 Heteronuclear Cross-Polarization

As an alternative to the previously described INEPT and DEPT techniques, heteronuclear cross-polarization (Figure 5) can also generate a coherence transfer (Zuiderweg, 1990), like its homonuclear equivalent TOCSY. Cross-polarization occurs with simultaneous spin-lock of two spins **I** and **S** under the condition of the so-called Hartmann-Hahn match (Hartmann and Hahn, 1962):

$$|\gamma_I B_1(\mathbf{I}) - \gamma_S B_1(\mathbf{S})| < J_{I,S}. \qquad (2.10)$$

This means that the spin-lock fields B_1 for the two nuclei ^{1}H and ^{13}C (typically several kilohertz to cover the desired chemical shift range) have to be matched within ~1% (for a $^1J_{H,C}$ of approximately 140 Hz). These spin-locks are (as in TOCSY experiments) usually performed with a pulse sequence like MLEV (Levitt and Freeman, 1979) or WALTZ (Allerhand et al., 1985). There is no need for J evolution periods, since the transfer proceeds from in-phase ^{1}H coherence directly to in-phase ^{13}C magnetization:

$$\begin{aligned}I_x \to\ & 0.5 I_x[1 + \cos(\pi J t)] + 0.5 S_x[1 - \cos(\pi J t)] \\ & + (I_z S_y - I_y S_z) \sin(\pi J t).\end{aligned} \qquad (2.11)$$

Neglecting the intermediate antiphase terms, this corresponds to an oscillation of in-phase coherence between the two spins with a period of $(^1J_{I,S})^{-1}$. This is the same duration as for a refocused INEPT transfer.

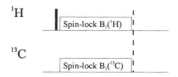

Figure 5. Pulse sequence for heteronuclear coherence transfer via cross-polarization (Zuiderweg, 1990). The two spin-lock fields have to be adjusted according to the Hartmann-Hahn condition $[\gamma(^{13}C)B_1(^{13}C) = \gamma(^1H)B_1(^1H)]$ and usually consist of composite pulse decoupling sequences.

Today, heteronuclear Hartmann-Hahn transfer has been successfully incorporated into a few multidimensional experiments (Majumdar et al., 1993; Richardson et al., 1993) and shown to be even slightly superior to INEPT steps in the case of large proteins. However, its application has been very limited so far, probably because the double spin-lock is much more demanding on the spectrometer hardware (especially on amplifiers and probes) than the INEPT or DEPT sequences.

2.4 Multidimensional Correlation Experiments

2.4.1 Natural ^{13}C Abundance

Because of the relatively low natural abundance of the ^{13}C isotope ($\sim 1.1\%$), NMR techniques relying on the presence of two or more ^{13}C spins in the same molecule are prohibitively insensitive, the only "practical" application being the INADEQUATE sequence (Bax et al., 1980a; with INEPT, Sørensen et al., 1982), which still requires sample concentrations greater than 100 mM and measuring times of days.

In this section we first describe the fundamental ^{1}H–^{13}C inverse correlation experiments Heteronuclear Single-Quantum Coherence (HSQC) (Bodenhausen and Ruben, 1980) and Heteronuclear Multiple-Quantum Coherence (HMQC) (Bax et al., 1983; Müller, 1979), as well as the long-range version Heteronuclear Multibond Correlation (HMBC) (Bax and Summers, 1986). Then we discuss some important 2D (or 3D) combinations of these sequences with other homonuclear experiments, such as TOCSY-HMQC (Lerner and Bax, 1986).

Direct detected techniques such as H,C-COSY, COLOC, and their combinations with homonuclear techniques will certainly be of limited use in the future because of the previously mentioned sensitivity problems. The interested reader is referred to the literature (Kessler et al., 1988; Martin and Zektzer, 1988).

2.4.1.1 Heteronuclear Single-Quantum and Multiple-Quantum Coherence

Both experiments, HSQC and HMQC, yield essentially the same information, namely, a one-bond correlation between protons and carbons. Yet they differ in the way the coherence transfers from ^{1}H onto ^{13}C and back to ^{1}H are accomplished.

HSQC (Bodenhausen and Ruben, 1980), in spite of the numerous pulses and delays, is the conceptually simpler experiment of the two. It consists of a nonrefocused INEPT transfer from proton to carbon, followed by the ^{13}C evolution period (t_1) with proton decoupling by the 180° ^{1}H pulse in its cen-

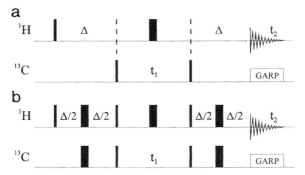

Figure 6. Pulse sequences of (a) the HMQC experiment (Bax et al., 1983; Müller, 1979) and (b) the HSQC experiment (Bodenhausen and Ruben, 1980). The delays Δ have to be set to Δ = $(2\ ^1J_{H,C})^{-1}$.

ter, and ends with a reverse INEPT transfer back to proton before signal acquisition under ^{13}C decoupling (Figure 6b). The coherence transfer pathway can be described as

$$-I_y \xrightarrow{\Delta} 2I_xS_z \xrightarrow{90°_y(I,S)} -2I_zS_x \xrightarrow{t_1} -2I_zS_x \cos(\omega_S t_1) \quad (2.12)$$
$$\xrightarrow{90°_y(I,S)} 2I_xS_z \cos(\omega_S t_1) \xrightarrow{\Delta} I_y \cos(\omega_S t_1).$$

It is obvious that there is no need for refocusing the carbon antiphase term $2I_zS_x$ created by the first INEPT transfer, because it is required again for the transfer back onto ^1H after the t_1 period. The appearance of the cross-signals in an HSQC spectrum (Figure 7) can be understood from the term that is finally detected (i.e., the last term in the given coherence transfer path), $I_y \cos(\omega_S t_1)$: It corresponds to a single signal in F_1 with the chemical shift ω_S of the carbon (neglecting ^{13}C–^{13}C coupling). In the proton dimension we start with in-phase magnetization that will develop the normal ^1H multiplet structure due to proton–proton coupling during t_2 (as in a normal 1D proton spectrum); the $^1J_{H,C}$ coupling is suppressed by the ^{13}C decoupling sequence.

The HMQC experiment (Bax et al., 1983; Müller, 1979) bears some resemblance with the DEPT sequence, because both make use of heteronuclear multiquantum coherence. As in DEPT, it is created by two 90° pulses on ^1H and ^{13}C, separated by a delay $\Delta = (2\ ^1J_{H,C})^{-1}$ for coupling evolution (Figure 6a):

$$I_z \xrightarrow{90°_x(I)} -I_y \xrightarrow{\Delta} 2I_xS_z \xrightarrow{90°_x(S)} -2I_xS_y, \quad (2.13)$$

2. Elucidation of Molecular Structure and Dynamics in the Liquid State / 21

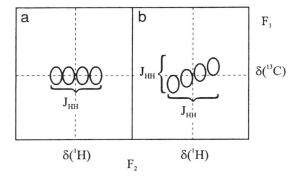

Figure 7. Schematic representation of a typical cross-peak fine structure in an (a) HSQC and (b) HMQC spectrum in natural abundance (i.e., neglecting ^{13}C–^{13}C coupling). The HMQC peaks are broadened in the ^{13}C dimension by ^{1}H–^{1}H couplings (shown here for a quartet), limiting the achievable resolution in this dimension.

neglecting chemical shift evolution. The term $-2I_xS_y$ represents simultaneous coherence on ^{1}H and ^{13}C, and its chemical shift evolution is described by a superposition of the chemical shifts of proton and carbon, i.e., it contains two components rotating with the sum and difference of ^{1}H and ^{13}C chemical shifts (since this term describes a state involving both nuclei, there is no evolution of coupling between the two):

$$-2I_xS_y \xrightarrow{t_1} (-I_xS_y - I_yS_x) \cos(\omega_S + \omega_I)t_1$$
$$+ (I_xS_x - I_yS_y) \sin(\omega_S + \omega_I)t_1 \quad (2.14)$$
$$(-I_xS_y + I_yS_x) \cos(\omega_I - \omega_S)t_1$$
$$- (I_xS_x + I_yS_y) \sin(\omega_I - \omega_S)t_1.$$

Now what we want during the t_1 interval is the sole evolution of the ^{13}C chemical shift, in such a way that we end up with a carbon dimension in F_1 after Fourier transformation. The 180° pulse on ^{1}H placed at the center of the t_1 evolution period has the effect of changing the sign of the ^{1}H contribution to the chemical shift, i.e., the term rotating with $(\omega_S + \omega_I)$ during the first half of t_1 will evolve with $(\omega_S - \omega_I)$ during the second half of t_1 (and vice versa). The result is a cancellation of the proton contribution to the chemical shift evolution of the multiquantum term:

$$-2I_xS_y \xrightarrow{[t_1/2,\ 180°(I),\ t_1/2]} -2I_xS_y \cos(\omega_S t_1) + 2I_xS_x \sin(\omega_S t_1). \quad (2.15)$$

In addition, since the two delays Δ are symmetrically placed around the 180° ^{1}H pulse, the proton chemical shift evolution during these two time periods is also refocused, so that the transfer pathway for a two-spin system is [with $\Delta = (2\,J_{I,S})^{-1}$]:

$$I_z \xrightarrow{90°_x(I)} -I_y \xrightarrow{\Delta} 2I_xS_z \xrightarrow{90°_x(S)} -2I_xS_y \xrightarrow{[t_1/2,\,180°(I),\,t_1/2]} \quad (2.16)$$

$$-2I_xS_y\cos(\omega_S t_1) \xrightarrow{90°_x(S)} -2I_xS_z\cos(\omega_S t_1) \xrightarrow{\Delta} -I_y\cos(\omega_S t_1).$$

This is exactly the same result as for the HSQC, so that one might expect two completely identical 2D spectra from both techniques. However, this will only be true for the simplified case described by the given formulas, i.e., an **I,S** two-spin system with neglectable relaxation.

The most obvious difference results from homonuclear $^1H-^1H$ coupling. In both experiments these couplings will become visible in the proton dimension F_2, since they evolve freely during the acquisition of the FID. In the HSQC experiment, only ^{13}C coherence exists during the t_1 period, so that $^1H-^1H$ coupling does not occur during this interval. Additionally, decoupling of ^{13}C from protons is accomplished via the 180° ^1H pulse at the center of t_1.

However, in the HMQC experiment we have $^1H,^{13}C$-multiquantum coherence during t_1. Indeed, the 180° proton pulse in the center can refocus the ^1H chemical shift contributions, but *not* the couplings between the proton involved in the multiquantum coherence and *other* protons. The reason for this is that the inversion of *all* proton spin states (by the 180° pulse) does not cause a change of the *relative* orientation between any two of them, which would be necessary for refocusing the $^1H-^1H$ couplings. As a result, in the HMQC the homonuclear proton couplings are visible not only in the proton dimension F_2, but also in the ^{13}C dimension, causing an additional broadening of the otherwise quite narrow ^{13}C linewidth (compare Figure 1). Furthermore, since the spin states (α or β) of the protons are not affected by any rf pulse between the t_1 and t_2 periods, the 2D multiplet structure caused by the $^1H-^1H$ couplings assumes a diagonal pattern rather than a rectangular one (compare Section 2.6.2.2), leading to a tilted appearance of the HMQC peaks (Figure 7b).

Another difference between HSQC and HMQC is the relaxation behavior during t_1, which is different for spin **S** antiphase coherence (HSQC) or **I-S** multiquantum coherence.

2.4.1.2 BIRD-Filtered HSQC and HMQC

Often t_1 noise is a serious problem with inverse experiments when run on samples with natural ^{13}C abundance. The reason for this is simple: In any given position, the chance of finding a ^{13}C spin is only 1.1%. Therefore, the great majority of all protons cannot contribute to a ^{13}C correlation spectrum and has to be eliminated by phase cycling. Although a factor of 100 is well within the dynamic range of modern receivers (unlike factors of 10,000 with protonated solvents), the cancellation is usually not perfect, leading to very unpleasant t_1 noise ridges at positions of strong ^{1}H resonances. The main reasons for this are, of course, instrumental instabilities, but also phase and amplitude errors in the transmitter and receiver.

Fortunately, there is a very convenient way of getting rid of these artifacts in HSQC and HMQC spectra: a BIRD (BIlinear Rotational Decoupling) pulse (Garbow et al., 1982) is inserted shortly before every scan with the actual sequence (Figure 8).

The BIRD pulse makes use of the fact that, after a delay $\Delta = (2 \ ^1J_{H,C})^{-1}$, the spins of protons bound to ^{13}C are 90° out of phase with respect to protons bound to ^{12}C. Thus it becomes possible to apply a 180° pulse only to ^{13}C-bound protons—or exclusively to protons bound to ^{12}C, depending on the relative pulse phases within the BIRD sequence (Kessler et al., 1988). With the following phase settings (so-called BIRD$_y$ pulse) we get inversion of the ^{12}C-bound protons only:

$$^1\text{H}-^{13}\text{C:}\quad I_z \xrightarrow{90^\circ_x(I)} -I_y \xrightarrow{\Delta} 2I_xS_z \xrightarrow{180^\circ_x(I,S)} -2I_xS_z \xrightarrow{\Delta} -I_y \xrightarrow{90^\circ_{-x}(I)} I_z \quad (2.17a)$$

$$^1\text{H}-^{12}\text{C:}\quad I_z \xrightarrow{90^\circ_x(I)} -I_y \xrightarrow{\Delta} -I_y \xrightarrow{180^\circ_x(I,S)} I_y \xrightarrow{\Delta} I_y \xrightarrow{90^\circ_{-x}(I)} -I_z \quad (2.17b)$$

with $\Delta = (2 \ ^1J_{H,C})^{-1}$.

As shown in this calculation, the BIRD pulse in front of any sequence delivers a 180°$_y$ pulse only to ^{1}H–^{12}C resonances, while ^{1}H–^{13}C pairs are left undisturbed. During the following short delay τ, the inverted ^{12}C-bound proton resonances are given time to relax back to their normal equilibrium state ($+z$); the actual experiment is started at the moment when most ^{1}H–^{12}C resonances are just going through zero (Figure 9). This leads to a drastic decrease in the intensity of the protons bound to ^{12}C—the rest is easily removed by the phase cycling of the experiment. In particular, the t_1 noise in HSQC or HMQC spectra acquired with a sequence equipped with a BIRD filter is comparable to ^{1}H homonuclear spectra.

Of course, the delay τ after the BIRD filter has to be optimized in order to reach a good null for the ^{12}C-bound protons. Theoretically, this is the case for a delay $\tau = T_1(^1\text{H}) \ln 2$. However, usually the proton T_1 times of a mole-

Figure 8. Pulse sequence of a BIRD$_y$ pulse (Garbow et al., 1982), acting like a 180°$_y$ pulse on ^{12}C-bound protons, but having no effect on ^{13}C-bound protons (see text). The delay Δ is set to $(2 \ ^1J_{H,C})^{-1}$.

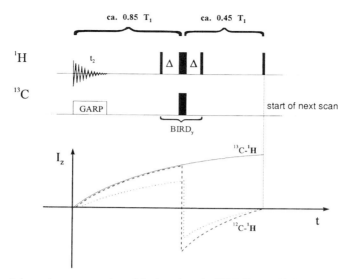

Figure 9. Schematic representation of the function of a BIRD filter, with $\Delta = (2\,{}^1J_{H,C})^{-1}$. While ^{13}C-bound protons are not affected by the BIRD$_y$ sequence (solid line), the already partially relaxed ^{12}C-bound protons are inverted by the BIRD pulse and go through zero at the end of the following delay (dashed line). ^{12}C-bound protons relaxing faster than the ones used for the calibration of the delays in the BIRD filter are also almost completely suppressed (dotted line).

cule cover a certain range, so that for any chosen delay τ only a few ^{12}C-bound protons will be completely suppressed. Fortunately, this problem can be easily solved by adjusting the whole recycling delay T_{recyc} (i.e., acquisition time *plus* relaxation delay *plus* τ) to approximately $1.3\,T_1$ of the fastest relaxing proton and setting $\tau = 0.35\,T_{recyc}$. After a few dummy scans, the *slowly* relaxing protons will reach a steady state where they *do not* come back to the full Boltzmann equilibrium z magnetization before the BIRD pulse (Figure 9). Therefore, their magnetization will not be completely inversed after the BIRD pulse, and it will take less time to go through zero than expected from their slow T_1 relaxation. Thus, even the slower relaxing protons will be effectively nulled, although the delay τ has been calibrated on the fastest relaxing protons.

As an additional benefit, the total recycling delay is now reduced to approximately $1.3\,T_1$, instead of ~3 to $5\,T_1$ as is usual. This means that in addition to suppressing the t_1 noise, the experiment runs almost three times faster than the conventional one without the BIRD filter, making BIRD-filtered techniques attractive even for rather dilute samples with natural ^{13}C abundance. Of course, a BIRD filter may not only be applied to HMQC or HSQC experiments, but also to techniques that involve an additional transfer step, e.g., HMQC-TOCSY and HMQC-NOESY. However, we should mention that with larger molecules dipolar relaxation between the ^{12}CH protons and the ^{13}CH protons leads to a significant deterioration of the performance of the BIRD filter.

2.4.1.3 Sensitivity-Enhanced HSQC

There is a nice improvement over the standard HSQC described earlier that can significantly increase the sensitivity of the experiment (Palmer et al., 1991). It makes use of the fact that HSQC, like most other multidimensional NMR experiments, acquires a signal that is amplitude-modulated in F_1, as can be seen from the transfer function given in Section 2.4.1.1. Although chemical shift evolution in t_1 leads to a phase modulation of the ^{13}C coherence in the x,y plane [$+2I_zS_y \cos(\omega_S t_1) - 2I_zS_x \sin(\omega_S t_1)$], only one of these components is used for the reverse transfer from ^{13}C to ^{1}H and contributes to the spectrum [e.g., with a $90°_x(^{13}C)$ pulse—and a $90°_y(^1H)$ pulse—only the *cosine* term can be converted back into proton SQC, $2I_yS_z \cos(\omega_S t_1)$]. The other half of the magnetization (i.e., the *sine* term) leads to indetectable MQC [$2I_yS_y \sin(\omega_S t_1)$] in a normal HSQC experiment.

However, a modification of the INEPT transfer step back to ^{1}H, resulting in the experiment called *Sensitivity-Enhanced HSQC* (Palmer et al., 1991), allows the conversion of the *sine* term into detectable proton magnetization as well (Figure 10a). During the following delay Δ the MQC term stays the same, since chemical shift is refocused and spins combined in an MQC do not show coupling to each other. The next two 90° pulses (on I and S) are phase-shifted by 90° with respect to the previous ones, therefore converting the MQC term into ^{1}H SQC [$2I_xS_z \sin(\omega_S t_1)$]. At the same time the already existing proton SQC term $I_x \cos(\omega_S t_1)$ is rotated into the z direction:

$$I_y \cos(\omega_S t_1) - 2I_xS_x \sin(\omega_S t_1) \xrightarrow{90°_x(I), 90°_y(S)} I_z \cos(\omega_S t_1) \quad (2.18)$$
$$+ 2I_xS_z \sin(\omega_S t_1)$$

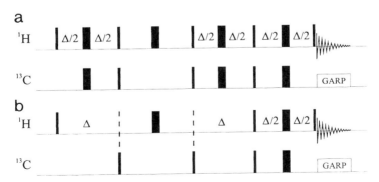

Figure 10. Pulse sequences for (a) the sensitivity-enhanced HSQC and (b) the sensitivity-enhanced HMQC experiment (for details, see Palmer et al., 1991); $\Delta = (2\ ^1J_{H,C})^{-1}$.

Now another delay Δ follows, leading to the refocusing of $2I_xS_z$ to detectable proton SQC, and the last $90°_y(^1H)$ pulse converts the *cosine* term that has been "waiting" in z back into the x,y plane:

$$I_z\cos(\omega_S t_1) + 2I_xS_z\sin(\omega_S t_1) \xrightarrow{\frac{\Delta}{2} - 180°_y(I,S) - \frac{\Delta}{2}}$$

$$- I_z\cos(\omega_S t_1) + I_y\sin(\omega_S t_1) \quad (2.19)$$

$$\xrightarrow{90°_y(I)} - I_x\cos(\omega_S t_1) + I_y\sin(\omega_S t_1)$$

At this point in time, *both* components have been converted back into detectable proton magnetization. However, there is a 90° phase shift between the two in *both* dimensions. This has to be compensated for by proper manipulation of the real and imaginary parts of the FID, containing the *sine* and *cosine* components, respectively, before the data set can be subjected to a 2D FT. The details of this processing scheme are not discussed here, and the interested reader is referred to the original publication (Palmer *et al.*, 1991), which also contains a similar enhancement scheme for the HMQC experiment (Figure 10b). As a result, we can acquire twice the signal intensity (in the ideal case) compared to the standard HSQC, but also pick up more noise by the square root of 2. This corresponds to a gain in SNR of 1.41, which would otherwise afford twice the measuring time.

Now, where are the drawbacks of this *Sensitivity-Enhanced* HSQC? First, the real gain in sensitivity might be lower than the theoretical prediction, e.g., due to relaxation. Then, the pulse sequence is quite lengthy compared to the HSQC, and the increased number of pulses can easily lead to a deterioration of the spectral quality (e.g., due to slightly "off" flip angles). Also the phase behavior of the resulting cross-peaks is usually not as clean as in a standard HSQC. However, the most important disadvantage of this technique is the quite cumbersome transformation of the raw data into the final 2D spectrum, due to the necessary rearrangement of real and imaginary data points. Therefore the best application of this sensitivity-increased HSQC might be the measurement of very dilute samples, where the saving of several hours of spectrometer time is more important than the inconveniences during data processing. Typically, this will be the case with natural abundance samples with a concentration of less than 10 mM, which will need an overnight run even on a high-field spectrometer to yield an adequate SNR.

2.4.1.4 Heteronuclear Multibond Correlation

The HMBC experiment (Bax and Summers, 1986) is the inverse analogue to the still widely used COLOC technique (Kessler *et al.*, 1984). Due to its much higher sensitivity we restrict ourselves to the HMBC.

As the name already implies, the HMBC correlates heteronuclei (e.g., ^{13}C

nuclei) with protons not only through one-bond, but also through long-range couplings (mostly ^2J and ^3J). In contrast to the HSQC/HMQC techniques, which are only applicable for studying single CH_n groups, the HMBC allows us to follow a CH_n-CH_n network, even including quaternary carbons (Figure 11). The pulse sequence (Figure 12a) is essentially an HMQC with a prolonged delay Δ (~40 to 100 ms) in the H,C transfer, so that small $^nJ_{H,C}$ couplings of the order of 1 to 10 Hz can also contribute to antiphase evolution and hence to coherence transfer onto ^{13}C. However, there is a consequence to this apparently minor change: Now the delay Δ is no longer short with respect to homonuclear H,H couplings (as it was in the HMQC). Therefore the proton resonances will undergo a generally nonuniform, unpredictable phase modulation, depending on their individual $^1H-^1H$ couplings. For just one additional proton spin I' we get

$$I_z \xrightarrow{90°_x(I)} -I_y \xrightarrow{\Delta} -I_y \cos(\Delta J_{IS}) \cos(\Delta J_{II'}) + 2I_x S_z \sin(\Delta J_{IS}) \cos(\Delta J_{II'})$$
$$+ 2I_x I'_z \cos(\Delta J_{IS}) \sin(\Delta J_{II'}) + 2I_y I'_z S_z \sin(\Delta J_{IS}) \sin(\Delta J_{II'})$$
(2.20)

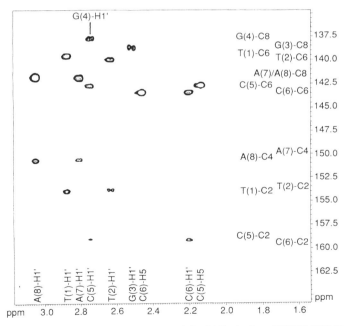

Figure 11. Region from the HMBC spectrum of the DNA duplex $d(TTGGCCAA)_2$ showing the correlations between the anomeric sugar protons H1' and the base protons H5 to the base carbons C2, C4, C6, and C8. (Reproduced with permission from Schmieder et al., 1992. Copyright 1992 Oxford University Press, Oxford.)

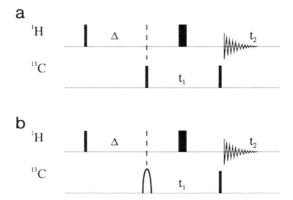

Figure 12. (a) Pulse sequence of the HMBC experiment (Bax and Summers, 1986). During the delay Δ, small heteronuclear long-range couplings evolve; typical settings are Δ = 60 to 80 ms. (b) HMBCS experiment yielding long-range correlations only to a small region of ^{13}C resonances excited by the selective first pulse on carbon (Kessler et al., 1990a). A 270° Gaussian pulse can be used advantageously for this purpose, yielding absorption mode spectra in F_1 due to neglectable phase distortions. The reduction of the spectral width in F_1 results in a high digital resolution in the carbon dimension.

prior to the first 90° (^{13}C) pulse. For this reason, the resulting HMBC spectrum does not show a uniform phase behavior in the proton dimension. For practical purposes, this means that a HMBC spectrum is usually shown in magnitude mode in F_2. But only in F_2!—there is *no* reason to sacrifice the nice absorptive lineshape (and therefore superior resolution) in the carbon dimension. Since the transformation is usually performed first along F_2, the whole 2D spectrum has to be 2D-transformed in the phase-sensitive mode first, neglecting the F_2 phases, and then converted to magnitude representation in F_2 only (an option that should be included in any good NMR processing program).

This "phase problem" of the HMBC is also responsible for the other small difference to the HMQC pulse sequence: There is no second delay Δ for refocusing of the proton antiphase coherence after the back-transfer from carbon. Instead, acquisition starts immediately after the second 90° (^{13}C) pulse, and the conversion of undetectable H,C-antiphase terms ($I_x S_z$) into detectable in-phase proton magnetization occurs during the acquisition period (no heteronuclear decoupling is performed!). This means that the signals will have antiphase structure in F_2—which does not really matter, since a magnitude calculation in F_2 is required anyway.

The HMBCS experiment (Figure 12b) differs from the standard HMBC sequence (Figure 12a) only in the substitution of the first 90° pulse on carbon by a shaped selective pulse (Kessler et al., 1990a). This feature allows us to restrict the spectral width in the carbon dimension to a small region, resulting in a much higher digital resolution. The foremost application of this technique is in the area of small and medium-sized peptides, where long-range correlations between an α–carbonyl carbon and the H$^\alpha$ of the *same* residue as well as the amide proton of the *sequentially following* amino acid can be used to establish the amino acid sequence of the molecule (Figure 13). In this case only the carbonyl carbons are of interest (i.e., a region of a few parts per million), but their signals have to be well resolved for an unambiguous assignment. Here the HMBCS technique is the experiment of choice, the selective pulse being advantageously performed as a 270° Gaussian pulse of a few milliseconds duration (Emsley and Bodenhausen, 1989). This yields high-resolution absorption mode spectra in the ^{13}C dimension, due to the neglectable phase distortions of the 270° Gaussian pulse (Kessler et al., 1991).

Another variant of the simple HMBC experiment includes a *low-pass J-filter* (Figure 14) (Kogler et al., 1983). In the HMBC spectrum not only do the desired long-range correlations occur, but the direct (^{1}J) connectivities as

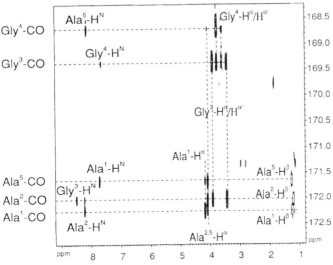

Figure 13. HMBCS of the carbonyl region of the cyclic peptide cyclo(–Ala^1–Ala^2–Gly^3–Gly^4–Ala^5–) (Kessler et al., 1990a). Each carbonyl carbon (in F$_1$) shows a long-range correlation to the H$^\alpha$ signals of its own and the sequentially following amino acid residue, as well as to its own H$^\beta$ (in the case of alanine) and to the sequentially following amide proton. This leads to the straightforward sequential assignment of all amino acid residues. (Courtesy of R. Haessner and M. Kurz.)

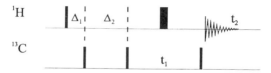

Figure 14. Pulse sequence of the HMBC experiment with low-pass J-filter (Kogler *et al.*, 1983). While the length of Δ_2 is 60 to 80 ms for the evolution of heteronuclear long-range couplings (as in the standard HMBC sequence of Figure 12), the delay Δ_1 is set to $(2\,^1J_{H,C})^{-1}$, so that protons directly coupled to ^{13}C nuclei are completely antiphase at the end of Δ_1 and are converted into undetectable multiquantum coherence by the first 90° pulse on ^{13}C.

well. Due to the length of the delay Δ (40 to 100 ms), the direct couplings (120 to 160 Hz) cause quite fast oscillations between in-phase and antiphase coherence. If, at the end of the period Δ, all proton magnetization is in the antiphase state with respect to the directly bound carbon, the heteronuclear transfer will go completely in this direction, leading to a very intense HMQC peak, but no other correlations. To circumvent this difficulty, the delay Δ should not be close to an odd multiple of $(2\,^1J_{H,C})^{-1}$, a condition that is practically impossible to be fulfilled due to the wide range of direct couplings.

A way out of this problem is offered by the *low-pass J-filter* (Kogler *et al.*, 1983): After a time $(2\,^1J_{H,C})^{-1}$, a 90° (^{13}C) pulse is applied, converting the evolved I_xS_z terms into MQC and suppressing them by appropriate phase cycling. This simple trick leads at least to a large reduction in intensity, if not to a complete removal of the direct HMQC peaks (due to nonuniform ^1J values).

Since long-range correlations always have an inherent low intensity compared to direct correlations, the sensitivity gain in switching from the forward COLOC (Kessler *et al.*, 1984) to the inverse HMBC experiment (Bax and Summers, 1986) must be greatly appreciated. Unfortunately, there is a drawback unique to the HMBC experiment that should be mentioned here: It is the experiment most prone to artifacts, specifically t_1 noise. The reason for this can be easily understood: Although the t_1 ridges in HSQC and HMQC spectra can be effectively reduced with a BIRD filter (compare Section 2.4.1.2), this trick does not work with long-range correlations such as HMBC. The BIRD filter only allows a distinction between ^{12}C and ^{13}C for *directly* bound carbons. Tuning its delays to long-range couplings is impractical due to the required length of these delays (^{1}H–^{1}H couplings and direct ^{1}H–^{13}C couplings will also evolve during that time) and the nonuniformity of long-range coupling constants, which cover more than one order of magnitude. Therefore, artifact suppression in an HMBC experiment completely relies on phase cycling, rendering it the most problematic one due to its sensitivity to spectrometer performance and stability. However, the use of pulsed field gradients for coherence selection (see Section 2.4.1.5) in modern spectrometers greatly helps to overcome some of the mentioned problems (e.g., solvent suppression and t_1 noise).

2.4.1.5 Gradient Selection

The HMBC experiment is a good opportunity to discuss the application of the novel *gradient-selected* techniques to heteronuclear NMR spectroscopy. The application of gradients to high-resolution NMR spectroscopy is not completely new (Bax *et al.*, 1980b; Maudsley *et al.*, 1978). However, only recent improvements in instrumental design (e.g., high-resolution probes with actively shielded gradient coils) allowed the use of the long ago proposed gradient coherence selection as a routine procedure.

Until recently only so-called *homospoil pulses* had been widely employed to destroy unwanted coherences, the most popular application of such a homospoil pulse being its insertion into the mixing period of a NOESY experiment (Boelens *et al.*, 1985; Sklenář and Bax, 1987). During this time interval, the desired state of the magnetization is in the form of a population along the z axis, because these I_z terms give rise to the NOE effect. Coherence selection has been accomplished so far with phase cycling only (Bain, 1984; Bodenhausen *et al.*, 1984; Piveteau *et al.*, 1985). However, during the mixing time of the NOESY experiment there are some quite unwanted terms of zero-quantum coherence (ZQC), which cannot be distinguished from z magnetization by phase cycling. Left alone, they lead to irritating antiphase cross-peaks between coupled spins in a NOESY spectrum. The best way to get rid of these coherences is through the use of the said homospoil pulse, i.e., a strong dc pulse several milliseconds in duration applied to the z shim coils. The result is a momentary destruction of magnetic field homogeneity (that has hopefully been achieved during the shimming process), which can be directly observed on the lock signal. Inhomogeneity means that the magnetic field strength—and hence the exact precession frequency of the spins—varies "largely" (i.e., more than ~1 in 10^9) within the sample volume. Any coherence, to some extent even the rather insensitive ZQC, is therefore quickly dephased within the x,y plane and thus effectively destroyed, while the z magnetization necessary for the NOE effect is left untouched (it does not have a *phase*!).

The gradients used in modern *gradient-selected* experiments are essentially homospoil pulses with a much improved performance. They are no longer applied to the z shim coils but to separate *actively shielded* gradient coils within the probe, thus allowing for a much higher gradient strength and faster recovery time of magnetic field homogeneity by suppression of eddy currents in the probe. Due to special gradient amplifiers under software control, these new gradients are highly reproducible and can be exactly calibrated in their individual strength. As a result, such a gradient cannot only be used to just *blow away* unwanted terms, but its dephasing action can also be compensated by another *refocusing* gradient under certain circumstances.

To understand this, we consider the factors that determine the *amount* of dephasing of a given coherence by a gradient pulse. The angle $\Delta\phi_G$ by which a coherence is rotated during the gradient pulse is given by the product of the gradient length τ_G and the angular velocity $\Delta\omega_G$ caused by the gradient field (*relative* to the normal precession with ω_0 under the static magnetic field B_0):

$$\Delta\phi_G = \tau_G \Delta\omega_G. \quad (2.21)$$

The angular velocity $\Delta\omega_G$ depends on the gradient strength G and a shape factor K_S (since the gradient pulse need not be *rectangular*, i.e., with constant amplitude during the whole duration). The angular velocity $\Delta\omega_G$ also depends on the *sensitivity* γ_C of the coherence to a change in magnetic field strength, which is proportional to the sum of the products of magnetogyric constants γ of the spins involved in the coherence and their *coherence order p*:

$$\gamma_C = \sum_i p_i \gamma_i. \quad (2.22)$$

For a detailed description of the concept of coherence order, the reader is referred to the literature (Ernst *et al.*, 1987). Essentially, this corresponds to the precession frequency ω_C of the coherence under consideration, which is the sum of the individual precession frequencies of all spins *i* contributing to the coherence, since the magnetogyric constant describes *per definition* the dependence of the precession frequency on the magnetic field strength

$$\omega_C = B_0 \gamma_C = B_0 \sum_i p_i \gamma_i. \quad (2.23)$$

Now the rotation angle caused by the gradient field is given by the product of gradient length (τ_G), gradient field strength (GK_S), and γ_C:

$$\Delta\phi_G = \tau_G G K_S \gamma_C. \quad (2.24)$$

If we use *two* gradient pulses, then the second one can be used to refocus the dephasing caused by the first gradient. The degree of dephasing during a given gradient pulse depends on the coherence order γ_C of the term under consideration; therefore, the second gradient can be calibrated to refocus only selected coherences.

For the effects of two gradients, 1 and 2, to cancel, the net effect has to be zero dephasing for any location in the sample:

$$\Delta\phi_{G1} + \Delta\phi_{G2} = 0 \quad (2.25)$$

$$\tau_{G1} G_1 K_{S1} \gamma_{C1} + \tau_{G2} G_2 K_{S2} \gamma_{C2} = 0. \quad (2.26)$$

If one chooses equal gradient length and shape for both, then the equation is reduced to:

$$G_1/G_2 = -\gamma_{C2}/\gamma_{C1}. \qquad (2.27)$$

Since we are essentially interested in *relative* gradient strength, we need only insert the *relative* values for γ, i.e., ± 4 for every proton, ± 1 for every carbon involved in the coherence (z operators are to be neglected). Thus, proton SQC (like I_x or $2I_yS_z$) counts as ± 4, proton DQC as ± 8, ^{13}C SQC as ± 1, heteronuclear MQC terms such as $2I_xS_x$ as either ± 3 ($+4 - 1$ or $-4 + 1$) or ± 5 ($+4 + 1$ or $-4 - 1$).

Let us consider the application of gradient selection to the HSQC experiment (Figure 15). Since we want to select ^{13}C coherence (based on the 1:4 ratio of the magnetogyric constants of ^{13}C and ^{1}H), we simply put two

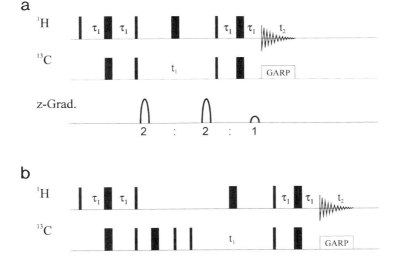

Figure 15. Pulse sequences for gradient-selected HSQC experiments. (a) A pair of gradients with a ratio of 4:1 selects for ^{13}C coherence during t_1 and ^{1}H coherence during t_2; the resulting 2D spectrum has to be processed in magnitude mode. *In praxi*, the first gradient is split into two equal parts in order to cancel artifacts from an imperfect 180° proton pulse, leading to a gradient ratio of 2:2:1 (Hurd and John, 1991). (b) Phase-sensitive gradient-selected HSQC (Willker et al., 1993). There is no gradient connected to the t_1 period; instead, after the first gradient ^{13}C coherence is converted back to z magnetization (the second gradient serves as a homospoil pulse, defocusing any residual coherences). Then ^{13}C coherence is created again by another 90° (^{13}C) pulse before the t_1 period, and the final gradient is applied on ^{1}H SQC after the reverse INEPT step.

gradients at two places in the sequence where we want to select ^{13}C SQC (during t_1) and ^{1}H SQC (directly before t_2). However, the gradients must usually have a duration of the order of a millisecond to be effective. During this time interval, unwanted chemical shift evolution will occur, leading to phase distortions (however, this can be refocused by a 180° pulse on the active spins after each gradient, followed by a delay of the same duration as the gradient pulse—or the other way around; compare Figure 15b). If we now choose a gradient ratio of 4:1 (in the order of occurrence in the pulse sequence), then only magnetization that is a ^{13}C coherence during t_1 and a ^{1}H coherence during t_2 will pass through the pair of gradients without any net defocusing. Any proton magnetization that has not been converted to ^{13}C SQC during t_1, e.g., protons bound to ^{12}C, will not be properly refocused by the second gradient and therefore be undetectable during t_2 (Hurd and John, 1991).

The advantage over phase cycling is obvious: With phase cycles, all the unwanted terms have to be detected in the first scan, but will be subtracted in the following ones, leaving only the desired components in the accumulating FID—*if* spectrometer stability is perfect and the dynamic range of the receiver sufficient. With gradients, we do not have to worry about proper subtraction of all artifacts, because they never reach the receiver in the first place. There is no more dynamic range problem with samples dissolved in H_2O, since the water signal is almost completely destroyed by the gradients (obviously there can't have been ^{13}C SQC during t_1). Additionally, we can theoretically run the gradient-selected experiment with just one scan per increment (sensitivity permitting), since we don't have to go through the minimum phase cycling for ^{13}C selection. Nevertheless, *some* phase cycling will usually still be maintained even with gradient selection, since there are other sources of artifacts that cannot be removed by these two gradients, e.g., signals arising from imperfect flip angles of the rf pulses, incomplete relaxation between scans, etc.

Unfortunately a few caveats must be mentioned when dealing with gradient selection in phase-sensitive experiments. The desired absorptive lineshape in multidimensional spectra can only be achieved by a superposition of N-type and P-type peaks, i.e., signals that come from terms with the *same absolute value*, but *opposite sign* of coherence order during the evolution times for the indirect dimensions. The reader is referred to the book by Ernst *et al.* (1987) for more details about the concepts of coherence order and of N- and P-type signals. In a greatly simplified way, the coherence pathway selection for phase-sensitive experiments *has* to be ambivalent to allow both pathways (with opposite signs during t_1) to contribute equally to the acquired spectrum. This means that the phase cycling schemes for these experiments are "incomplete" in this respect, making no distinction between the two required pathways. However, this *is* a problem when using gradient selection: A pair of gradients strictly selects for a *single* ratio of coherence orders (γ_1/γ_2), including signs of γ_1 and γ_2! Therefore, a gradient during t_1 will generally select only *one* of the two coherences necessary for a phase-sensitive experiment; the other coherence has to be recorded in a separate experiment (same pulse sequence, but different gradient ratio), which is then added to the first one. In addition to this rather awkward way of acquiring a simple 2D spectrum, it also means a 50% reduction of SNR when compared to the phase-cycling version (where both components can be acquired in a single run).

However, in many experiments gradients during t_1 can be avoided by slightly modifying the pulse sequence. As an example, we discuss the gradient-selected HSQC sequence shown in Figure 15b (Willker *et al.*, 1993). Here the magnetization is—as usual—initially transferred from ^{1}H to ^{13}C via an INEPT step, and then the first gradient is applied on the ^{13}C single quantum coherence, removing one of the two possible coherence paths. Before the beginning of the t_1 evolution period, a z filter is employed to convert the remaining ^{13}C coherence into a z population and then back to ^{13}C coherence, thus creating anew both coherence types required for phase-sensitive spectra. The rest is identical to the standard HSQC sequence, with the insertion of another gradient pulse into the REVINEPT refocusing period on ^{1}H, directly before the start of the acquisition. To avoid the introduction of artifacts with the z filter, an additional gradient is applied within this period to remove any remaining coherences, thus serving as a homospoil pulse. The gradient ratios for the first and last gradients (on ^{13}C and ^{1}H coherence, respectively) are 4:1, whereas the second *homospoil* gradient can have any amplitude "incommensurable" with the other two gradients to avoid accidental refocusing of unwanted coherences. For example, a gradient ratio of 8:3:2 performs well in this experiment. This sequence still yields a 50% lower SNR, because the first gradient (on ^{13}C) destroys exactly one-half of the ^{13}C coherence created by the INEPT transfer. After the z filter, only the remaining half of the magnetization is then equally distributed among the two coherence types needed in t_1. But at least this pulse sequence yields a phase-sensitive HSQC spectrum after normal 2D Fourier transformation, and it shows very good suppression of H_2O and ^{12}C-bound protons, thanks to the gradients (Figure 16).

We should mention that the positive features of gradient selection may well overcompensate the disadvantage of a 50% loss of sensitivity in most cases. The loss in SNR is often outweighed by a significant gain in the *signal-to-artifact ratio*, since gradient-selected experiments don't have to rely on instrument stability over a lengthy phase cycle for proper subtraction of artifacts. In addition, the suppression of very intense unwanted signals (^{12}C-

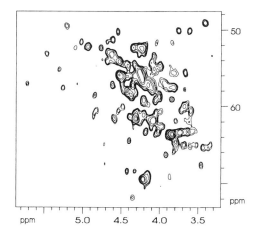

Figure 16. Phase-sensitive ^{13}C-HSQC with gradient selection (for pulse sequence, see Figure 15b) of the uniformly ^{15}N and ^{13}C labeled protein IIB^{Glc} (Golic Grdadolnik et al., 1994); 101 amino acid residues, 1.2 mM solution in H$_2$O/D$_2$O = 9:1, with a measuring time of ~1 h. The H$^\alpha$–C$^\alpha$ region is shown to demonstrate the complete suppression of the water resonance (at 4.7 ppm) by the gradient-selection scheme.

bound protons in natural abundance samples, solvent signals) even in a *single scan* usually allows a large increase in receiver gain and is thus a better use of the available dynamic range of the receiver. This is especially important in the case of very dilute samples in nondeuterated solvents (e.g., natural products in aqueous solution, biological fluids), where the molar concentration of the solvent may be more than 10^5 times higher than the molecule under investigation.

2.4.1.6 Inverse Correlation Experiments with Additional Transfer Steps

Inverse heteronuclear ^{1}J correlation techniques (i.e., HMQC or HSQC) can be readily combined with ^{1}H homonuclear transfer steps, such as in COSY, TOCSY, NOESY, or ROESY (see Figure 17). As a result, the 2D spectrum will contain essentially the same connectivity information as the homonuclear version (e.g., ^{1}H,^{1}H-TOCSY). But all signals will be labeled with their ^{13}C chemical shift in F$_1$ instead of the ^{1}H shift; this can be quite advantageous when there is severe overlap in the ^{1}H spectrum, but not in ^{13}C (see Figures 18 and 19). For example, in carbohydrates a ^{13}C-edited TOCSY [2D HMQC-TOCSY (Lerner and Bax, 1986)] is usually much easier to interpret than a conventional ^{1}H,^{1}H-TOCSY, where most of the cross-peaks occur in a very crowded area. A simple HMQC spectrum should also be run,

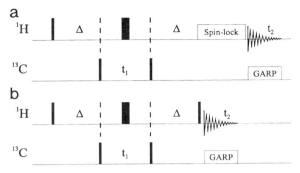

Figure 17. (a) Pulse sequence for the 2D HMQC-TOCSY experiment with ^{13}C chemical shift in F_1. The TOCSY mixing process appears during the spin-lock period. (b) 2D HMQC-COSY pulse sequence. In both sequences Δ has to be set to $(2\ ^1J_{H,C})^{-1}$ (Lerner and Bax, 1986).

so that the direct ^{1}H,^{13}C-correlation signals (corresponding to the diagonal signals in a ^{1}H,^{1}H-TOCSY) can be readily identified in the HMQC-TOCSY spectrum.

The disadvantages of these techniques in comparison with their homonuclear version are, of course, the lower sensitivity (theoretically 1.1%, although some additional sensitivity can be gained by shorter recycle delays together with BIRD filters; see Section 2.4.1.2) and the collapse of signals

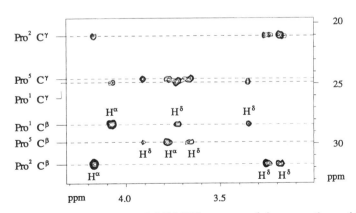

Figure 18. Region from the 2D HMQC-TOCSY spectrum of the cytostatic peptide hymenistatin, cyclo(–Pro^1–Pro^2–Tyr^3–Val^4–Pro^5–Leu^6–Ile^7–Ile^8–) (Konat *et al.*, 1993). The three proline spin systems can be clearly distinguished due to the good separation of the ^{13}C chemical shifts. (Courtesy of R. Konat.)

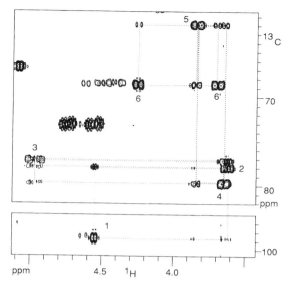

Figure 19. Region from the 2D HMQC-COSY of a glucose derivative. The HSQC signals of the glucose moiety are labeled with their position number, and the COSY correlations between them are indicated by the dashed lines. In spite of severe overlap in the proton spectrum, the superior ^{13}C dispersion leads to the unambiguous assignment of all signals. (Courtesy of S. Zimmer.)

from diastereotopic methylene protons in F_1, appearing inevitably at the same ^{13}C frequency in the carbon-edited spectra.

Naturally, the sensitivity problems are much more severe in the case of ROESY or NOESY transfers, where the cross-peak intensity is usually already low in homonuclear spectra. On the other hand, the gain in resolution in the ^{13}C dimension achieved with HMQC-COSY and HMQC-TOCSY makes these techniques interesting alternatives to homonuclear 2D methods.

2.4.1.7 *Heteronuclear 3D NMR in Natural Abundance*

The experiments including two transfer steps (one homo- and one heteronuclear) can be easily expanded into 3D techniques by inserting a ^{1}H evolution period before the homonuclear transfer (either directly before or after the ^{13}C evolution time). The resulting 3D spectrum will consist of a 2D ^{1}H,^{1}H spectrum (COSY, TOCSY, etc.) edited according to the carbon chemical shift in the third dimension (Kessler *et al.*, 1990b; Schmieder *et al.*, 1990). The evaluation of such a ^{13}C-edited 3D NMR experiment is relatively straightforward and explained in detail in the literature (e.g., Clore and Gronenborn, 1991; Wagner, 1990). Although these techniques are already used for the study of highly complex molecules, such as biopolymers (proteins, nucleic acids, etc.), their widespread application has been hindered by the requirements of data storage, processing, and visualisation, which are admittedly much higher for a 3D than for a 2D spectrum.

The sensitivity of these 3D experiments is inherently comparable to the 2D homonuclear versions with a ^{13}C filter, because the coherence transfer steps involved are the same (excitation of ^{1}H, transfer to ^{13}C and back to ^{1}H, homonuclear proton transfer, detection of ^{1}H). The minimum amount of measuring time for a 3D spectrum is determined by the relatively large number of increments (since there are *two* indirect dimensions now) and the required phase cycle for each of them (which can be reduced when gradient selection is employed).

While analog frequency filters can be used to reduce the bandwidth in the direct dimension, this possibility is excluded for the indirect dimensions. However, reduction of the spectral range to increase the resolution can be accomplished in three ways:

1. *Use of frequency selective pulses*: Because the carbon chemical shifts in proteins cluster in rather well-defined, well-separated regions, selective excitation can be advantageous. Typical frequency-selective pulses used in high-resolution ^{13}C NMR are Gaussian-shaped and related pulses (for a review, see Kessler *et al.*, 1991), the Gaussian cascades G3 and G4 (Emsley and Bodenhausen, 1990), and the BURP (Band-Selective, Uniform Response, Pure-Phase) pulse family (Geen and Freeman, 1991). In addition, band-selective decoupling sequences containing selective pulses have been designed that can be used for selective decoupling of, e.g., the ^{13}C-carbonyl region (McCoy and Mueller, 1992, 1993; Zuiderweg and Fesik, 1991).

2. *Multiplicity selection techniques*: These techniques allow us to select only carbon atoms with a certain number of directly bound protons (CH/CH$_2$/CH$_3$) and thus to reduce the number of signals and in many cases also the required spectral width (Kessler and Schmieder, 1991).

3. *Folding in the indirect dimensions*: This refers to the use of a narrower spectral width in the indirect dimension than required by the signal dispersion. Signals that were originally outside the chosen spectral window are not clipped, but "folded" into the available spectral window (Schmieder *et al.*, 1991).

Multiplicity selection has been successfully employed in several cases, e.g., for the selection of only the methyl resonances in cyclosporin derivatives with a Heteronuclear Quadruple Quantum Correlation (HQQC) experiment (Kessler *et al.*, 1991). Since at least four coupled spins are required to create quadruple-quantum coherence, only CH$_3$ signals can pass such a filter, leading to a reduction of the spectral width to the narrow methyl region.

Folding is certainly the most generally applicable method; appropriate folding of the signals, e.g., in the ^{13}C dimension, allows a considerable reduction in sweep width in this dimension and hence a lower number of increments for achieving sufficient resolution. The resolution required in the indirect dimensions is significantly lower for a 3D than for a 2D spectrum, because the signals are now spread in three dimensions, reducing the chance of accidental overlap.

Thus a typical 3D spectrum, for example, a HMQC-TOCSY, can be run with 32(^{13}C)*64(^{1}H) complex increments, which amounts to a spectrometer time of approximately 13 h (8 scans, 0.7-s recycle delay with BIRD filter), a measurement time comparable to that of a well-resolved 2D spectrum. With the availability of cheap data storage and processing equipment as well as commercial 3D processing software, the use of heteronuclear NMR spectroscopy with nonenriched samples will soon become more widespread as a last resort in demanding routine cases. In addition, the use of gradient-selection techniques can improve the quality of natural abundance 3D spectra by reducing the minimal phase cycle and suppressing artifacts from ^{12}C-bound protons (compare Section 2.4.1.5).

As an example of a natural abundance heteronuclear 3D spectrum, Figure 20 shows a cross section through a 3D ^{13}C-HQQC-TOCSY of thiocyclosporin A (Seebach et al., 1991).

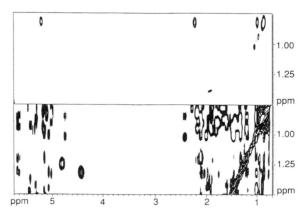

Figure 20. (Top) ^{1}H,^{1}H plane at δ(^{13}C) = 20.35 ppm from a 3D HQQC-TOCSY spectrum of thiocyclosporin A (Seebach et al., 1991). A comparison with the conventional 2D TOCSY spectrum (bottom) demonstrates the complete removal of severe signal overlap in the methyl region by editing according to the ^{13}C frequency in the third dimension. (Courtesy of P. Schmieder.)

2.4.2 ^{13}C-Enriched Samples

In recent years the availability of isotope-enriched samples has been advanced by molecular biology and biotechnology. Today most proteins under NMR investigation have been synthesized by genetically modified strains of bacteria or mammalian cells (Hansen *et al.*, 1992), and isotope labeling can be readily carried out (although it is still costly) by supplementing isotope-enriched nutrients in the culture broth (for ^{13}C enrichment, mostly labeled glucose, acetate, or amino acid mixtures from algal hydrolysate are employed). Methods for the convenient enrichment of other biologically important substance classes, e.g., nucleic acids (Nikonowicz and Pardi, 1992), are currently under development.

The use of isotopic enrichment has several advantages over natural abundance ^{13}C NMR. The most obvious one is clearly the sensitivity factor. Compared to the natural ^{13}C abundance of 1.1%, a 10% enrichment leads to a ninefold gain in the SNR, corresponding to a more than eightyfold reduction in measuring time. However, for most biosynthetically produced proteins, an enrichment of more than 95% is standard, meaning a reduction in spectrometer time by more than a factor of 7000! These fantastic gains in sensitivity have to be set off against the costs for ^{13}C enrichment; however, the solubility of most proteins is rather limited, so that the use of a highly concentrated nonenriched sample instead is no valid alternative.

Another important advantage of highly enriched molecules is the possibility of using ^{13}C–^{13}C connectivities for coherence transfer (see also in Chapter 5, Sections 5.2 and 5.3). This requires that *two adjacent* carbon positions contain the ^{13}C isotope, and the probability for this is given by the *square* of the (uniform) isotopic abundance. The resulting gain in sensitivity for these experiments is so large that experiments such as ^{13}C,^{13}C-INADEQUATE (requiring a weekend of spectrometer time with a several hundred millimolar nonenriched sample) can be easily conducted on very dilute (often just 1 mM), but highly ^{13}C-enriched protein samples in a few hours. Other experiments requiring still more adjacent ^{13}C spins, e.g., ^{13}C-TOCSY, are completely impossible with nonenriched samples, but widely used with labeled compounds. The advantage of such techniques for the study of highly complex molecules is based on the better dispersion of the ^{13}C spectrum, the possibility of combining the ^{1}H and ^{13}C dispersions in separate spectral dimensions to further reduce overlap, and the efficient coherence transfer through the relatively large $^{1}J_{C,C}$ couplings (~30 to 60 Hz). Considering the fast relaxation in large proteins, the faster coherence transfer along a ^{13}C chain (when compared to protons) leads to a considerable gain in sensitivity. Additionally, quaternary carbons, especially the carbonyl carbons, can also be employed for coherence propagation, e.g., along the peptide chain.

2.4.2.1 Isotope Filters

Another advantage of ^{13}C labeling becomes obvious when one considers *nonuniform* enrichment. Different kinds of *selective* labels have been used in the study of large molecules, allowing even in very complex spectra the selective observation of signals from the labeled positions in so-called *isotope-filtered* NMR spectra (Otting et al., 1986; Otting and Wüthrich, 1989). For instance, the incorporation of certain ^{13}C-labeled amino acids (e.g., phenylalanine) in biotechnologically produced proteins leads to a quick identification of the corresponding signals in the often overcrowded spectra. Additionally, the study of supramolecular complexes (e.g., enzyme-inhibitor) can be greatly simplified by using selective labeling: If only one of the components of the complex has been enriched, then an isotope-filtered spectrum solely will show this specific component (Fesik et al., 1988). Thus the influence of the complex partner on the labeled component can be easily studied by just comparing its spectra in the free and bound states, without being confused by additional signals from the other component.

The principle of an isotope filter can be easily explained with Cartesian product operators (Figure 21): First we have to create a coherence that involves the selected isotope (e.g., ^{13}C). This is done by letting the direct heteronuclear coupling evolve into antiphase during a delay $\Delta = (2\ J_{H,C})^{-1}$. If we now apply a 0° [Eq. (28a)] and a 180° pulse [Eq. (28b)] on ^{13}C in subsequent scans and afterward refocus again, we end up with in-phase proton magnetization *with opposite signs* in the two scans:

$$I_z \xrightarrow{90^\circ_x(I)} -I_y \xrightarrow{\Delta} 2I_xS_z \xrightarrow{0^\circ_x(S)} 2I_xS_z \xrightarrow{\Delta} I_y \qquad (2.28a)$$

$$I_z \xrightarrow{90^\circ_x(I)} -I_y \xrightarrow{\Delta} 2I_xS_z \xrightarrow{180^\circ_x(S)} -2I_xS_z \xrightarrow{\Delta} -I_y \qquad (2.28b)$$

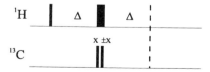

Figure 21. Pulse sequence of an isotope half-filter, $\Delta = (2\ ^1J_{H,C})^{-1}$ (Otting et al., 1986; Otting and Wüthrich, 1989). In odd-numbered scans the two 90° pulses on ^{13}C add up to a 180° ^{13}C pulse ($90^\circ_x + 90^\circ_x$), while they cancel each other in even-numbered scans ($90^\circ_x + 90^\circ_{-x}$). Thus all ^{13}C-bound protons undergo a sign flip; all other protons are not affected. Subtraction of two subsequent scans therefore selects for ^{13}C-bound protons; addition suppresses them selectively.

[The effect of ^{1}H chemical shift evolution during Δ is canceled by the 180°(^{1}H) pulse in the center of the sequence.]

While all this happens to protons bound to ^{13}C, the magnetization of ^{12}C-bound protons is not affected by the ^{13}C pulse:

$$I_z \xrightarrow{90°_x(I)} -I_y \xrightarrow{\Delta} -I_y \xrightarrow{0°_x(S)} -I_y \xrightarrow{\Delta} -I_y \quad (2.29a)$$

$$I_z \xrightarrow{90°_x(I)} -I_y \xrightarrow{\Delta} -I_y \xrightarrow{180°_x(S)} -I_y \xrightarrow{\Delta} -I_y. \quad (2.29b)$$

Therefore subtraction of the signals from these two scans leads to the cancelation of ^{12}C-bound protons, while the ^{13}C-labeled protons are kept. To avoid the need for two separate pulse sequences with and without the 180°(^{13}C) pulse, one uses a pair of 90°(^{13}C) pulses instead that add up to 0° or 180°, depending on their relative phase ($90°_x + 90°_{-x} = 0°_x$, $90°_x + 90°_x = 180°_x$).

Such a filter can be inserted, e.g., in a standard 2D pulse sequence to yield the corresponding isotope-filtered spectrum. As an example, let us consider an isotope-filtered 2D-NOESY sequence (Bax and Weiss, 1987): There are two possibilities for the placement of the filter, either *before* (Figure 22a) or *after* the NOESY mixing time (Figure 22b). In the first case, the filter selects for ^{13}C-bound protons during t_1, and magnetization can then be transferred via NOE to *any* proton (labeled or not) during t_2 (Figure 23a). The result will be an unsymmetric 2D-NOESY spectrum that contains NOE signals be-

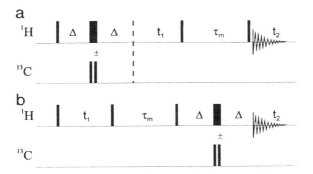

Figure 22. Two isotope-filtered 2D-NOESY sequences (Bax and Weiss, 1987). (a) X$_1$-filtered NOESY; the half-filter at the beginning of the t_1 evolution period selects for ^{13}C-bound protons in F$_1$, while in F$_2$—after the NOE mixing time τ_m—*all protons* can be observed that show an NOE to any ^{13}C-bound proton. (b) X$_2$-filtered NOESY; no selection occurs for the t_1 period, but the half-filter *after* the NOE mixing time τ_m allows only ^{13}C-bound protons to be observed in F$_2$.

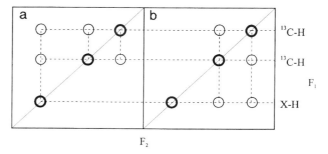

Figure 23. Schematic illustration of isotope-filtered NOESY spectra (compare pulse sequences in Figure 22): (a) X_1-filtered NOESY; signals can be observed between all proton signals in F_2 and exclusively the ^{13}C-bound protons in F_1; (b) X_2-filtered NOESY; here the ^{13}C selection occurs in the F_2 dimension.

tween ^{13}C-bound protons only (in F_1) and any proton resonance (in F_2). If one puts the isotope filter *after* the mixing time, then the result will be a selection for NOEs between *any* proton in F_1 and only the ^{13}C-bound protons in F_2 (Figure 23b). A combination of both filters, i.e., isotope filtering *before and after* the NOE transfer step will obviously result in signals only between two ^{13}C-bound protons. Since each one of these filters selects for ^{13}C-bound protons in just one dimension of a 2D spectrum, they are also called *X-half-filters*.

Often one is primarily interested in the structure of a complex component that *cannot* be easily labeled. A typical example would be a protein-inhibitor complex, where one wants to determine the *bioactive* conformation (i.e., the bound state) of a potent inhibitor as a lead structure for the design of improved drugs (Fesik, 1993). If this inhibitor is a synthetic drug that cannot be biotechnologically manufactured from readily available labeled precursors, then complete isotopic labeling will become prohibitively expensive. However, in most cases the protein involved can be labeled by using an appropriate expression system.

In this case there is the possibility that filtering techniques can be used that *select* for the *unlabeled* component and *suppress* the whole bulk of signals from the *labeled protein* (Gemmecker et al., 1992; Ikura and Bax, 1992). Since these synthetic drugs are usually relatively small molecules compared to the protein, the application of such filters renders it possible to determine their structure with a few simple 2D NMR experiments—even in large protein complexes with molecular weights of several kDa (Ikura and Bax, 1992; Petros *et al.*, 1992).

The principle of these filters (Figure 24) is easily understood: During the

2. Elucidation of Molecular Structure and Dynamics in the Liquid State / 45

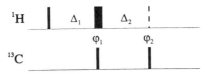

Figure 24. Pulse sequence of the double ^{12}C isotope filter (Gemmecker *et al.*, 1992; Ikura and Bax, 1992). The delays Δ_1 and Δ_2 have to be set to $(2\ ^1J_{H,C})^{-1}$ and should be tuned to two different $^1J_{H,C}$ values for optimum suppression. The 90°(^{13}C) pulses are phase cycled independently ($\phi_1 = x, -x; \phi_2 = x, x, -x, -x$) and four subsequent scans have to be added for complete suppression of all ^{13}C-bound protons.

---

delay $\Delta = (2\ J_{H,C})^{-1}$, the ^{13}C-bound protons evolve into heteronuclear antiphase coherence and are subsequently converted into undetectable multiquantum coherence; in addition, a two-step phase cycle on the 90°(^{13}C) pulse leads to a sign flip and hence to the cancelation of these terms after addition of two subsequent scans:

$$I_z \xrightarrow{90°_x(I)} -I_y \xrightarrow{\Delta} 2I_xS_z \xrightarrow{90°_{+x}(S)} -2I_xS_y \quad (2.30a)$$

$$I_z \xrightarrow{90°_x(I)} -I_y \xrightarrow{\Delta} 2I_xS_z \xrightarrow{90°_{-x}(S)} 2I_xS_y. \quad (2.30b)$$

This step is repeated once more in the second half of the filter sequence. The ^{12}C-bound protons are not affected by the whole sequence—they do not evolve into heteronuclear antiphase coherence, and chemical shift evolution is refocused by the 180°(^{1}H) pulse in the center.

There is a reason for using such a *double* filter (Gemmecker *et al.*, 1992): As mentioned earlier, the delay Δ has to be tuned to $(2\ ^1J_{H,C})^{-1}$ to quantitatively yield antiphase coherence, since only these terms are subtracted. Now the values for $^1J_{H,C}$ usually cover a certain range, e.g., ~125 Hz for methyl groups, ~140 Hz for aliphatic methin signals, and up to 160 Hz and more for aromatic groups. Therefore it is impossible to completely remove all ^{13}C-bound protons with a single filter—the remaining leakage of up to 10% renders the resulting spectrum quite useless. In the case of a double filter, the leakage from the first filter is subject to another filtering procedure, reducing the total leakage to less than 1%. In addition, the two defocusing delays Δ_1 and Δ_2 in a double filter can be tuned to different coupling constants, so that a good suppression over a broad range of $^1J_{H,C}$ coupling constants results. An application of such a double filter is the MEXICO (Measurement of EXchange rates in Isotopically labeled COmpounds) experiment (Gemmecker *et al.*, 1993) for the measurement of fast exchange rates between water and amide protons in labeled proteins.

2.4.2.2 2D ^{13}C-TOCSY

As an example of NMR experiments that rely on high ^{13}C enrichment, we would like to briefly discuss the 2D ^{13}C-TOCSY technique. The expansion of this 2D sequence into a 3D or even 4D experiment is discussed in Section 2.4.2.3.

^{1}H,^{1}H-TOCSY (TOtal Correlation SpectroscopY) is among the most popular 2D techniques, since it yields the connectivities between all protons belonging to the same spin system. During the TOCSY mixing time (a spin-lock usually performed by composite pulse sequences), in-phase magnetization is transferred along the J-coupling networks. Transfers can occur not only among neighboring protons, but also as multistep transfers, depending on the length of the mixing process (usually ~10 to 100 ms). With larger molecules, however, the sensitivity of the TOCSY experiment is drastically reduced, mainly because relaxation becomes more severe during the quite lengthy mixing periods. For proteins of 20 kDa and more, the proton TOCSY transfer is so inefficient that it might even become difficult to observe a one-step transfer.

However, there is an alternative in the case of ^{13}C-labeled molecules: We perform the TOCSY transfer not between ^{1}H spins, but among the ^{13}C nuclei. The principle of the experiment is quite easy (Figure 25a): After the ^{13}C-decoupled ^{1}H evolution period t_1, proton magnetization is transferred onto ^{13}C via an INEPT step and refocused to in-phase ^{13}C coherence. Now the spin-lock is performed at the ^{13}C frequency; to cover the whole range of ^{13}C chemical shifts sufficiently (e.g., 60 to 80 ppm for the aliphatic signals), spe-

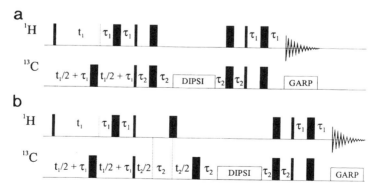

Figure 25. (a) Pulse sequence of the 2D ^{13}C-TOCSY experiment (τ_1 and τ_2 as in Figure 2b). During the ^{13}C spin-lock period, a TOCSY transfer occurs between coupled ^{13}C nuclei. (b) 3D version of the same experiment; an additional ^{13}C evolution time results in separating the cross-peaks in the third dimension according to their ^{13}C chemical shifts. The values for τ_1 and τ_2 are the same as in Figure 2b (Fesik et al., 1990).

cifically designed pulse sequences should be used for this purpose, for example, DIPSI (Shaka et al., 1988). After the ^{13}C-TOCSY mixing period, magnetization is transferred back onto ^{1}H via INEPT and then recorded under ^{13}C decoupling.

Why should this technique present an advantage over conventional ^{1}H-TOCSY experiments? First, the transfer is actually *faster*: Although each of the two refocused INEPT transfers requires $(^1J_{H,C})^{-1} \approx$ about 7 ms, the ^{13}C-TOCSY transfer is about four times faster than the ^{1}H-TOCSY because of the larger coupling constants $^1J_{C,C} \approx$ 30 to 60 Hz, whereas the proton–proton coupling used in the ^{1}H-TOCSY transfer is much smaller ($^3J_{H,H} \approx$ 1 to 10 Hz). In addition, ^{13}C relaxation is slower than for protons. Therefore a ^{13}C-TOCSY with a 20-ms mixing time will still give good results (Figure 26), whereas in an equivalent ^{1}H-TOCSY experiment (with an 80-ms mixing time) relaxation may reduce the sensitivity below a reasonable level.

2.4.2.3 3D and 4D Experiments

With ^{13}C enrichment, the use of carbon chemical shift labeling in a separate dimension becomes obvious. Because practically all multidimensional heteronuclear experiments start and end with ^{1}H magnetization for sensitivity reasons, the resulting transfer ^{1}H–^{13}C–(COSY or TOCSY)–^{13}C–^{1}H lends itself easily to the acquisition of 3D and even 4D spectra. As an example, we discuss 3D and 4D ^{13}C-TOCSY in more detail.

Furthermore, the combination of ^{13}C labeling with other isotopic enrichments (^{15}N in particular) leads to an even greater variety of possible magne-

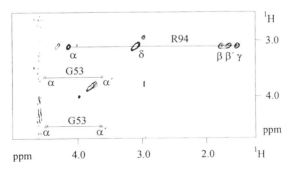

Figure 26. 2D slice at $\delta(^{13}C = 43.22$ ppm) from a 600-MHz 3D ^{13}C-TOCSY spectrum of the IIB^{Glc} domain of the glucose phosphotransferase system of *E. coli* (1.2-mM solution of the uniformly ^{13}C- and ^{15}N-enriched protein in H$_2$O:D$_2$O = 9:1, 101 amino acid residues). The chosen ^{13}C plane corresponds to the chemical shifts of G53–C$^\alpha$ and R94–C$^\delta$, therefore the spin systems of these two amino acids can be seen in this slice. Water suppression was achieved with pulsed field gradients (Golic Grdadolnik et al., 1994).

tization transfers in proteins. In this case the protein backbone is completely labeled with NMR-active nuclei, and coherence transfers can be conducted among them almost as desired. The names of these techniques symbolize the transfer path of the magnetization, with relay spins given in parentheses. Thus, HNCA refers to a 3D experiment connecting an amide proton, amide nitrogen, and C^α shifts in separate dimensions; HN(CA)CO yields connectivities between an amide proton, amide nitrogen, and carbonyl carbon in the three dimensions, and C^α is used as a relay spin for the step from ^{15}N to CO. The same experiment can also be conducted as a 4D version and is then referred to as HNCACO, with the C^α chemical shift in the additional dimension. Other techniques include sidechain resonances, such as the CBCA(CO)NH experiment connecting C^β and C^α chemical shifts to the amide ^{1}H and ^{15}N frequencies of the following residue, using CO for the relay spin. (Here the naming becomes slightly inconsequential, because the name could also suggest a 4D spectrum with C^β, C^α, ^{15}N, and ^{1}H in different dimensions.) As an example, we discuss the so-called 3D HNCA technique in more detail.

3D and 4D HCCH-TOCSY The 2D ^{13}C-TOCSY (see Section 2.4.2.2) can be readily expanded into a 3D HCCH-TOCSY. During the transfer process

$$^1\text{H}-^{13}\text{C}-(^{13}\text{C}-)_{\text{TOCSY}}{}^{13}\text{C}-^1\text{H} \qquad (2.31)$$

four nuclei are available for chemical shift labeling in an evolution time: the two ^{1}H–^{13}C pairs before and after the mixing step. In the 2D version of the ^{13}C-TOCSY described above the two dimensions were used for the two protons involved:

$$^1\text{H}(t_1)-^{13}\text{C}-(^{13}\text{C}-)_{\text{TOCSY}}{}^{13}\text{C}-^1\text{H}(t_2). \qquad (2.32)$$

In a 3D experiment, the additional dimension will show the chemical shifts of one of the two carbons to reduce ambiguities from overlap in the proton dimensions, e.g.:

$$^1\text{H}(t_1)-^{13}\text{C}(t_2)-(^{13}\text{C}-)_{\text{TOCSY}}{}^{13}\text{C}-^1\text{H}(t_3). \qquad (2.33)$$

In this experiment the protons in F_1 are defined not only by their own chemical shift, but can be further discriminated by the chemical shift of the directly bound carbon in F_2. The corresponding pulse sequence is shown in Figure 25b. The only difference with respect to the 2D ^{13}C-TOCSY sequence (Figure 25a) is the insertion of a carbon evolution time before the TOCSY transfer, including a slight rearrangement of the ^{1}H 180° pulse to allow for proton decoupling during t_2.

Obviously this 3D experiment can be further expanded into a 4D experiment by running another evolution on carbon *after* the TOCSY step:

$$^1H(t_1)-^{13}C(t_2)-(^{13}C-)_{TOCSY}{}^{13}C(t_3)-^1H(t_4). \qquad (2.34)$$

This experiment will contain carbon shift information for *both* proton dimensions, thereby minimizing the chance of misassignment due to degenerate ^{1}H resonances. However, the time requirements are even higher than for a 3D spectrum to allow for reasonable digitization in the *three* indirect dimensions, although a slightly lower digital resolution can be used than in a 3D experiment (because the additional dimension helps separate neighboring peaks). Also visualization of the data in the form of 2D slices is now even less adequate than for a 3D spectrum, and the data evaluation has to rely much more on appropriate software. Nevertheless, most commercial processing software does already include 4D (or higher dimensional) capabilities.

3D HNCA   As an example of a *triple resonance* experiment (i.e., one involving pulses on *three different* spin species; here ^{1}H, ^{13}C, and ^{15}N), we now briefly discuss the HNCA technique (Farmer *et al.*, 1992; Kay *et al.*, 1990). The HNCA experiment yields a correlation between amide protons, amide nitrogens, and alpha carbons in a peptide backbone (hence its name). Essentially magnetization is transferred from ^{1}H to ^{15}N via INEPT, then on to ^{13}C$^\alpha$ with another INEPT step. After a chemical shift evolution period, the magnetization is transferred back to ^{15}N and finally to ^{1}H, where the signal is detected under ^{15}N decoupling. However, one often gets correlations from a ^{15}N–^{1}H pair to the intraresidual α–carbon (via the $^1J_{N,C^\alpha}$ coupling) *and* to the preceding C$^\alpha$, via the only slightly smaller $^2J_{N,C^\alpha}$ coupling (Kay *et al.*, 1990).

Figure 27 shows two slightly different pulse sequences for the HNCA experiment. In the original sequence (Figure 27a), the ^{15}N evolution time t_1 is followed by a delay τ_2 long enough (~33 ms) to develop ^{15}N antiphase magnetization with respect to ^{13}C$^\alpha$ for the subsequent transfer onto C$^\alpha$. In addition τ_2 is set to an integral multiple of $(^1J_{H,N})^{-1}$ in order to avoid refocusing of ^{15}N antiphase coherence with respect to ^{1}H which remains from the first INEPT transfer. In the next step, the double antiphase ^{15}N magnetization is now converted into three-spin coherence (^{1}H, ^{15}N, ^{13}C$^\alpha$). During t_2, the proton and nitrogen chemical shift evolutions are each refocused with a 180° pulse, so that we end up with pure ^{13}C$^\alpha$ chemical shift evolution in t_2. The rest of the sequence is practically symmetric to the beginning (without ^{15}N evolution): The three-spin coherence is converted back to double ^{15}N anti-

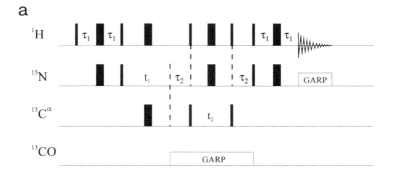

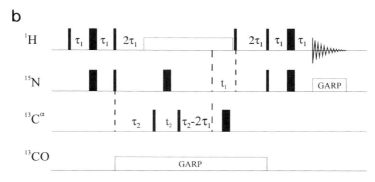

Figure 27. Two HNCA pulse sequences: (a) (Kay et al., 1990) and (b) (Farmer et al., 1992); for details see text. The delays are $\tau_1 = (4\ ^1J_{H,NC})^{-1}$, $\tau_2 = 22$ or 33 ms.

phase magnetization. The antiphase magnetization with respect to $^{13}C^\alpha$ is then refocused during τ_2, and the sequence is finished with an INEPT step back to ^1H. The ^{15}N decoupling is performed during the acquisition time, as is selective decoupling of the carbonyl carbons during the times when the magnetization is on ^{15}N or $^{13}C^\alpha$ (since these nuclei show significant couplings to the carbonyl carbon).

The use of three-spin coherence during t_2 seems to be somewhat unconventional, but is justified by the authors with the smaller number of rf pulses in comparison to alternative versions. For example, ^{15}N magnetization is kept antiphase to ^1H all the time by properly setting τ_2, so the final transfer back from ^{15}N to ^1H can proceed immediately after the refocusing of the ^{15}N–$^{13}C^\alpha$ coupling.

In contrast to the sequence in Figure 27a, in the more recent sequence introduced by Farmer et al. (1992) (Figure 27b), ^{15}N antiphase magnetization with respect to ^{1}H resulting from the first INEPT step is immediately refocused during the delay $2\tau_1 = (2\ ^1J_{H,N})^{-1}$. Then ^{1}H decoupling is applied during the longest part of the sequence to prevent the creation of new antiphase terms. Not until shortly before the final INEPT transfer back from ^{15}N to ^{1}H is proton decoupling turned off so that the required ^{15}N–^{1}H antiphase can build up during the delay $2\tau_1$. Again, ^{15}N decoupling during the acquisition period and CO decoupling are performed as in Figure 27a.

From a detailed analysis of the relaxation pathways relevant for the HNCA experiment, the avoidance of ^{15}N–^{1}H antiphase magnetization in Figure 27b should be very advantageous in the case of fast relaxing molecules. Indeed, it has been shown that the sequence of Figure 27b is clearly superior to that of Figure 27a when applied to large proteins (Figure 28). In spite of the rather long delays Δ, the sequence of Figure 27b can be used to study proteins of molecular weights up to 30 kDa with tolerable relaxation losses. For a more modern version to avoid saturation effects via chemical exchange, see Jahnke and Kessler (1995).

2.5 Determination of ^{13}C Relaxation Times

Carbon T_1 relaxation times are widely used to determine the local mobility of molecules (Lyerla and Levy, 1974; Wehrli and Wirthlin, 1974). Ei-

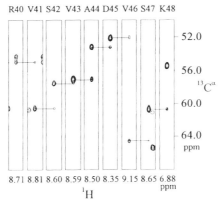

Figure 28. 3D HNCA spectrum of uniformly ^{13}C- and ^{15}N-enriched IIB^{Glc} protein at 600 MHz, sample conditions as in Figure 26 (Golic Grdadolnik *et al.*, 1994). 2D slices at the amide ^{15}N frequencies of residues 40–48 show cross-peaks correlating the amide protons with the C$^\alpha$ resonances of their own and the preceding amino acid. This leads to the sequential assignment of the amide signals, as indicated by the horizontal lines.

ther T_1 times are used for a qualitative picture of molecular flexibility, or more sophisticated procedures can be applied to separate between global and local modes of reorientation (Lipari and Szabo, 1982a,b; Woessner et al., 1969).

Obviously the easiest way of determining ^{13}C T_1 times is the use of one-dimensional ^{13}C-detected techniques such as inversion–recovery (Vold et al., 1968), fast inversion–recovery (Candet et al., 1975), saturation–recovery (Freeman and Hill, 1971a), or the so-called "Freeman–Hill" approach (Freeman and Hill, 1971b). All these techniques are well established, therefore we do not discuss them here, but refer the interested reader to a detailed review article (Weiss and Ferretti, 1988).

Due to the inherent narrow lineshape and large spectral dispersion of carbon spectra, these methods can be applied to many small- and medium-sized molecules without risking signal overlap. However, the relatively low sensitivity of direct ^{13}C detection represents a major drawback of such techniques.

Advantageous in terms of SNR (and also spectral resolution) is the use of inverse 2D techniques for the determination of T_1. As an example, we show in Figure 29 a sequence containing an INEPT transfer from proton to carbon and back (Kay et al., 1989). Similar pulse sequences have also been published with slightly modified ^1H decoupling (Peng and Wagner, 1992) and with DEPT transfers instead of INEPT (Nirmala and Wagner, 1988).

In the 2D T_1 experiment of Figure 29, magnetization is first transferred from the protons onto carbon, leading to a sensitivity gain because of the larger magnetization to start with *and* because of the shorter recycle times (since ^1H magnetization usually relaxes faster than ^{13}C). After refocusing, the ^{13}C in-phase magnetization is turned onto the $-z$ axis and given time to relax comparable to an inversion–recovery experiment. A carbon evolution time t_1 follows to yield the indirect dimension of the 2D spectrum. Finally, magnetization is transferred back from ^{13}C to ^1H and detected under ^{13}C decoupling for the highest sensitivity. To suppress spurious $^{12}C-^1H$ signals when

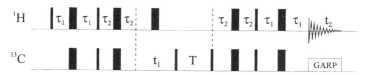

Figure 29. 2D pulse sequence for the measurement of ^{13}C T_1 times, employing an INEPT transfer (Kay et al., 1989); τ_1 and τ_2 are, respectively, slightly less than and equal to $(4\ ^1J_{H,C})^{-1}$; T is the delay for T_1 relaxation.

working with unlabeled compounds, a BIRD filter can precede the whole sequence (as described in Section 2.4.1.2). The result is a series of 2D ^{1}H,^{13}C HSQC-like spectra in which the intensity of every cross-peak is modulated with an exponential decay corresponding to the carbon T_1 relaxation rate.

Despite the substantial advantages of indirect detection methods in terms of sensitivity, significant errors in the measured ^{13}C relaxation rates can occur, resulting in part from the creation of multispin terms by the ^{1}H–^{13}C transfer. These multispin terms are converted into observable magnetization later in the sequence. Measurements can be additionally complicated by the interference of dipolar ^{1}H–^{13}C interactions and, particularly in large molecules, by the nonexponential relaxation of the methyl protons.

Recently pulse schemes have been described for the measurement of ^{13}C T_2 relaxation times in AX$_3$ spin systems (Kay *et al.*, 1992), based on INEPT (Freeman *et al.*, 1981) and DEPT transfer schemes (Bendall *et al.*, 1983). A comparison between relaxation rates obtained from direct ^{13}C observations and from magnetization-transfer experiments showed that they are identical only if magnetization from *all* ^{13}C transitions is equally transferred to the detected ^{1}H spins, with the consequence that flip angles and delays have to be carefully adjusted in the heteronuclear transfer steps. For applications to macromolecules, experimental and theoretical results suggest that magnetization transfer schemes based on the INEPT technique are superior to DEPT-derived sequences (Kay *et al.*, 1992).

2.6 Heteronuclear Coupling Constants

The application of heteronuclear scalar J-coupling constants for structural studies has recently seen a certain renaissance. Coupling data are usually used in addition to NOE data to further refine the structure. They contain valuable information about dihedral angles that helps in defining, e.g., the ϕ and ψ angles in peptide backbones or the conformation of protein sidechains.

Theoretically, coupling constants could be determined by measuring the splitting of resonance lines in a 1D spectrum. However, in almost all relevant cases this is impractical because there are generally quite a few homonuclear and heteronuclear couplings present in every signal that cannot be properly assigned nor even resolved in a 1D spectrum. This is especially true for the very interesting *long-range* heteronuclear coupling constants, which are usually smaller than the linewidth for most larger molecules. Therefore, the accurate determination of J values usually requires the acquisition of additional NMR experiments specifically designed to allow the measurement of one or several types of coupling constants. After a short glance at $^1J_{H,C}$ couplings, we will briefly explain the different approaches to the measurement of long-range $^nJ_{H,C}$ ($n > 1$) coupling constants.

2.6.1 $^1J_{H,C}$ Coupling Constants

Direct heteronuclear couplings depend on the electronic configuration of the nuclei involved, and therefore contain potential information about the local conformation. However, this dependence is usually small and difficult to translate into conformational data; therefore, their application to structure refinement is limited to a few special cases.

The most prominent application is certainly the use of $^1H^\alpha-^{13}C^\alpha$ coupling constants for the determination of the ψ angle in amino acid residues. Although the size of this coupling depends on both the ψ and the ϕ angles, its variation as a function of ψ is large enough (~5 Hz) (Egli and Philipsborn, 1981) to allow the use of $^1J_{H,C^\alpha}$ as a constraint in structure calculations of peptides (Mierke et al., 1992) and proteins (Vuister et al., 1993).

The size of the direct $^1H^\alpha-^{13}C^\alpha$ couplings (ranging primarily from ~120 to 200 Hz) is at least one order of magnitude larger than all other couplings to either the $^{13}C^\alpha$ or $^1H^\alpha$ nucleus. Therefore, they can be easily determined by acquiring a 2D HMQC or HSQC spectrum *without* heteronuclear decoupling during t_2. Although all other couplings are usually not resolved (especially with larger molecules), the direct coupling can be readily extracted from the large doublet splitting in F_2. Keep in mind, however, that the spectrum has to be recorded and processed with a sufficiently high resolution in F_2.

In the case of ^{13}C-enriched proteins it is advisable to run the HSQC with decoupling of the $^{13}C^\alpha$ spin from the neighboring $^{13}C=O$ (and also ^{15}N in the case of a $^{13}C,^{15}N$-labeled protein) during t_1. Otherwise these relatively large couplings (approximately 55 and 15 Hz, respectively) lead to a considerable broadening or even splitting of the cross-peaks in F_1, decreasing the sensitivity and resolution of the HSQC spectrum. For larger proteins where the $^1H^\alpha-^{13}C^\alpha$ region becomes quite crowded in a 2D HSQC, one has also the choice of adding a third dimension (e.g., $^{13}C=O$) in order to reduce signal overlap (Vuister et al., 1993).

2.6.2 Heteronuclear Long-Range Coupling Constants

Due to the small size of heteronuclear long-range couplings (in the range of ~10 Hz down to less than 1 Hz), their determination is by far not as straightforward as that of large direct couplings. A direct measurement, e.g., from the fine structure of the resonance lines in a 1D ^{13}C (or ^1H) spectrum, is usually complicated by several factors. Among these is the large number of long-range couplings for every carbon resonance, which makes their unambiguous assignment impossible. In addition, for many large- or medium-sized molecules, the linewidth of the individual resonances comes close to or even exceeds the size of these small couplings, leading to excessive overlap of the multiplet lines caused by heteronuclear long-range couplings.

To overcome these problems, many different methods have been proposed, which can be divided into three different general strategies:

1. analysis of the lineshape of signals containing the desired coupling (Section 2.6.2.1)
2. simplification of multiplet patterns (Section 2.6.2.2)
3. modulation of cross-peak intensities by the long-range coupling (Section 2.6.2.3).

Next we briefly describe the background of each of them and explain their application in a specific pulse sequence.

2.6.2.1 Lineshape Analysis

This approach is based on the fact that the overall multiplet lineshape of a signal is a result of all of its inherent couplings, even if they are not resolved (as is the case for small heteronuclear long-range couplings). Therefore, these small couplings can in principle be obtained by a computational analysis of the multiplet lineshape. However, there are some requirements: The amount of information that can be reliably extracted from such a lineshape is limited by its quality, i.e., its digital resolution and noise content. To obtain meaningful results, the approach should use high-quality spectra (the mere presence of a signal is not enough!), and the number of parameters to be fitted should be as small as possible. Therefore, the analysis is usually based on a reference lineshape that already contains additional information.

This principle can be best explained with a technique for the determination of heteronuclear long-range couplings from ^{1}H,^{13}C-HMBC spectra introduced by Keeler and Neuhaus (Keeler *et al.*, 1988, 1989; Keeler and Titman, 1990). In an HMBC cross-peak, the multiplet pattern in F_2 (i.e., the proton dimension) consists of all homonuclear proton–proton couplings *in phase* and the *antiphase* ^{1}H,^{13}C heteronuclear long-range coupling (which caused the cross-peak). However, a similar pattern can be seen in many homonuclear ^{1}H spectra (1D, 2D TOCSY/ROESY/NOESY): Here the fine structure of a signal consists of all of its homonuclear ^{1}H–^{1}H couplings *in phase*. Therefore, if this pattern is used as a reference lineshape for the HMBC multiplet structure, there are only two parameters to be fitted: the desired (antiphase) heteronuclear proton–carbon coupling constant and a scaling factor between the two lineshapes. Figure 30 shows an application of the Keeler/Neuhaus technique. For further experimental details, the reader is referred to the original literature.

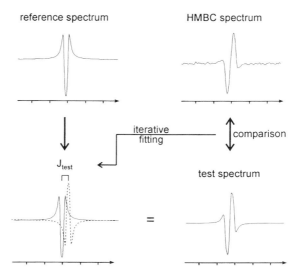

Figure 30. Illustration of the Keeler/Neuhaus technique for the measurement of ^{1}H,^{13}C long-range couplings (Keeler and Titman, 1990). Starting from a homonuclear reference spectrum, the lineshape of the proton under consideration is extracted from a suitable signal and used for the simulation of an antiphase pattern (dashed line) caused by the desired long-range coupling. Iterative fitting to the real lineshape from an HMBC spectrum yields the value of this coupling constant.

2.6.2.2 Simplification of Multiplet Patterns (E.COSY)

The so-called E.COSY techniques (Griesinger et al., 1987) make use of a large, well-resolved splitting in one dimension of a 2D (or 3D) cross-peak to allow the measurement of a small, otherwise not resolved ^{1}H,^{13}C long-range coupling in another dimension (Figure 31).

The requirements for the generation of E.COSY patterns for the determination of a certain long-range ^{1}H,^{13}C coupling are the following:

1. The *carbon* under consideration has to exhibit a large, easily resolvable direct coupling (either $^1J_{H,C}$ or $^1J_{C,C}$).
2. Magnetization from the (^{1}H or ^{13}C) spin directly coupled to the carbon of interest has to be somehow transferred to the *proton* under consideration.
3. During this transfer process the spin states of the *carbon* must not be mixed by rf pulses (i.e., with flip angles that are not multiples of 180°).

Condition 3 may cause problems in some experiments involving ^{13}C-enriched molecules, e.g., if the necessary transfer steps involve other carbon

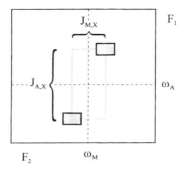

Figure 31. Basic E.COSY pattern. In a spin system of three mutually coupled spins, A, M, and X, the A,M cross-peak displays a general pattern of two vertically and horizontally displaced elements (themselves more or less complex multiplet patterns, here represented by shaded squares). The orthogonal displacements directly yield the two passive couplings to the X spin, as indicated. Because of the two-dimensional nature of the displacement, *one* large passive coupling is sufficient to resolve the E.COSY pattern and allow the measurement of *both* couplings (Griesinger *et al.*, 1987).

spins, or if the signals are to be decoupled from other ^{13}C nuclei. In these cases one can either use selective pulses, which affect only limited regions of the ^{13}C spectrum, or substitute 90° pulses by small flip angle pulses (causing only neglectable mixing of the spin states, but concomitantly reducing the transfer efficiency).

Several E.COSY-type techniques have been applied to the determination of different ^{1}H,^{13}C long-range couplings in ^{13}C labeled proteins, e.g., of $^2J_{H\alpha,C'}$ (Vuister and Bax, 1992), $^3J_{H\beta,C'}$ (Eggenberger *et al.*, 1992), $^3J_{H^N,C'}$ and $^3J_{H^N,C\beta}$ (Seip *et al.*, 1994).

We now explain the mechanism of these experiments with an example, the HETLOC technique (Figure 32) (Kurz *et al.*, 1991). At first glance, HETLOC is a 2D ^{1}H,^{1}H-TOCSY experiment with a ^{13}C filter at the beginning (see Section 2.4.1.2). This means we only observe TOCSY peaks start-

Figure 32. HETLOC pulse sequence (Kurz *et al.*, 1991): an X_1 half-filter selecting for ^{13}C-bound protons precedes a homonuclear 2D-TOCSY experiment. No ^{13}C decoupling is performed during t_1 and t_2, resulting in an E.COSY-type cross-peak pattern (compare with Figure 21).

ing from ^{13}C-bound protons (in F$_2$). However, there is no ^{13}C decoupling during t_1 or t_2. Since we select only ^{13}C-bound protons in t_1, all signals will show the large $^1J_{H,C}$ splitting in F$_1$ in the processed TOCSY spectrum. However, during the TOCSY mixing time, magnetization will be partially transferred to neighboring protons, which are generally *not* bound to ^{13}C (for ^{13}C in natural abundance). When we detect the magnetization on these protons during t_2, they will show only a small long-range ^{1}H,^{13}C coupling.

Now the molecules we observe in this experiment can be divided into two groups: One half contains a ^{13}C spin in the α state, the other half a ^{13}C spin in the β state. Remember that, in a doublet, the two lines correspond to the two species of molecules with the coupling nucleus in the α or β state, respectively. However, the state of each ^{13}C spin is maintained between t_1 and t_2, because we do not apply any pulses on carbon that could induce a transition. This means that, out of the four possible combinations ($t_1 : t_2 = \alpha : \alpha$, $\alpha : \beta$, $\beta : \alpha$, $\beta : \beta$), we will only see the ones with *identical* spin states in t_1 and t_2 (i.e., $\alpha : \alpha$ and $\beta : \beta$), corresponding to only *half* of the possible multiplet lines in the 2D cross-peak (Figure 33). These two remaining lines are diagonally shifted, with the displacement in F$_1$ and F$_2$ corresponding to the large $^1J_{H,C}$ and small long-range ^{1}H,^{13}C coupling, respectively. Due to the size of the direct ^1J coupling, the two lines are well separated, and the small heteronuclear long-range coupling can be easily extracted from the multiplet pattern (Figure 34).

2.6.2.3 Long-Range Couplings from Cross-Peak Intensities

Signal intensities have been used for a qualitative determination of long-range couplings for many years, e.g., from COLOC (Anders *et al.*, 1987;

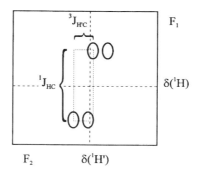

Figure 33. HETLOC multiplet scheme (Kurz *et al.*, 1991): the H,H' cross-peak (here shown as a doublet) is split by the heteronuclear couplings $^1J_{H,C}$ (large, in F$_1$) and $^3J_{H',C}$ (small, in F$_2$).

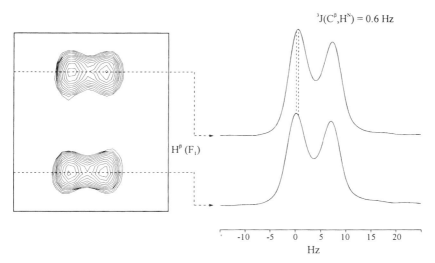

Figure 34. HETLOC cross-peak caused by a heteronuclear long-range coupling of only 0.6 Hz. Although this is far less than the actual linewidth, the coupling constant can be reliably extracted from the horizontal displacement between the upper and lower part of the cross-peak. (Courtesy of M. Eberstadt.)

Kessler *et al.*, 1987) or HMBC spectra (Grzesiek *et al.*, 1992). If one compares, e.g., the two H^β–C' HMBC cross-peaks of an aromatic amino acid, their relative intensities are proportional to $\sin(\pi\,{}^3J_{H^\beta,C'}\Delta)\sin(\pi\,{}^3J_{H^\beta,H^\alpha}\Delta)$ (see pulse sequences in Figure 12). With the knowledge of the values for ${}^3J_{H^\alpha,H^\beta}$ from homonuclear spectra, one can estimate the relative size of the two H^β–C' couplings and use this information for their diastereotopic assignment.

The most recent techniques for the determination of heteronuclear long-range couplings rely on intensity modulations of cross-peaks. Typically, a 2D spectrum is acquired (usually of the HMQC/HSQC type) with an additional fixed delay in the pulse sequence. During this delay the coupling of interest leads to the partial development of antiphase terms, which are then removed, resulting in an intensity loss in the final spectrum. If one now varies the time $t_{\text{eff}}(J)$ during which coupling evolution is effective (e.g., by shifting a refocusing 180° pulse within the fixed delay), then one gets a modulation of the cross-peak intensities, in the easiest case with $\cos[\pi J t_{\text{eff}}(J)]$:

$$I_y \xrightarrow{t_{\text{eff}}} I_y \cos\pi J_{I,S} t_{\text{eff}}(J) - 2I_x S_z \sin\pi J_{I,S} t_{\text{eff}}(J). \qquad (2.35)$$

Of course, chemical shift evolution has to be refocused, and the overall length of the pulse sequence has to be constant, independent of t_{eff}, to elimi-

TABLE 2
Techniques of Modern NMR Spectroscopy

Name	Information content	Enrichment required	Typical molecules	Typical sample concentrations	Typical time
2D-HMQC, HSQC	Direct ^{1}H, ^{13}C correlation	No	All molecules	>10 mM	2 h
2D-HMBC	Long-range ^{1}H, ^{13}C correlations	No	<2 kDa	>20 mM	12 h
2D-HMQC-COSY	COSY with ^{13}C dispersion	No	<2 kDa	>20 mM	12 h
2D-HMQC-TOCSY	TOCSY with ^{13}C dispersion	No	<2 kDa	>20 mM	12 h
3D-HMQC-TOCSY	TOCSY with ^{13}C dispersion	Yes	<10 kDa	2 mM	2 d
3D-HCCH-COSY	^{1}H-detected ^{13}C-^{13}C COSY	Yes	<30 kDa	2 mM	2 d
3D-HCCH-TOCSY	^{1}H-detected ^{13}C-^{13}C TOCSY	Yes	<30 kDa	2 mM	2 d
3D-HNCA	^{1}H$_N$-^{15}N-^{13}C$^\alpha$ correlation	^{13}C and ^{15}N	Proteins <30 kDa	2 mM	2 d
2D-HETLOC	$^3J_{H,C}$ coupling constants	No	<2 kDa	>20 mM	1 d

nate contributions from relaxation effects. Finally, only *one* coupling must evolve during t_{eff} (which can be done by using, e.g., selective 180° pulses). Then the interesting heteronuclear long-range coupling can either be calculated from the intensities in two spectra with different t_{eff}, or from a fitting procedure to the cross-peak intensities from a whole series of such spectra.

2.7 Concluding Remarks

The intention of this chapter was to present an overview of the use of heteronuclear and multidimensional ^{13}C techniques in modern NMR spectroscopy. Table 2 sums up the techniques mentioned and gives estimated values for typical sample concentrations and measuring times. Of course, these numbers may vary considerably from case to case, since they depend on quite a few sample-specific factors (such as relaxation behavior, solubility, spectral dispersion). However, these data might be useful in estimating *relative* sensitivities of the different NMR techniques.

References

Allerhand, A., Addleman, R. E., and Osman, D. (1985). *J. Am. Chem. Soc.* **107**, 5809.
Anders, U., Gemmecker, G., Kessler, H., and Griesinger, C. (1987). *Fresenius Z. Anal. Chem.* **327** (1), 72.
Aue, W. P., Bartholdi, E., and Ernst, R. R. (1976). *J. Chem. Phys.* **64**, 2229.
Bain, A. (1984). *J. Magn. Reson.* **56**, 418.
Bax, A., and Pochapski, S. S. (1992). *J. Magn. Reson.* **99**, 638.
Bax, A., and Summers, M.F. (1986). *J. Am. Chem. Soc.* **108**, 2093.
Bax, A., and Weiss, M. A. (1987). *J. Magn. Reson.* **71**, 571.
Bax, A., Freeman, R., and Kempsell, S. P. (1980a). *J. Am. Chem. Soc.* **102**, 4849.
Bax, A., de Jong, P. G., Mehlkopf, A. F., and Smidt, J. (1980b). *Chem. Phys. Lett.* **69**, 567.
Bax, A., Griffey, R. H., and Hawkins, B. L. (1983). *J. Magn. Reson.* **55**, 301.
Bax, A., Clore, G. M., Driscoll, P. C., Gronenborn, A. M., Ikura, M., and Kay, L. E. (1990a). *J. Magn. Reson.* **87**, 620.
Bax, A., Clore, G. M., and Gronenborn, A. M. (1990b). *J. Magn. Reson.* **88**, 425.
Bendall, M.R., Doddrell, D.M., and Pegg, D.T. (1981). *J. Am. Chem. Soc.* **103**, 4603.
Bendall, M. R., Pegg, D. T., Doddrell, D. M., and Field, J. (1983). *J. Magn. Reson.* **51**, 520.
Bodenhausen, G., and Ruben, D. J. (1980). *Chem. Phys. Lett.* **69**, 185.
Bodenhausen, G., Kogler, H., and Ernst, R. R. (1984). *J. Magn. Reson.* **58**, 370.
Boelens, R., Scheek, R., Dijkstra, K., and Kaptein, R. (1985). *J. Magn. Reson.* **62**, 378.
Braunschweiler, L., and Ernst, R. R. (1983). *J. Magn. Reson.* **53**, 521.
Byrd, R., Summer, M., Jon, H., Spellmeyer-Fonts, C., and Mavilli, L. (1986). *J. Am. Chem. Soc.* **108**, 504.
Candet, D., Levy, G. C., and Peat, I. (1975). *J. Magn. Reson.* **18**, 199.
Canet, D., Brondeau, J., Mischler, E., and Humbert, F. (1993). *J. Magn. Reson. A* **105**, 239.
Clore, G. M., and Gronenborn, A. M. (1991). *Prog. Nucl. Magn. Reson. Spectrosc.* **23**, 43.

Clubb, R. T., Thanabal, V., and Wagner, G. (1992). *J. Biomol. NMR* **2**, 203.
Eggenberger, U., Karimi Nejad, Y., Thüring, H., Rüterjans, H., and Griesinger, C. (1992). *J. Biomol. NMR* **2**, 583.
Egli, H., and Philipsborn, W. (1981). *Helv. Chim. Acta* **64**, 976.
Emsley, L., and Bodenhausen, G. (1989). *J. Magn. Reson.* **82**, 211.
Emsley, L., and Bodenhausen, G. (1990). *Chem. Phys. Lett.* **165**, 469.
Ernst, R. R., Bodenhausen, G., and Wokaun, A. (1987). "Principles of Nuclear Magnetic Resonance in One and Two Dimensions." Clarendon Press, Oxford.
Farmer, B. T., II, Venters, R. A., Spicer, L. D., Wittekind, M. G., and Müller, L. (1992). *J. Biomol. NMR* **2**, 195.
Fesik, S. W. (1993). *J. Biomol. NMR* **3**, 261.
Fesik, S. W., Luly, J. R., Erickson, J. W., and Abad-Zapatero, C. (1988). *Biochemistry* **27**(22), 8297.
Fesik, S. W., Eaton, H. L., Olejniczak, E. T., and Zuiderweg, E. R. P., McIntosh, L. D., and Dahlquist, F. W. (1990). *J. Am. Chem. Soc.* **112**, 886.
Freeman, R., and Hill, H. (1971a). *J. Chem. Phys.* **54**, 3367.
Freeman, R., and Hill, H. (1971b). *J. Chem. Phys.* **55**, 1985.
Freeman, R., Mareci, T. H., and Morris, G. (1981). *J. Magn. Reson.* **42**, 341.
Garbow, J. R., Weitekamp, D. P., and Pines, A. (1982). *Chem. Phys. Lett.* **93**, 504.
Geen, H., and Freeman, R. (1991). *J. Magn. Reson.* **93**, 93.
Gemmecker, G., Olejniczak, E. T., and Fesik, S. W. (1992). *J. Magn. Reson.* **96**, 199.
Gemmecker, G., Jahnke, W., and Kessler, H. (1993). *J. Am. Chem. Soc.* **115**, 11620.
Golic Grdadolnik, S., Eberstadt, M., Gemmecker, G., Kessler, H., Buhr, A., and Erni, B. (1994). *Eur. J. Biochem.* **219**, 945.
Griesinger, C., Sørensen, O. W., and Ernst, R. R. (1987). *J. Magn. Reson.* **75**, 474.
Grzesiek, S., Ikura, M., Clore, G. M., Gronenborn, A. M., and Bax, A. (1992). *J. Magn. Reson.* **96**, 215.
Guéron, M., Plateau, P., and Decorps, M. (1991). *Prog. Nucl. Magn. Reson.* **23**, 137.
Hansen, A. P., Petros, A. M., Mazar, A. P., Pederson, T. M., Rueter, A., and Fesik, S. W. (1992). *Biochemistry* **31**, 12713.
Hartmann, S. R., and Hahn, E. L. (1962). *Phys. Rev.* **128**, 2042.
Hurd, R. E., and John, B. K. (1991). *J. Magn. Reson.* **91**, 648.
Ikura, M., and Bax, A. (1992). *J. Am. Chem. Soc.* **114**, 2433.
Ikura, M., Kay, L. E., and Bax, A. (1990). *Biochemistry* **29**, 4659.
Jahnke, W., and Kessler, H. (1995). *Angew. Chem., Int. Ed. Engl.* (in press).
Kalinowski, H.-O., Berger, S., and Braun, S. (1984). "^{13}C-NMR-Spektroskopie." Thieme, Stuttgart.
Kay, L. E., Torchia, D., and Bax, A. (1989). *Biochemistry* **28**, 8972.
Kay, L. E., Ikura, M., Tschudin, R., and Bax, A. (1990). *J. Magn. Reson.* **89**, 496.
Kay, L. E., Bull, T. E., Nicholson, L. K., Griesinger, C., Schwalbe, H., Bax, A., and Torchia, D. A. (1992). *J. Magn. Reson.* **100**, 538.
Keeler, J., and Titman, J. J. (1990). *J. Magn. Reson.* **89**, 640.
Keeler, J., Neuhaus, D., and Titman, J. J. (1988). *Chem. Phys. Lett.* **146**, 545.
Keeler, J., Neuhaus, D., and Titman, J. J. (1989). *J. Magn. Reson.* **85**, 111.
Kessler, H. (1982). *Angew. Chem., Int. Ed. Engl.* **21**, 512.
Kessler, H., and Bermel, W. (1986). In "Applications of NMR Spectroscopy to Problems in Stereochemistry and Conformational Analysis" (Y. Takeuchi and A. P. Marchand, eds.), MSA 6, pp. 179–205, VCH, Deerfield Beach, FL.
Kessler, H., and Schmieder, P. (1991). *Biopolymers* **31**, 621.
Kessler, H., Griesinger, C., Zarbock, J., and Loosli, H. R. (1984). *J. Magn. Reson.* **57**, 331.

Kessler, H., Griesinger, C., and Wagner, K. (1987). *J. Am. Chem. Soc.* **109**, 6927.
Kessler, H., Gehrke, M., and Griesinger, C. (1988). *Angew. Chem., Int. Ed. Engl.* **27**, 447.
Kessler, H., Schmieder, P., Köck, M., and Kurz, M. (1990a). *J. Magn. Reson.* **88**, 615.
Kessler, H., Schmieder, P., and Oschkinat, H. (1990b). *J. Am. Chem. Soc.* **112**, 8599.
Kessler, H., Mronga, S., and Gemmecker, G. (1991). *Magn. Reson. Chem.* **29**, 527.
Kogler, H., Sørensen, O. W., Bodenhausen, G., and Ernst, R. R. (1983). *J. Magn. Reson.* **55**, 157.
Konat, R. K., Mierke, D. F., Kessler, H., Kutscher, B., Bernd, M., and Voegeli, R. (1993). *Helv. Chim. Acta* **76**, 1649.
Kumar, A., Wagner, G., Ernst, R. R., and Wüthrich, K. (1980). *Biochem. Biophys. Res. Commun.* **96**, 1156.
Kurz, M., Schmieder, P., and Kessler, H. (1991). *Angew. Chem., Int. Ed. Engl.* **30**, 1329.
Lerner, L., and Bax, A. (1986). *J. Magn. Reson.* **69**, 375.
Levitt, M., and Freeman, R. (1979). *J. Magn. Reson.* **33**, 473.
Lipari, G., and Szabo, A. (1982a). *J. Am. Chem. Soc.* **104**, 4546.
Lipari, G., and Szabo, A. (1982b). *J. Am. Chem. Soc.* **104**, 4559.
Lyerla, J. R., and Levy, G. C. (1974). *Top. Carbon-13 NMR Spectrosc.* **1**, 79.
Majumdar, A., Wang, H., Morshauser, R. C., and Zuiderweg, E. R. P. (1993). *J. Biomol. NMR* **3**, 387.
Marshall, J. L. (1983). "Carbon-Carbon and Carbon-Proton NMR Couplings: Application to Organic Stereochemistry and Conformational Analysis," MSA 2. VCH, Deerfield Beach, FL.
Martin, G. E., and Zektzer, A. S. (1988). "Two-Dimensional NMR Methods for Establishing Molecular Connectivity." VCH, Weinheim.
Maudsley, A. A., Wokaun, A., and Ernst, R. R. (1978). *Chem. Phys. Lett.* **55**, 9.
McCoy, M. A., and Mueller, L. (1992). *J. Am. Chem. Soc.* **114**, 2108.
McCoy, M. A., and Mueller, L. (1993). *J. Magn. Reson.* **101**, 122.
Messerle, B. A., Wider, G., Otting, G., Weber, C., and Wüthrich, K. (1989). *J. Magn. Reson.* **85**, 608.
Mierke, D. F., Golic Grdadolnik, S., and Kessler, H. (1992). *J. Am. Chem. Soc.* **114**, 8283.
Morris, G. A., and Freeman, R. (1979). *J. Am. Chem. Soc.* **101**, 760.
Müller, L. (1979). *J. Am. Chem. Soc.* **101**, 4481.
Nikonowicz, E. P., and Pardi, A. (1992). *Nature (London)* **355**, 184.
Nirmala, N. R., and Wagner, G. (1988). *J. Am. Chem. Soc.* **110**, 7557.
Otting, G., and Wüthrich, K. (1989). *J. Magn. Reson.* **85**, 586.
Otting, G., Senn, H., Wagner, G., and Wüthrich, K. (1986). *J. Magn. Reson.* **70**, 500.
Palmer, A. G., III, Cavanagh, J., Wright, P. F., and Rance, M. (1991). *J. Magn. Reson.* **93**, 151.
Pavia, A., and Ferrari, B. (1983). *Int. J. Pept. Protein Res.* **22**, 549.
Peng, J. W., and Wagner, G. (1992). *J. Magn. Reson.* **98**, 308.
Petros, A. M., Kawai, M., Luly, J. R., and Fesik, S. W. (1992). *FEBS Lett.* **308**, 309.
Piveteau, D., Delsuc, M.-A., and Lallemand, J.-Y. (1985). *J. Magn. Reson.* **63**, 255.
Plateau, P., and Guéron, M. (1982). *J. Am. Chem. Soc.* **104**, 7310.
Richardson, J. M., Clowes, R. T., Boucher, W., Domaille, P. J., Hardman, C. H., Keeler, J., and Laue, E. D. (1993). *J. Magn. Reson. B* **101**, 223.
Schmieder, P., Kessler, H., and Oschkinat, H. (1990). *Angew. Chem., Int. Ed. Engl.* **29**, 546.
Schmieder, P., Zimmer, S., and Kessler, H. (1991). *Magn. Reson. Chem.* **29**, 375.
Schmieder, P., Ippel, J. H., van den Elst, H., van der Marel, G. A., van Boom, J. H., Altona, C., and Kessler, H. (1992). *Nucleic Acids Res.* **20**(18), 4747.
Seebach, D., Ko, S. Y., Kessler, H., Köck, M., Reggelin, M., Schmieder, P., Walkinshaw, M. D., Bölsterli, J. J., and Bevec, D. (1991). *Helv. Chim. Acta* **74**, 1953.

Seip, S., Balbach, J., and Kessler, H. (1992). *J. Magn. Reson.* **100**, 406.
Seip, S., Balbach, J., and Kessler, H. (1994). *J. Magn. Reson.* B **104**, 172.
Shaka, A. J., Lee, C. J., and Pines, A. (1988). *J. Magn. Reson.* **77**, 274.
Sklenář, V., and Bax, A. (1987). *J. Am. Chem. Soc.* **74**, 469.
Sørensen, O. W., Freeman, R., Frenkiel, T. A., Mareci, T. H., and Schuck, R. (1982). *J. Magn. Reson.* **46**, 180.
Sørensen, O. W., Eich, G. W., Levitt, M. H., Bodenhausen, G., and Ernst, R. R. (1983). *Prog. NMR Spectrosc.* **16**, 163.
Tyburn, J.-M., Brereton, I. M., and Doddrell, D. M. (1992). *J. Magn. Reson.* **97**, 305.
Vold, R., Waugh, J., Klein, M., and Phelps, D. (1968). *J. Chem. Phys.* **48**, 3831.
Vuister, G. W., and Bax, A. (1992). *J. Biomol. NMR* **2**, 401.
Vuister, G. W., Delaglio, F., and Bax, A. (1993). *J. Biomol. NMR* **3**, 67.
Wagner, G. (1990). *Prog. Nucl. Magn. Reson. Spectrosc.* **22**(2), 101.
Wehrli, F. W., and Wirthlin, T. (1974). "Interpretation of Carbon-13 NMR-Spectra." Heyden, London.
Weiss, G., and Ferretti, J. (1988). *Prog. Nucl. Magn. Reson. Spectrosc.* **20**, 317.
Willker, W., Leibfritz, D., Kerssebaum, R., and Bermel, W. (1993). *Magn. Reson. Chem.* **31**, 287.
Woessner, D. E., Snowden, B. S., Jr., and Meyer, G. H. (1969). *J. Chem. Phys.* **50**, 719.
Zuiderweg, E. R. P. (1990). *J. Magn. Reson.* **89**, 533.
Zuiderweg, E. R. P., and Fesik, S. W. (1991). *J. Magn. Reson.* **93**, 653.

CHAPTER 3

Measurement of Internuclear Distances in Biological Solids by Magic-Angle Spinning ^{13}C NMR

Joel R. Garbow
Monsanto Corporate Research
Monsanto Company
St. Louis, Missouri 63198

Terry Gullion
Department of Chemistry
Florida State University
Tallahassee, Florida 32306

3.1 Introduction

Elucidating the structure of macromolecules is of fundamental importance in a number of disciplines, including chemistry, structural biology, and polymer science. A variety of techniques have been developed for providing this structural information. Among the most important and versatile of these techniques are X-ray crystallography and nuclear magnetic resonance (NMR) spectroscopy. X-ray crystallography may provide the complete molecular structure of macromolecules for which suitable single crystals can be grown (Banner et al., 1993; Chen et al., 1994; Lima et al., 1994; Zanotti, 1992). This requirement for high-quality single crystals limits the general applicability of crystallography in biological systems and completely precludes its use in amorphous systems such as glassy polymers or inorganic glasses. Liquid-state NMR has been used to map the covalent structure of a wide variety of molecules, providing conformational information about proteins and other biological macromolecules (Clore and Gronenborn, 1991; Wagner, 1990) and also about the microstructure of polymers (Ando et al., 1990; Tonelli, 1989). As molecular weights increase, however, solubilities and spectral crowding combine to limit the applicability of liquid-state techniques.

Solid-state NMR can provide important structural information without limits imposed by either molecular weight or solubility and can be applied equally well to microcrystalline and amorphous materials. As such, the solid-state NMR technique is complementary to both X-ray crystallography and liquid-state NMR. By combining high-power ^{1}H decoupling and magic-angle spinning (MAS) (Andrew et al., 1958; Lowe, 1959; Schaefer and Stejskal, 1976), high-resolution, *single-site* spectra can be obtained of a number of important spin-$\frac{1}{2}$ nuclei, including ^{13}C, ^{15}N, ^{31}P, and ^{19}F. The chemical-shift values of observed signals, their linewidths, and their relaxation properties can reveal much about the structure of macromolecules in the solid state. One particularly important element of solid-state structure characterization, which might provide *detailed* information about individual regions of macromolecules, is the measurement of distances between selected spin-$\frac{1}{2}$ sites through the measurement of through-space, dipolar couplings (Griffiths and Griffin, 1993).

In this article, we review several MAS solid-state NMR experiments for measuring distances between pairs of spin-$\frac{1}{2}$ nuclei. In particular, we focus on methods for determining internuclear separations between dilute spin pairs that can be incorporated into samples by stable isotopic enrichment. For example, the low natural abundance of ^{13}C and ^{15}N (1.11 and 0.37 at.%, respectively) makes these spins ideal candidates for site-specific isotopic labeling. In addition, distance measurements can include spin-$\frac{1}{2}$ species such as ^{31}P and ^{19}F, which are 100% abundant, in molecules containing single phosphorus or fluorine sites.

We start our description of these MAS experiments with the internal spin Hamiltonian, $H_{int}(t)$, for an isolated pair of spin-$\frac{1}{2}$ nuclei labeled I and S. Ignoring indirect spin couplings (J couplings), the internal Hamiltonian is composed of the chemical-shift interaction, H_{CSA}, and either the heteronuclear dipolar interaction, H_D^{he}, or the homonuclear dipolar interaction, H_D^{ho}. Because of the magic-angle sample rotation, each of the terms in the internal spin Hamiltonian is time dependent. The internal Hamiltonian H_{int} is

$$H_{int}(t) = H_{CSA}(t) + H_D^{he}(t) + H_D^{ho}(t). \quad (3.1)$$

In frequency units, the chemical-shift interaction for the I spin is given by

$$H_{CSA}(t) = [\delta_I + \eta_I \xi(t)] I_z, \quad (3.2)$$

where the isotropic chemical shift is

$$\delta_I = \gamma_I H_0 \bar{\sigma}_I = \frac{\gamma_I H_0}{3} (\sigma_{I,xx} + \sigma_{I,yy} + \sigma_{I,zz}) \quad (3.3)$$

and the chemical-shift asymmetry is

$$\eta_I = \sigma_{I,zz} - \bar{\sigma}_I. \quad (3.4)$$

The $\sigma_{I,ii}$'s in these equations are the principal values of the I-spin chemical-shift tensor, γ_I is the magnetogyric ratio of spin I, and H_0 is the strength of the external magnetic field. Similar expressions describe the S-spin chemical-shift term. In frequency units, the dipolar interactions under MAS conditions between an I-S spin pair, separated by a distance r, in a high magnetic field are (Maricq and Waugh, 1979; Munowitz and Griffin, 1982) as follows:

$$H_D^{he}(t) = 2\frac{\gamma_I \gamma_S \hbar}{r^3} \xi(t) I_z S_z \qquad (3.5)$$

and

$$H_D^{ho}(t) = \frac{\gamma_I \gamma_S \hbar}{r^3} \xi(t)(3I_z S_z - \bar{I} \cdot \bar{S}). \qquad (3.6)$$

The time dependence of the interactions caused by the sample rotation is contained in $\xi(t)$, which can be written as

$$\xi(t) = C_1 \cos(\omega_r t) + S_1 \sin(\omega_r t) + C_2 \cos(2\omega_r t) + S_2 \sin(2\omega_r t). \qquad (3.7)$$

The C's and S's are specific to each interaction (Maricq and Waugh, 1979) and their detailed forms need not be specified here. From Eq. (3.7), the time dependence of $H_{int}(t)$ is governed by the spinning rate ω_r and is periodic with period $T_r (= 2\pi/\omega_r)$. In addition to possessing the property $\xi(t + T_r) = \xi(t)$, the average of $\xi(t)$ over one complete rotor period is zero. These properties of $\xi(t)$ lead to the formation of rotational echoes at integral multiples of the rotor period following excitation of the spin system.

Internuclear distances between spin pairs I and S are easily determined from the $1/r^3$ distance dependence of the dipolar interactions [Eqs. (3.5) and (3.6)], assuming these interactions can be measured. Several factors, however, complicate the accurate measurement of dipolar couplings. As described earlier, MAS coherently averages both the chemical-shift anisotropy (CSA) and dipolar interactions to zero, producing a train of rotational echoes at integral multiples of the spinning period. Fourier transformation of this echo train produces a spectrum consisting of centerbands at the isotropic chemical shifts and a series of spinning sidebands located at multiples of the spinning frequency away from the centerbands. In principle, the CSA or dipolar interactions can be determined from a detailed analysis of the spinning-sideband patterns (Herzfeld and Berger, 1980). However, under the fast-spinning conditions typically employed to obtain high-resolution solid-state spectra, sideband intensities are small (this is what is meant by high resolution!), making such analyses impossible. Slowing the rotor speed to increase spinning-sideband intensities leads to severe overlap of sidebands from different chemically distinct spin sites. Thus, the goals of collecting high-resolution solid-state spectra and of measuring anisotropic interactions appear to be mutually exclusive.

A further complication arises in measuring dipolar interactions because, at typical magnetic field strengths, the CSA dominates the dipolar interaction. For example, in a peptide or protein, the heteronuclear dipolar coupling between a backbone ^{13}C-enriched carbonyl carbon and a ^{15}N-enriched amide nitrogen four positions from the labeled ^{13}C site along the primary sequence ranges from 20 to 300 Hz, depending on the peptide conformation (Garbow and McWherter, 1993). By contrast, a typical chemical-shift asymmetry, δ, for such a carbonyl carbon is several kilohertz. Thus, weak dipolar interactions must usually be measured in the presence of more dominant chemical-shift anisotropies.

A wide variety of techniques have been developed to recover the dipolar and/or CSA interactions under fast MAS conditions. For example, the CSA can be recovered by changing the orientation of the sample spinning axis (Bax et al., 1983) or by applying suitable pulse trains synchronously with the sample rotation (Gan, 1992; Gullion, 1989; Hu et al., 1993; Tycko et al., 1989; Yarim-Agaev et al., 1982). In a similar manner, dipolar interactions can be recovered in MAS experiments by manipulating the spinning axis (Gullion et al., 1991; Terao et al., 1986; Tycko, 1994), by applying rotor-synchronous pulses, or through a judicious choice of spinning speeds. The latter two methods for dipolar recovery are the subject of this chapter. We focus on two specific MAS NMR experiments that can be used to measure accurately the distances between labeled spin pairs. One of these, Rotational-Echo, Double-Resonance (REDOR) measures heteronuclear dipolar interactions (e.g., ^{13}C–^{15}N), the other, Rotational Resonance (R^2), measures homonuclear interactions (e.g., ^{13}C–^{13}C). In addition to our discussions of REDOR and R^2, we will briefly describe a number of other recently developed experiments for determining internuclear distances.

Although the REDOR and R^2 experiments are relatively new, a number of interesting and diverse applications have already been described. Both the REDOR and R^2 methods were initially developed for characterizing biological solids and the majority of applications of these techniques published to date have been devoted to studies of biological systems. Table 1 summarizes the wide range of biological solids studied by means of REDOR and R^2. Inspection of this table shows that distances ranging from 1.3 to 11 Å have been measured. In addition, these techniques have been applied to a variety of nonbiological systems, including glassy polymers (Pavlovskaya et al., 1993; Schmidt et al., 1993) and zeolites (Fyfe et al., 1992; Murray et al., 1993). In the following two sections of this review, we first examine REDOR and then R^2 and highlight in greater detail several of the studies of biological systems listed in Table 1.

TABLE 1
Distance Measurements in Biological Solids by REDOR and R² NMR

System and spin pair	MW (kDa)	r (Å)	Technique	Reference
D-Ala-D-Ala ligase/inhibitor complex $^{31}P-^{31}P$	39	2.7	R²	McDermott et al., (1990)
Amyloid-protein fragment $^{13}C-^{13}C$	1	3.9	R²	Spencer et al., (1991)
Bacteriorhodopsin-bound retinal $^{13}C-^{13}C$	26	3.0–4.2	R²	Creuzet et al. (1991); Thompson et al. (1992); Lakshmi et al. (1993)
α-helical undecapeptide $^{13}C-^{13}C$	1	3.7–6.8	R²	Peersen et al. (1992)
GpATM model peptide $^{13}C-^{13}C$	1	4.5–5.0	R²	Peersen and Smith (1993)
Dipalmitoylphosphatidylcholine $^{13}C-^{13}C$	0.2	4.0–4.5	R²	Peersen and Smith (1993)
Thermolysin/inhibitor complex $^{13}C-^{31}P$	35	1.6–1.7	2D REDOR	Copie et al., (1990)
Emerimicin fragment $^{13}C-^{15}N$	1	4.0	REDOR	Marshall et al. (1990)
Emerimicin fragment $^{13}C-^{19}F$	1	7.8	REDOR	Holl et al. (1992)
EPSPS/S3P/glyphosate complex $^{13}C-^{31}P$	46	5.6–7.2	REDOR	Christensen and Schaefer (1993)
Melanostatin $^{13}C-^{15}N$	0.3	2.4–4.8	REDOR	Garbow and McWherter (1993)
Gramacidin $^{13}C-^{2}H$	2	1.3	2D REDOR	Hing and Schaefer (1993)
Rat cellular retinol binding protein II $^{13}C-^{19}F$	21	7–11	REDOR	McDowell et al. (1993)

3.2 Measuring Distances between Isolated Heteronuclear Spin Pairs by MAS NMR

In Section 3.1, we discussed the motivation behind the development of solid-state NMR experiments for measuring dipolar couplings between isolated spin pairs under fast MAS conditions. Ideally, such experiments should measure dipolar interactions independently of all other spin parameters, including CSA and relaxation. For heteronuclear spin pairs, *decoupling* of the dipolar interaction from the CSA can be easily achieved since the heteronuclear dipolar interaction commutes with the chemical-shift interaction. Relaxation effects can be eliminated by performing a difference experiment similar to that suggested in Section 3.2.1.

In this section, we focus on rotational-echo, double-resonance (REDOR) NMR as a technique for accurately measuring heteronuclear dipolar couplings (Gullion and Schaefer, 1989a,b). REDOR recovers the heteronuclear dipolar interaction by applying trains of rotor-synchronous π pulses. These pulse trains cause a net dipolar dephasing, which leads to a loss of signal intensity. After describing the REDOR experiment and two applications thereof, we examine two other experiments for recovering heteronuclear dipolar interactions. The first, Transfer-Echo, Double-Resonance (TEDOR) NMR (Hing *et al.*, 1992, 1993), also uses rotor-synchronous π pulse trains. However, TEDOR differs from REDOR in that it involves polarization transfer of magnetization between the coupled spins. Finally, we will briefly describe Rotary-Resonance Recoupling (R^3) (Levitt *et al.*, 1988; Oas *et al.*, 1988; Raleigh *et al.*, 1988a), an experiment that recovers the heteronuclear dipolar interaction between a coupled spin pair by applying a weak, continuous rf field to one spin while acquiring a signal from the other. Throughout this section we emphasize the measurement of $^{13}C-^{15}N$ and $^{13}C-^{31}P$ dipolar couplings, although the methodology described can be applied equally well to other isolated heteronuclear spin pairs, as alluded to in Section 3.1.

3.2.1 A Simple Analogy

As an introduction to REDOR, we begin by discussing a simple liquid-state NMR analogy. Consider an isolated spin at exact resonance in a strong, static Zeeman field. At time t_0, we subject this spin to a low-amplitude, oscillatory magnetic field, H_L, of amplitude D and angular frequency ω_r, applied parallel to the Zeeman field. The question we ask is this: Given this field's angular frequency, how can we determine D?

One simple solution to this question is suggested in Figure 1a, in which the spin system is prepared in an initial state of magnetization M_x by a $(\pi/2)_y$ pulse before the oscillatory field is turned on. The field then acts on the spin for a time T_r ($= 2\pi/\omega_r$) after which data acquisition begins. The x component of magnetization at T_r is

$$M_x(T_r) = M_0 \cos \phi(T_r) e^{-T_r/T_2}, \quad (3.8)$$

where M_0 is the equilibrium magnetization, T_2 is the transverse relaxation

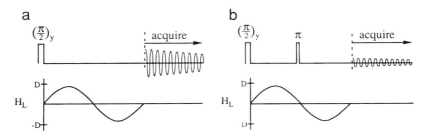

Figure 1. Measuring the strength of a weak, periodic interaction. (a) A periodic interaction is applied during the evolution period and the magnetization is sampled afterward. Since the applied interaction averages to zero, it has no effect on the magnetization. (b) The interaction can be recovered by applying a π pulse at the halfway point of the evolution period. The π pulse interferes with the averaging of the periodic interaction, causing a decrease in signal amplitude.

time, and $\phi(T_r)$ is the accumulated phase caused by the oscillatory field during the time between the $(\pi/2)_y$ pulse and data acquisition. For the sequence shown in Figure 1a, $\phi(T_r)$ is zero since the oscillatory field, on average, is equally positive and negative, and the signal intensity at the point of acquisition is simply

$$M_{x,1}(T_r) = M_0 e^{-T_r/T_2}. \qquad (3.9)$$

Thus, signal loss is due only to relaxation and not to the effects of the small oscillatory field.

Now consider the pulse sequence in Figure 1b, which contains a π pulse at $T_r/2$. The function of this additional pulse is to prevent the refocusing of the dephased magnetization during the time interval from $T_r/2$ to T_r. For this pulse sequence, the x component of magnetization at T_r is

$$M_{x,2}(T_r) = M_0 \cos\left(\frac{4\gamma D}{\omega_r}\right) e^{-T_r/T_2}, \qquad (3.10)$$

where γ is the magnetogyric ratio of the nuclear spin. Although this equation contains the strength of the oscillating magnetic field D, it also contains M_0 and T_2. This is undesirable because determination of D by this sequence alone requires prior knowledge of both M_0 and T_2. However, if we perform both experiments sketched in Figure 1, D can be determined without knowing either M_0 or T_2. For example, the ratio of the *difference* signal $M_{x,1}(T_r) - M_{x,2}(T_r)$ [defined as $\Delta M_x(T_r)$] to the *full* signal $M_{x,1}(T_r)$ is

$$\frac{\Delta M_x(T_r)}{M_{x,1}(T_r)} = 1 - \cos\left(\frac{4\gamma D}{\omega_r}\right). \qquad (3.11)$$

Thus, this measured ratio eliminates both M_0 and T_2 and contains D as the only unknown parameter. A similar normalized difference is used when determining heteronuclear dipolar couplings in REDOR experiments.

3.2.2 Rotational-Echo, Double-Resonance NMR

As described previously, the heteronuclear dipolar coupling between two spin-$\frac{1}{2}$ nuclei (e.g., ^{13}C and ^{15}N) is spatially averaged to zero by MAS. However, as illustrated by the simple experiment described in Section 3.2.1, interactions that are coherently averaged to zero can be recovered by defeating the averaging process with rf irradiation. There, the oscillating field strength D was recovered by the application of a single π pulse. We now describe an experiment to recover the heteronuclear dipolar interaction in solids. This experiment, REDOR NMR (Gullion and Schaefer, 1989a,b), uses rotor-synchronous π pulse trains to recover the dipolar interaction.

The specific form for the dipolar Hamiltonian given by Eq. (3.5) for an isolated heteronuclear spin pair spinning about the magic angle at a rate ω_r is (Maricq and Waugh, 1979)

$$H_D(t) = \frac{\gamma_I \gamma_S \hbar^2}{r^3}[\sqrt{2}\sin2\beta\cos(\alpha + \omega_r t) - \sin^2\beta\cos2(\alpha + \omega_r t)]I_z S_z. \tag{3.12}$$

The polar angle β and azimuthal angle α define the internuclear vector with respect to the sample spinning axis. Over one rotor cycle, the *spatial* part of the interaction is averaged to zero by MAS. To recover and measure the ^{13}C–^{15}N distance, a method must be found to interfere with the spatial averaging of the dipolar interaction. By applying rotor-synchronous rf pulses, REDOR manipulates the *spin* coordinates, which defeats the averaging caused by sample rotation.

The pulse sequences shown in Figure 2 illustrate how the spatial averaging of the dipolar couplings can be partly defeated. In analyzing these pulse sequences, we assume a pair of spin-$\frac{1}{2}$ nuclei, I and S, and initially consider only the heteronuclear dipolar interaction between them. The pulse sequence in Figure 2a prepares S-spin transverse magnetization with a $(\pi/2)_y$ pulse. This magnetization evolves for one rotor cycle and the signal is acquired after this evolution period. During the one-rotor-cycle evolution period, the magnetization evolves in the transverse plane according to the equation of motion defined by Eq. (3.12). [An illustrative spin trajectory (Olejniczak *et al.*, 1984) for a single molecular orientation is shown in Figure 2b.] Immediately after the pulse, the spin is oriented along the x axis. During the evolution period, it dephases. However, the net dephasing after one rotor cycle is zero because of the spatial averaging caused by the sample spinning. Thus, as predicted by

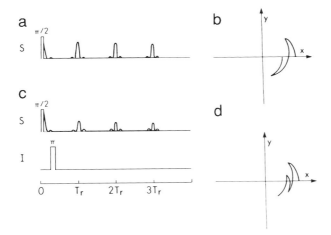

Figure 2. Illustration of the effect of a single I-spin π pulse on the S-spin magnetization in a heteronuclear I-S dipolar-coupled spin pair. (a) A train of rotational echoes, separated by T_r, following an excitation pulse. The rotational echoes result from the spatial averaging of the dipolar coupling caused by MAS. (b) A representative S-spin magnetization trajectory. The dipolar coupling causes dephasing as the sample rotates. However, no *net* dipolar evolution occurs and, consequently, the net dephasing after one rotor cycle is zero. (c) Application of an I-spin π pulse interrupts the normal dipolar evolution, producing rotational echoes having diminished amplitudes. (d) Following the I-spin π pulse, the trajectory of the S spin is changed relative to the trajectory shown in Part (b). (Reproduced with permission from Gullion and Schaefer, 1989b. Copyright © 1989 Academic Press.)

Eq. (3.12), a *rotational echo* forms at the end of the one-rotor-cycle evolution period and after each subsequent rotor period. Each rotational echo is less intense than its predecessor due to T_2 processes.

Now consider the application of a π pulse on the I channel a time t_1 after the S-spin $(\pi/2)_y$ pulse, as illustrated in Figure 2c. Up to the point of this π pulse, the evolution of S magnetization is the same as that described earlier. However, by flipping the orientation of the I spin, the π pulse changes the sign of the local dipolar magnetic field experienced by the S spin. Thus, after the pulse, the S-spin evolution is opposite to that occurring when no pulse is applied. The spin trajectory for the S spin under these conditions is shown in Figure 2d. In addition to losses due to T_2 relaxation, the amplitude of the rotational echo at T_r will be further reduced because of a net dephasing of magnetization due to the dipolar interaction. This reduction due to dipolar dephasing clearly depends on the placement of the I-channel π pulse. For example, if t_1 is zero or T_r, no net dipolar dephasing occurs.

It is straightforward to show that when the I-channel π pulse is placed t_1 from the beginning of a rotor cycle, the average S-spin precessional frequency during that rotor cycle is (Gullion and Schaefer, 1989b)

$$\bar{\omega}_D = \pm \frac{\gamma_I \gamma_S \hbar}{4\pi r^3} \{2\sqrt{2} \sin 2\beta [\sin(\alpha + \omega_r t_1) - \sin\alpha] \\ - \sin^2\beta [\sin 2(\alpha + \omega_r t_1) - \sin 2\alpha]\}. \quad (3.13)$$

For the one-rotor-cycle experiment shown in Figure 2c, the dephasing due to the application of the π pulse is $\phi(\alpha,\beta;t_1) = \bar{\omega}_D(\alpha,\beta;t_1)T_r$. This accumulated phase can be increased by simply adding more rotor cycles with I-channel π pulses during the evolution period. For an N_c rotor cycle experiment, the net dipolar dephasing is $\Delta\Phi(\alpha,\beta;t_1) = \bar{\omega}_D(\alpha,\beta;t_1)N_cT_r$.

Figure 3a shows the pulse sequence for the original REDOR experi-

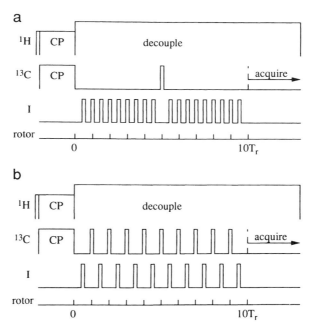

Figure 3. Two versions of the REDOR pulse sequence. In each, a dipolar evolution period of 10 rotor cycles ($N_c = 10$) is shown. The REDOR experiment requires two π pulses per rotor cycle during the evolution period. The control experiment producing the *full* REDOR spectrum is performed without I-channel π pulses, whereas the *reduced* REDOR spectrum is obtained by applying the I-channel pulses. (a) Original REDOR pulse sequence using a single ^{13}C refocusing π pulse. (b) Alternative version of the REDOR experiment with one [13]C π pulse per rotor cycle. This sequence refocuses carbon magnetization better than in the original REDOR experiment. In both experiments, xy-8 phase cycling of the π pulse trains compensates for off-resonance effects. Typical π pulse lengths are 10 to 15 μs and a strong proton decoupling field [H_1(^{1}H)$_{DCP}$ > 90 kHz] is applied during the evolution period to minimize signal losses due to [1]H–[13]C dipolar couplings. Because REDOR experiments depend on the formation of rotational echoes, careful control of the spinning speed is required. Specially designed electronic circuitry can control spinning speeds to ±3 Hz of their nominal values.

ment. This experiment begins with cross-polarization of the ^{13}C spins via the protons. After cross-polarization, a strong rf decoupling field removes the proton–carbon dipolar interaction throughout the remainder of the experiment. Typical proton decoupling levels are 100 kHz during the dipolar evolution period and 80 kHz during data acquisition. The cross-polarization preparation is followed by a dipolar evolution period consisting of N_c rotor cycles (here $N_c = 10$). A π pulse is applied on the ^{13}C channel in the middle of the dipolar evolution period to refocus isotropic chemical shifts prior to data acquisition, which immediately follows the evolution period. The *full* spectrum is obtained with no π pulses on the I channel. In this experiment, the dipolar interaction is averaged to zero by the sample rotation. However, when a train of π pulses is applied on the I channel, the ^{13}C magnetization is dephased by the ^{13}C–I dipolar interaction. Two π pulses are applied per rotor cycle to ensure that dipolar dephasing is additive. This dephased signal is called the *reduced* signal.

An alternative version of the REDOR experiment is shown in Figure 3b (Gullion *et al.*, 1990a). In principle, there is no difference between this pulse sequence and the one shown in Figure 3a. However, experience shows that this alternative sequence refocuses the carbon magnetization better than the experiment with a single carbon π pulse. We believe this is due to the effects of resonance offsets, which cause the effective flip angle of a single π pulse to deviate from 180°. However, by using a compensated π pulse train (see Figure 7), flip angle errors due to resonance offset can be minimized (Gullion *et al.* 1990a,b; Gullion and Schaefer, 1991; Gullion, 1993).

Figure 4 shows ^{13}C–^{15}N REDOR spectra of [1–^{13}C, ^{15}N]glycine diluted 1:10 with natural-abundance glycine for N_c equal to 10, 18, 26, and 34. Dilution helps to minimize intermolecular ^{13}C–^{15}N dipolar interactions between labeled glycine molecules in the sample. Clearly, significant carbon dephasing occurs when the ^{15}N pulses are present (Figure 4b) and the amount of dipolar dephasing is sensitive to the number of rotor cycles present during the evolution period.

Figure 5 shows the dependence of $\Delta S/S_0$ on sample spinning rate for a sample of [2–^{13}C, ^{15}N]alanine diluted 1:10 with natural-abundance alanine (Gullion and Schaefer, 1989a). The term $\Delta S/S_0$ is simply the *difference* signal, ΔS, normalized by the *full* signal, S_0, after accounting for natural-abundance ^{13}C spins not coupled to a ^{15}N spin (see Section 3.2.3.3) and is given by (Gullion and Schaefer, 1989b)

$$\Delta S(t_1)/S_0 = 1 - \frac{1}{2\pi} \int_0^{2\pi} d\alpha \int_0^{\pi/2} \cos\Delta\Phi(\alpha, \beta; t_1) \sin\beta \, d\beta. \quad (3.14)$$

Since the dipolar evolution develops according to Eq. (3.13), the signal loss due to dipolar dephasing is described by the cosine projection, and spatial integration of this equation is necessary only over a hemisphere. As sug-

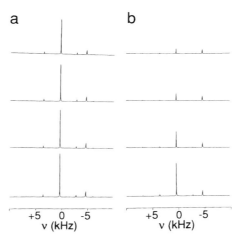

Figure 4. REDOR ^{13}C NMR spectra of [1–^{13}C, ^{15}N]glycine diluted 1 part in 10 with natural-abundance glycine. These spectra were collected using the pulse sequence of Figure 3b. (a) *Full* spectra, collected in the absence of any ^{15}N π pulses. (b) *Reduced* spectra, collected with one ^{15}N π pulse per rotor period. In a and b N_c equals 10, 18, 26, and 34 from bottom to top, respectively. The decrease in signal intensity of the *full* spectra as N_c increases is due to T_2 effects (e.g., incomplete proton decoupling). Experimental: H_0 = 3.55 T; H_1(^{13}C, ^{15}N) = 40 kHz; v_r = 3.2 kHz; proton decoupling field, H_1(^{1}H)$_{DCP}$, = 110 kHz.

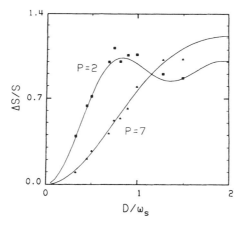

Figure 5. Dependence of the normalized REDOR difference signal $\Delta S/S$ on the spinning rate, ω_s, for a sample of [2–^{13}C,^{15}N]alanine recrystallized 1 part in 10 in natural-abundance alanine. The value $\Delta S/S$ differs from $\Delta S/S_0$ in that the former is the normalized REDOR difference signal *before* correcting for natural-abundance spins (Section 3.2.2.3). In this isotopically enriched alanine sample, the difference between the values of $\Delta S/S$ and $\Delta S/S_0$ is negligible. These results were generated using the pulse sequence of Figure 3a with N_c = 2. The P = 2 (squares) data were obtained with ^{15}N pulses at integer multiples of $T_r/2$ and the P = 7 (triangles) data were obtained with ^{15}N pulses at integer multiples of $T_r/7$ and T_r. The solid lines were calculated using Eq. (3.14) assuming a dipolar coupling D of 895 Hz. Experimental: H_0 = 4.7 T; H_1(^{13}C, ^{15}N) = 38 kHz; v_r = 1 kHz; H_1(^{1}H)$_{DCP}$ = 110 kHz. (Reproduced with permission from Gullion and Schaefer, 1989a. Copyright © 1989 *Academic Press*.)

gested by our analysis of the experiment described in Figure 1, this ratio is independent of T_2. Since D, the dipolar coupling constant, is fixed for this sample, these data were collected by changing the spinning speed ω_s to vary the ratio D/ω_s. The $P = 2$ (squares) and $P = 7$ (triangles) data of Figure 5 are for t_1 equal to $T_r/2$ and $T_r/7$, respectively. The solid lines are simulations using Eq. (3.14) and a dipolar coupling constant of 895 Hz, corresponding to a $^{13}C-^{15}N$ distance of 1.50 Å. The experimental data and simulations are in excellent agreement; they also agree well with x-ray results reporting a $^{13}C-^{15}N$ distance of 1.49 Å (Wyckoff, 1966). The small difference in the $^{13}C-^{15}N$ distance between the REDOR and x-ray measurements is probably due to ultra-high-frequency molecular motions (Henry and Szabo, 1985; Torchia and Szabo, 1982). Measurements on fixed-distance model compounds consistently show heteronuclear dipolar couplings measured by REDOR to be ~10% smaller than the rigid-lattice limit. Similar reductions in coupling have also been noted in the measurement of $^1H-^{13}C$ dipolar couplings and were attributed to high-frequency molecular vibrations and librations (Munowitz and Griffin, 1982; Schaefer et al., 1983). Several features of this plot are worth noting. First, the $P = 2$ data rise much more rapidly than the $P = 7$ data. This demonstrates that the $P = 2$ experiment requires a shorter dipolar evolution time to achieve the same $\Delta S/S_0$ value. Second, dipolar oscillations, reminiscent of a free-induction decay of a dipolar coupled spin system, are evident for large values of D/ω_s in the $P = 2$ case.

For weak dipolar couplings, REDOR data are best characterized by computing values of $\Delta S/S_0$. For stronger couplings, dipolar powder patterns can be generated in 2D REDOR experiments (Gullion et al., 1988; Gullion and Schaefer, 1989b; Schmidt et al., 1992; Schmidt and Vega, 1992; Weintraub and Vega, 1993). One version of 2D REDOR involves fixing the total evolution period and incrementing t_1 from 0 to T_r. This procedure modulates the signal intensity and produces, upon Fourier transformation, a dipolar pattern defined by a series of sidebands. Figure 6a shows such a pattern, which has an envelope characteristic of the familiar Pake powder pattern (Pake, 1948). Simulation of this pattern yields the dipolar coupling constant. In processing data from this experiment, it is important to add the data obtained at times t_1 and $T_r - t_1$, since this creates a data set of pure absorption mode spectra (Gullion and Schaefer, 1989b; Gullion and Conradi, 1990; Levitt, 1989). Another method for generating dipolar powder patterns is to fix t_1 and increment the number of rotor cycles of dipolar evolution, N_c (Gullion and Schaefer, 1989b). Fourier transformation of these data with respect to the dipolar evolution time produces a Pake-spun pattern such as that shown in Figure 6b. In this experiment, the dipolar coupling can be determined from the separation of the edges of the spectrum. It is important to

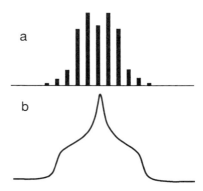

Figure 6. Simulated 2D REDOR powder patterns. (a) Sideband pattern resulting from incrementing t_1 from 0 to T_r and Fourier transforming the data with respect to t_1. The sideband pattern has an envelope characterized by the traditional Pake powder pattern. (b) Pake-spun pattern. This pattern can be produced in either of two ways: (1) Fix the time between I-spin π pulses to $T_r/2$ and increment the number of rotor cycles or (2) place the I-spin π pulses at integer multiples of $T_r/4$ and $3T_r/4$ and acquire the data synchronously every rotor cycle. Fourier transformation of the acquired data produces a Pake-spun spectrum. The dipolar coupling is $\sqrt{8}/\pi$ times the difference between the shoulders of this pattern.

emphasize that the generation and analysis of such 2D powder patterns is only practical for large dipolar couplings. For weak couplings, REDOR data are analyzed by measuring $\Delta S/S_0$.

Since REDOR experiments consist of rotor-synchronous pulse trains, it is important to understand the types of errors that may arise in such experiments. First, we consider errors arising from the rf pulses. Figure 7 shows $\Delta S/S$ values for a series of ^{13}C-observe REDOR experiments collected as a function of ^{15}N carrier offset (Gullion and Schaefer, 1991). The filled circles show results of experiments in which all ^{15}N pulses had the same phase. Clearly, the data are severely dependent on resonance offset. This resonance-offset dependence can be eliminated by phase-alternating the ^{15}N pulses according to the xy-4 (xyxy) or xy-8 (xyxyyxyx) phase cycling schemes (Gullion et al., 1990a,b; Gullion, 1993; Weintraub and Vega, 1993). The $N_c =$ 32 and 48 data illustrate the benefit of this phase cycling. REDOR data collected using xy-8 phase cycling are nearly offset-independent over a large range of resonance offsets.

Another concern of any experiment involving a synchronously applied pulse train is sensitivity to the stability of the spinning speed. Typically, spinning speeds in REDOR experiments are controlled to within ± 3 Hz. The pulse sequences sketched in Figure 3 are tolerant to minor rotor-speed variations (Garbow and Gullion, 1992).

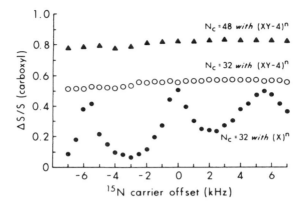

Figure 7. Dependence of the ^{13}C REDOR $\Delta S/S$ on the ^{15}N carrier frequency for the carboxyl carbon of a recrystallized equimolar mixture of L-[2-^{13}C, ^{15}N]alanine and L-[1-^{13}C]alanine. The carrier offset is referenced to the isotropic ^{15}N chemical shift of alanine. The solid circles are for an N_c = 32 REDOR experiment using the pulse sequence of Figure 3a with all ^{15}N pulses having the same phase. The open circles are for the same N_c = 32 REDOR experiment with xy-4 phase cycling of the ^{15}N pulses. The closed triangles are for a 48-rotor-cycle REDOR experiment with xy-4 phasing of the ^{15}N pulses. The experiments were performed with ^{15}N pulses spaced by half the rotor period. Experimental: H_0 = 4.7 T; H_1(^{13}C, ^{15}N) = 38 kHz; v_r = 2.7 kHz. (Reproduced with permission from Gullion and Schaefer, 1991. Copyright © 1991 J. Mag. Res.)

3.2.3 A General REDOR Strategy for the Elucidation of Protein Backbone Conformations: ^{13}C–^{15}N REDOR of Melanostatin

As an application of the REDOR technique, we summarize a recently reported study (Garbow and McWherter, 1993) of the tripeptide melanostatin, Pro–Leu–Gly–NH$_2$, a neurohormone responsible for modulating dopamine receptors in the central nervous system that might also inhibit the release of melanin-stimulating hormone (Mishra et al., 1983). In that study, a series of ^{13}C,^{15}N-labeled tripeptides were prepared and studied by ^{13}C-detected REDOR. From the REDOR data, several ^{13}C–^{15}N dipolar coupling constants were determined, allowing the distances between the ^{13}C- and ^{15}N-labeled sites to be calculated. The REDOR distances effectively constrained the backbone dihedral angles of melanostatin, thereby providing a direct determination of its solid-state conformation.

Figure 8 shows the REDOR ^{13}C NMR spectra of a sample of [1-^{13}C]Pro–Leu–[^{15}N]Gly–NH$_2$, diluted 1:9 in natural-abundance melanostatin, collected with N_c = 32. The internuclear distance between the labeled

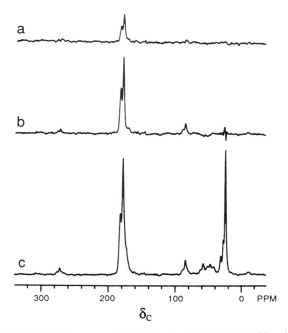

Figure 8. REDOR ^{13}C NMR spectra of [1-^{13}C]Pro-Leu-[^{15}N]Gly-NH$_2$ diluted 1 part in 9 with natural-abundance tripeptide. (a) REDOR *difference* spectrum. (b) Spectrum showing signal due to [1-^{13}C]leucyl carbon only. (c) *Full* spectrum collected in the absence of all ^{15}N π pulses. Signals in this spectrum arise from the ^{13}C-enriched and natural-abundance carbonyl carbons (δ_C, 172 to 178 ppm) and natural-abundance methyl (δ_C, 23ppm), methylene (δ_C, 27, 31 ppm), other sidechain aliphatic (δ_C, 40 to 50 ppm), and C$_\alpha$ (δ_C, 50 to 60 ppm) in the tripeptide. Signals near 80 and 270 ppm are spinning sidebands. Experimental: H$_0$ = 3 T; H$_1$(^{13}C) = 50 kHz; H$_1$(^{15}N) = 38 kHz; v_r = 3 kHz; N$_c$ = 32; H$_1$(^{1}H)$_{DCP}$ = 110 kHz during the dipolar evolution period and 80 kHz during data acquisition. (Reprinted with permission from Garbow and McWherter, 1993. Copyright © 1993 American Chemical Society.)

sites in this molecule depends on the dihedral angles ϕ_{Leu} and ψ_{Leu}. The spectrum of Figure 8c is that of the *full* sample collected in the absence of all ^{15}N π pulses and shows signals arising from both the ^{13}C-enriched leucyl carbonyl carbon and of all the natural-abundance carbons in the tripeptide. Figure 8b shows the ^{13}C signal due to the labeled leucyl carbonyl carbon only, obtained by subtracting the spectrum of a natural-abundance melanostatin sample collected under identical experimental conditions. The spectrum of this labeled carbon is an asymmetric doublet, with the splitting due to the dipolar coupling between the carbonyl carbon and directly bonded ^{14}N spin in the peptide bond. The REDOR *difference* signal, ΔS, is shown in Figure 8a. As discussed later, the experimentally measured $\Delta S/S_0$ value (0.374) must be corrected for natural-abundance contributions before a dipolar coupling can be determined (see Section 3.2.3.3).

The determination of dipolar couplings requires being able to account accurately for natural-abundance contributions to both the REDOR echo and difference signals. As illustrated in Figure 8, contributions to the echo signal from natural-abundance spins are generally removed experimentally by acquiring the spectrum of an unlabeled sample and subtracting it from the spectrum of the labeled sample. Often, natural-abundance contributions to the REDOR *difference* signal can be calculated directly, as described in Section 3.2.3.3. However, for situations in which such calculations are not possible, a *double-difference* REDOR method, illustrated in Figure 9, was developed (Garbow and McWherter, 1993). Figure 9 shows REDOR ^{13}C NMR spectra of two different [^{15}N]Leu-labeled melanostatins: Pro–[^{15}N]Leu–[1–^{13}C]Gly–NH$_2$ and Pro–[^{15}N]Leu–[2–^{13}C]Gly–NH$_2$, each diluted 1:49 in natural-abundance tripeptide. Because the [1–^{13}C]Gly-enriched melanostatin sample is natural abundance in the C$_\alpha$ position of

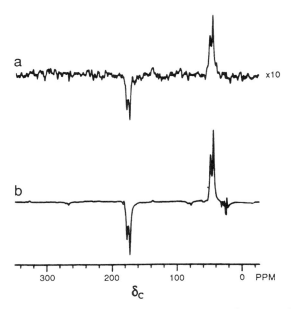

Figure 9. REDOR ^{13}C NMR spectra of Pro–[^{15}N]Leu–[1–^{13}C]Gly–NH$_2$ and Pro–[^{15}N]Leu–[2–^{13}C]Gly–NH$_2$, each diluted 1:49 with natural-abundance tripeptide. (a) Difference of REDOR *difference* spectra. Experimental conditions are the same as given in Figure 8. (b) Difference of *full* spectra of these two samples, showing signals from labeled carbons only. (Reprinted with permission from Garbow and McWherter, 1993. Copyright © 1993 American Chemical Society.)

Gly and vice versa for the [2-^{13}C]Gly-enriched sample, these two samples can serve to experimentally correct each other's natural-abundance ^{13}C contributions to both the full-echo signal and $\Delta S/S_0$. This experimental self-correction for natural-abundance REDOR *difference* signals is necessary, because there are contributions to ΔS from natural-abundance sidechain leucyl carbons that are not chemical-shift resolved from the labeled [2-^{13}C]Gly carbon. The partial overlap of carbon signals makes these contributions to $\Delta S/S_0$ difficult to calculate.

Figure 9a shows the *difference* spectrum of the REDOR *difference* spectra for these two samples. Figure 9b shows the *difference* spectrum formed by subtracting the *full* spectrum of the [1-^{13}C]Gly/[^{15}N]Leu sample from that of the [2-^{13}C]Gly/[^{15}N]Leu sample. The signal from the enriched [2-^{13}C]Gly appears as a positive-going doublet in this spectrum; the signal from the enriched [1-^{13}C]Gly position as a negative-going signal.

The *double-difference* REDOR method is useful anytime contributions to REDOR *difference* signals arising from natural-abundance carbons cannot be easily calculated. Examples include this melanostatin case, in which the natural-abundance signals are only partially chemical-shift resolved, and cases where natural-abundance contributions are unknown, which might occur for a bound, labeled inhibitor that is near naturally occurring ^{13}C spins in the active site of an enzyme.

The last step in the processing of REDOR data is to convert the measured dipolar couplings to REDOR distances according to Eq. (3.14). In converting these couplings to distances, the aforementioned reduction in dipolar coupling due to ultra-high-frequency motion (i.e., 875 Hz ↔ 1.49 Å) was assumed. Results of the REDOR experiments on melanostatin are summarized in Table 2. The near exact agreement in the [1-^{13}C,^{15}N]Leu sample between the distance measured in REDOR and the distance known from the

TABLE 2
REDOR ^{13}C NMR of ^{13}C, ^{15}N-Enriched Melanostatins

					r_{CN}	
^{13}C Site	^{15}N Site	N_c	$\Delta S/S_0^{a,b}$	D_{CN} (Hz)	REDOR (Å)	X-ray[c] (Å)
[1-^{13}C]Leu	[^{15}N]Leu	16	0.777	199	2.43	2.42
[1-^{13}C]Pro	[α-^{15}N]Gly	32	0.346	57	3.69	3.70
[2-^{13}C]Gly	[^{15}N]Leu	32	0.081	26	4.78	4.68
[1-^{13}C]Gly	[^{15}N]Leu	32	0.073	25	4.87	4.97

[a]Computed as the ratio of measured peak heights.
[b]Includes correction for natural-abundance terms, as described under Section 3.2.3.3.
[c]C. A. McWherter, J. R. Garbow, H.-S. Shieh, and M. Chiang, unpublished results (1993).

leucyl geometry validates the assumption that the averaging of dipolar couplings by high-frequency motions in melanostatin is similar to that in crystalline glycine. The three conformationally dependent REDOR distances measured are in excellent agreement with those determined in an X-ray study of melanostatin (Reed and Johnson, 1973).

3.2.3.1 Conformational Analysis

Having established that carbon–nitrogen distances can be accurately measured in this tripeptide via REDOR, the problem becomes one of deriving constraints on molecular conformation from these distances. Planar peptide bonds, standard amino acid bond geometries and bond lengths, and an error of ± 0.1 Å in all REDOR determined distances are assumed throughout the analysis. The result of combining the conformational constraints imposed by REDOR experiments on [1-^{13}C]Pro–Leu–[^{15}N]Gly–NH$_2$ (ϕ_{Leu}, ψ_{Leu}) and Pro–[^{15}N]Leu–[2-^{13}C]Gly–NH$_2$ (ψ_{Leu} only) is shown in black in Figure 10. The two black patches, describing those dihedral angle values that simultaneously satisfy both distance constraints, represent only 2% of the

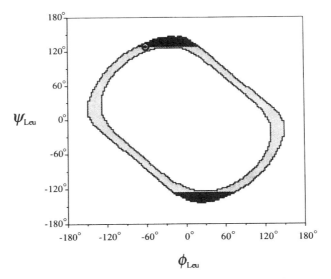

Figure 10. Conformational map showing how the REDOR-measured distances constrain the allowed values of ϕ_{Leu} and ψ_{Leu} in melanostatin. The shaded annular region show the conformations consistent with the REDOR-determined ^{13}C–^{15}N distance of 3.69 Å between [1-^{13}C]Pro and [^{15}N]Gly. Further constraints (shown in black) are provided by the 4.78-Å ^{13}C–^{15}N distance between [^{15}N]Leu and [2-^{13}C]Gly. The lower black patch can be discarded from conservative energy considerations derived from conformational energy maps for the nonglycine amino acids. (Reprinted with permission from Garbow and McWherter, 1993. Copyright © 1993 American Chemical Society.)

total (ϕ_{Leu}, ψ_{Leu}) dihedral space. Conservative energy arguments permit the lower patch to be rejected, leaving the upper patch, which includes the X-ray determined distance. A similar approach was followed to establish REDOR constraints on other pairs of dihedral angles. Considering only those portions of the molecule constrained by the REDOR measurements, a family of backbone structures consistent with the REDOR data was generated. These 14-atom backbone fragments are all characterized by a turn throughout the central portion of the tripeptide. The pair-wise, minimum RMSD between the family of REDOR structures and the X-ray conformation is 0.437 ± 0.224 Å (average ± standard deviation).

3.2.3.2 General Labeling Strategy

The results of the melanostatin work suggest a general strategy for mapping the backbone of peptides and proteins using REDOR-derived distances (Garbow and McWherter, 1993). This strategy is summarized in Figure 11, which indicates an approach for selecting sites of ^{13}C and ^{15}N isotopic enrichment. Designating the site of ^{15}N enrichment as residue i, ^{13}C labels are placed in the carbonyl carbon of residue $i - 2$ and C_α of $i + 1$. The distances between the labeled carbonyl carbon and nitrogen sites depend on the dihedral angles ϕ_{i-1} and ψ_{i-1}, while the $^{13}C_\alpha$–^{15}N distance depends on ψ_i. In addition, the entire range of spanned distances (2.2 to 5.1 Å for the carbonyl; 4.1 to 5.0 Å for C_α) can be measured by ^{13}C–^{15}N REDOR. Thus, when labeling sites are selected according to this scheme, REDOR experiments are guaranteed to provide two measurable, nontrivial carbon–nitrogen distances, which furnish constraints on backbone dihedral angles. As illustrated in the figure, subsequent samples can be prepared by shifting the triplet of labels, residue-by-residue along the peptide/protein chain to generate a series of overlapping dihedral constraints.

From the point of view of the synthesis, the labeling strategy outlined in Figure 11 can be achieved by preparing either triple-labeled (^{13}C,^{15}N,^{13}C) samples or pairs of double-labeled (^{13}C,^{15}N) samples. The synthesis of triple-labeled samples has the advantage of minimizing the total number of samples that must be prepared. (In this regard, the need to prepare natural-abundance samples must not be overlooked.) One obvious disadvantage of preparing triple-labeled samples is that these samples cannot be studied by ^{15}N-observe REDOR, since signal from the ^{15}N spin will be simultaneously dephased by two different ^{13}C sites.

Note that dihedral-angle mapping of melanostatin, illustrated in Figure 10, is a somewhat special case because only two regions of the ϕ_{Leu}/ψ_{Leu} space satisfy the REDOR-measured distances. In the more general case, up to four patches of conformational space may be consistent with the measured distances. A similar observation was made for ^{13}C–^{13}C distance con-

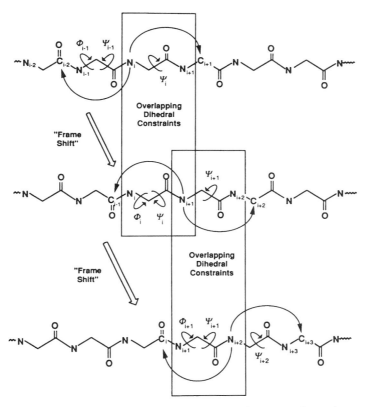

Figure 11. General strategy for mapping the backbone conformation of peptides or proteins using ^{13}C, ^{15}N REDOR NMR. Labels are placed in the [α–N] position of residue i, the carbonyl carbon of residue $i - 2$, and C_α of residue $i + 1$. REDOR experiments provide constraints on the angles ϕ_{i-1}, ψ_{i-1}, and ψ_i. A frame shift of the labeled triple along the backbone leads to overlapping dihedral constraints, which can be used to map the conformation of the backbone. (Reprinted with permission from Garbow and McWherter, 1993. Copyright © 1993 American Chemical Society.)

straints in a synthetic amyloid-protein fragment (Spencer et al., 1991), where homonuclear backbone distances alone failed to uniquely determine peptide dihedrals. In these situations, other experiments, including the introduction of sidechain labels and the measurement of ^{13}C–^{13}C distances using techniques described in the next section, can be used to supplement the REDOR measurements.

We make one additional note about conformational analysis using REDOR data. In designing experiments to map peptide and protein backbones, it was instructive to analyze the REDOR data by building up a series of sequential dihedral-angle constraints, and it was this overlapping constraint approach that led to the development of the labeling strategy outlined earlier. From a practical point of view, however, it is probably easier and more efficient to analyze the data by considering all of the REDOR-generated distances simultaneously and generating families of structures that satisfy the whole of the REDOR data.

3.2.3.3 Natural-Abundance Corrections

As noted, before heteronuclear dipolar couplings can be accurately determined, REDOR NMR data must be corrected for contributions due to natural-abundance ^{13}C and ^{15}N spins. Experimental approaches have been described that employ a fourth spin species, X (e.g., ^{31}P, ^{19}F), to select the ^{13}C signal from labeled carbons only. One such experiment, RETRO (Holl et al., 1990), selects the labeled site through REDOR-like dephasing of the ^{13}C signal due to the X spin. A second method, which combines the TEDOR (see Section 3.2.5) and REDOR experiments, selects the labeled ^{13}C by polarization transfer from the X spin. Once ^{13}C from the labeled site is selected, REDOR experiments are then performed to measure the dipolar coupling between this site and a labeled I spin. While RETRO and TEDOR/REDOR are elegant solutions to the problem of natural-abundance background, they lack generality because they require both a fourth spin species in the sample and a four-channel solids spectrometer. In addition, REDOR *difference* signals from carbons selected via either RETRO or TEDOR/REDOR still receive contributions from natural-abundance I spins near the labeled site. Clearly, other solutions to this problem are required.

Since the dipolar coupling is determined from the $\Delta S/S_0$ ratio, natural-abundance terms affecting either the numerator or denominator will contribute to this ratio. Natural-abundance corrections to $\Delta S/S_0$ can be divided into two broad categories:

1. natural-abundance contributions to the *full* echo signal, S (i.e., the difference between $\Delta S/S$ and $\Delta S/S_0$)
2. natural-abundance contributions to the REDOR *difference* signal, ΔS.

In the case of the *full* echo signal, we must accurately determine the intensity of the signal due to the labeled position. To do this, signals due to natural-abundance ^{13}C spins occurring at the same frequency as the label must be subtracted. For simple molecular systems, such as a small, labeled molecule diluted in a natural-abundance host, it might be possible to calculate the relative contributions of label and natural-abundance signals. An adjustment to S can then be made to accurately determine S_0. For more complex systems, the natural-abundance contributions must be experimentally determined by collecting data on a natural-abundance sample under identical REDOR conditions as those applied to the labeled sample and then subtracting these spectra, one from the other. This has the clear drawback that the *difference* spectrum, resulting from the subtraction of signals from different samples, is never completely clean and often displays minor subtraction artifacts. Nonetheless, the excellent agreement between NMR and X-ray results for melanostatin provides strong evidence that the use of a second sample to estimate natural-abundance contributions to the *full* echo signal is a legitimate approach.

Since REDOR ^{13}C *difference* spectra contain signals from carbons coupled to ^{15}N spins only, it might seem that no correction for natural-abundance contributions to ΔS need be made. However, there are two different classes of molecules that can make natural-abundance contributions to REDOR *difference* spectra:

1. natural-abundance ^{13}C spins having the same resonance frequency as the labeled ^{13}C spin and which are near the labeled ^{15}N position
2. natural-abundance ^{15}N spins near the labeled ^{13}C position.

One strategy for experimentally determining the natural-abundance contribution to ΔS is the *double-difference* REDOR method described previously. Alternatively, corrections to ΔS can be made experimentally by collecting REDOR data on ^{13}C- and ^{15}N-only-labeled samples. Preparation of ^{15}N-only-labeled samples has the advantage that these samples can also serve as the natural-abundance ^{13}C control samples necessary to measure accurately natural-abundance contributions to the *full* echo signal. Often, the necessary natural-abundance corrections to ΔS can be directly calculated. For peptide and protein molecules, as well as others whose covalent molecular structure is known, it is straightforward to enumerate the major natural-abundance contributors to ΔS. These will occur at fixed, short (one- and two-bond) distances. The contribution of these short distances to $\Delta S/S_0$ can be calculated analytically and subtracted from experimentally measured values prior to their conversion into dipolar couplings. This calculational approach, which was used together with the *double-difference* REDOR method in the melanostatin work, has been successfully employed in a variety of different REDOR studies.

3.2.4 ^{31}P–^{13}C REDOR of the EPSPS/S3P/Glyphosate Complex

5-enolpyruvylshikimate-3-phosphate (EPSP), the product of the reversible condensation of shikimate 3-phosphate (S3P) and phosphoenolpyruvate (PEP), is an important intermediate in the biosynthesis of aromatic amino acids. This reaction is catalyzed by EPSP synthase (EPSPS), a 46-kDa enzyme whose structure has been determined previously by X-ray analysis (Stallings et al., 1991). The formation of EPSP is inhibited by the herbicide N-(phosphonomethyl)glycine (glyphosate), which binds to the enzyme in the presence of S3P (Steinrucken and Amrhein, 1980). The structure of the enzyme/S3P/glyphosate ternary complex is unknown because good crystals of this complex are not available. The molecular weight of the complex precludes its analysis by liquid-state NMR. However, important structural information about the complex has been determined from $^{31}P-^{13}C$ REDOR experiments (Christensen and Schaefer, 1993).

Structures of glyphosate and S3P are shown in Figure 12. Each molecule contains one ^{31}P spin, present at 100% natural abundance. If a ^{13}C label is placed on one of the molecules, two $^{31}P-^{13}C$ distance constraints can be generated (one intramolecular, one intermolecular). Although, in principle, REDOR experiments could be performed by observing the ^{13}C spin, data analysis would be complicated because the ^{13}C label is simultaneously coupled to two ^{31}P spins. Instead, the ^{31}P signal was collected because, in the absence of natural abundance ^{13}C, each phosphorus is coupled to a single ^{13}C spin. Initially, the ^{13}C spin label was placed in the carbonyl position of glyphosate, with the intramolecular $^{31}P-^{13}C$ distance probing the extension of the glyphosate molecule and the intermolecular $^{31}P-^{13}C$ distance probing the proximity of S3P to the glyphosate.

REDOR ^{31}P NMR spectra of EPSPS/S3P/[1-^{13}C]glyphosate ternary complex are shown in Figure 13a. Since there are only two phosphorus spins, the spectra consist of the two well-resolved ^{31}P resonances and their associated spinning sidebands. The lower spectrum is the ^{31}P *full* spectrum obtained in the absence of ^{13}C pulses, and the upper spectrum is the REDOR

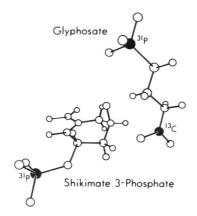

Figure 12. Structures of shikimate-3-phosphate and [1-^{13}C]glyphosate. (Reprinted with permission from Christensen and Schaefer, 1993. Copyright © 1993 American Chemical Society.)

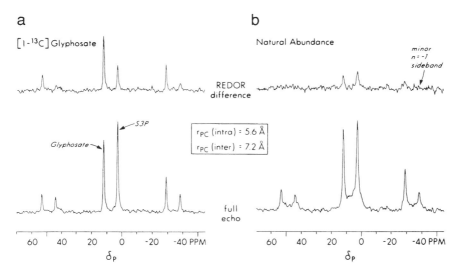

Figure 13. REDOR ^{31}P NMR spectra of S3P and glyphosate bound to EPSP synthase. (a) [1-^{13}C]glyphosate. (b) Natural-abundance glyphosate. Shown at top, REDOR *difference* spectra; shown at bottom, *full* spectra. Experimental: H_0 = 7 T; $H_1(^{31}P)$ = 50 kHz; v_r = 5 kHz; N_c = 64; $H_1(^1H)_{DCP}$ = 70 kHz throughout. (Reprinted with permission from Christensen and Schaefer, 1993. Copyright © 1993 American Chemical Society.)

difference spectrum. As described earlier, the $\Delta S/S_0$ values measured for these spectra must be corrected for dephasing due to natural-abundance ^{13}C spins located near the phosphorus nuclei. The REDOR spectra of an EPSPS/S3P/natural-abundance glyphosate complex, shown in Figure 13b provide an experimental measure of this natural-abundance dephasing. After correcting for these natural-abundance ^{13}C contributions to the REDOR experiment, two ^{31}P–^{13}C dipolar couplings were measured.

Analysis of the REDOR data of the EPSPS/S3P/glyphosate complexes yielded a ^{31}P–^{13}C (phosphonate) distance of 5.6 Å ($\pm$0.2 Å), indicating that the glyphosate molecule is nearly fully extended. The measured ^{31}P–^{13}C (phosphate) distance was 7.2 Å ($\pm$0.4 Å), demonstrating that S3P is close to glyphosate in the complex. Using these two distances and assuming a hydrogen bond between the glyphosate NH proton and the C5 oxygen of S3P, the structural relationship between S3P and glyphosate shown in Figure 12 was proposed. This structure can be confirmed and further refined by measuring other ^{31}P–^{13}C and ^{15}N–^{13}C distances between labeled sites in the S3P and glyphosate molecules.

The ^{31}P–^{13}C distances measured in the EPSPS complex are considerably longer than any reported REDOR measurements for ^{13}C–^{15}N spin pairs. This increased distance range is due to the larger magnetogyric ratio of ^{31}P compared to ^{15}N, since the dipolar coupling between a pair of spins is directly proportional to the product of their magnetogyric ratios. Experience has shown that dipolar couplings as small as 25 Hz can be reliably measured in REDOR experiments. From Eq. (3.5), a 25-Hz dipolar coupling translates to theoretical internuclear distances of 4.96, 7.87, and 10.42 Å for ^{13}C–^{15}N, ^{13}C–^{31}P, and ^{13}C–^{19}F spin pairs, respectively. Examples in this section have illustrated REDOR measurement of ^{13}C–^{15}N and ^{13}C–^{31}P distances. Measurement of ^{13}C–^{19}F distances have also been demonstrated in biological systems in which a specific proton was selectively replaced with a fluorine (Holl et al., 1992; McDowell et al., 1993).

3.2.5 Transfer-Echo, Double-Resonance NMR

As described in preceding sections, natural-abundance contributions to both echo and difference signals can be a major limitation on the accuracy of REDOR-determined distances. Various strategies were discussed for measuring and correcting for these natural-abundance spins. A different approach is provided by another recently developed method, transfer-echo double-resonance (TEDOR) NMR (Hing et al., 1992). In TEDOR, polarization is transferred from one rare spin to another (e.g., ^{15}N → ^{13}C) prior to data acquisition. Thus TEDOR filters out natural-abundance contributions, because only signal from molecules containing *both* rare spins is acquired.

The pulse sequence for ^{13}C-detected TEDOR is shown in Figure 14. The first step in the TEDOR experiment is cross-polarization of the ^{15}N spins

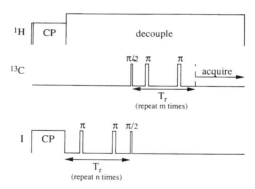

Figure 14. A ^{13}C-detected TEDOR pulse sequence. The ^{13}C- and I-channel $\pi/2$ pulses are applied simultaneously and separate the n- and m-rotor-cycle evolution periods. After cross-polarization from protons, I-spin magnetization evolves into antiphase magnetization under the action of the heteronuclear dipolar coupling. The simultaneous $\pi/2$ pulses produce antiphase ^{13}C magnetization, which evolves into an observable ^{13}C signal.

from the protons. Following this cross-polarization step, polarization is transferred to the ^{13}C spins under the influence of the ^{13}C–^{15}N dipolar coupling. As illustrated in the pulse sequence of Figure 14, π pulses, applied twice per rotor period on the ^{15}N rf channel, interfere with averaging of the heteronuclear dipolar coupling. This effect is the same as that created by the π pulses in the evolution period of REDOR experiments. Although the two π pulses may be placed anywhere during the rotor period, they are most often placed at $T_r/4$ and $3T_r/4$. This placement has the advantage of refocusing the heteronuclear scalar coupling, so that magnetization dephasing is due to the dipolar coupling only.

The nonzero dipolar interaction permits the development of ^{15}N antiphase magnetization described by the density operator $I_y S_z$ (I = ^{15}N, S = ^{13}C) for ^{15}N spins coupled to ^{13}C. This term is allowed to develop for a total of n rotor periods, after which $\pi/2$ pulses are simultaneously applied to the ^{13}C and ^{15}N spins. These pulses generate antiphase ^{13}C magnetization described by the density operator $I_z S_y$. This transfer is reminiscent of INEPT-like transfers in liquids (Morris and Freeman, 1978). However, unlike INEPT polarization transfer, which occurs under the action of scalar coupling, it is the dipolar interaction that drives the polarization transfer in TEDOR experiments in solids. Since the ^{13}C magnetization arises from ^{15}N magnetization transferred to carbon via the dipolar coupling, only those ^{13}C spins that experience a nonzero dipolar coupling are polarized. Thus, TEDOR is selective for ^{13}C spins coupled to ^{15}N spins, with the background signal from isolated carbons cleanly removed. Following the simultaneous $\pi/2$ pulses, the ^{13}C antiphase magnetization is refocused into an observable ^{13}C signal under the action of the carbon–nitrogen dipolar coupling. The approach is the same as in the first half of the TEDOR experiment, except that now the two π pulses per rotor period are applied on the ^{13}C channel.

The selectivity and sensitivity of TEDOR are illustrated in Figure 15, which shows results for a mixed sample of sodium–[1–^{13}C]propionate and L–[4–^{13}C, amide–^{15}N]asparagine. Figure 15a shows a Hahn-echo spectrum of this sample, with signals due to the labeled carbons in both propionate and asparagine; Figure 15b shows the TEDOR result, in which only signal from the double-labeled asparagine is observed.

The magnitude of the observed signal following an m-rotor-cycle refocusing period depends on the rotation period T_r, the dipolar coupling D, and the values m and n. Computer simulations of the observable magnetization, M_S, as a function of the dimensionless parameters mDT_r and nDT_r are shown in Figure 16. The signal is zero immediately following the simultaneous $\pi/2$ pulses, builds to a maximum, and then decays in an oscillatory manner. Maximum TEDOR magnetization is achieved when $|nDT_r| = 0.84$.

A number of 2D TEDOR NMR approaches have been described for de-

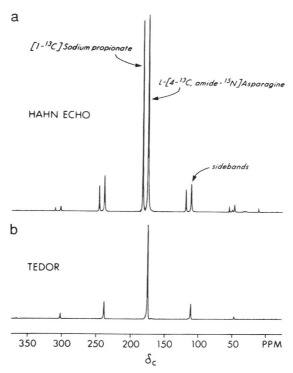

Figure 15. Hahn-echo and TEDOR ^{13}C NMR spectra of a mixture of L–[4–^{13}C, amide–^{15}N] asparagine and [1–^{13}C]sodium propionate. (a) Hahn-echo MAS spectrum obtained by acquiring data two rotor cycles after cross-polarization with a single π pulse on the ^{13}C channel at T_r. Strong resonances from the labeled sites in asparagine and sodium propionate are observed. (b) TEDOR ^{13}C spectrum acquired using the pulse sequence of Figure 14 with $n = m = 2$. Only the ^{13}C–4 carbons contribute to the TEDOR spectrum. Experimental: $H_0 = 4.7$ T; $H_1(^{13}C, ^{15}N)$ = 38 kHz; $v_r = 3.2$ kHz. (Reproduced with permission from Hing et al., 1993. Copyright © 1993 J. Mag. Res.)

termining carbon–nitrogen dipolar couplings (Hing et al., 1993). For example, a series of TEDOR spectra can be collected as a function of m for a fixed value of n (an optimal value for n is chosen by estimating D), the amplitudes of the observed signals measured, and the results fitted to curves such as those of Figure 16. However, the requirement of performing 2D experiments limits the applicability of TEDOR as a distance-measuring method. For systems with a limited signal-to-noise ratio, REDOR, in which $\Delta S/S_0$ is measured for a single set of experimental conditions, is a more practical distance-measuring technique. A TEDOR preparation period may also serve as

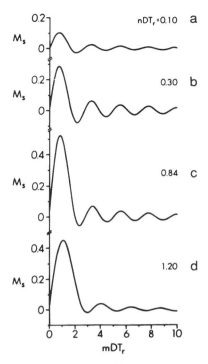

Figure 16. Computer simulations of the observable magnetization M_S for the TEDOR experiment described in Figure 14, where M_S is plotted as a function of the product of the dipolar coupling D and evolution period mT_r for nDT_r values of (a) 0.10, (b) 0.30, (c) 0.84, and (d) 1.20. (Reproduced with permission from Hing et al., 1993. Copyright © 1993 J. Mag. Res.)

the front-end for a TEDOR/REDOR four-channel experiment (Holl et al., 1992) to suppress natural-abundance signals. Under the influence of the ^{15}N–^{13}C dipolar coupling, TEDOR is first used to select a specific subset of ^{13}C spins, as described earlier. Distances between these carbons and other NMR-active nuclei, such as ^{31}P or ^{19}F, are then measured in conventional REDOR experiments as described in Section 3.2.2.

3.2.6 Rotary-Resonance Recoupling

As described in Section 3.1, the dipolar interaction between a heteronuclear I-S spin pair is effectively removed by high-speed MAS. This dipolar coupling can also be removed by applying a strong, resonant rf field to the non-observe I spin, whether or not the sample is spinning. However, when the nutation frequency of this resonant decoupling field, ω_1, matches an integral multiple of the spinning speed, ω_r, the decoupling is partially defeated and significant changes occur in the observed NMR lineshape (Levitt et al., 1988; Oas et al., 1988; Raleigh et al., 1988a). This *recoupling* of the I-S dipolar interaction through the application of a continuous, resonant I-spin rf field matched to an integral multiple of the spinning speed is referred to as *Rotary-Resonance Recoupling* (R^3) NMR. The pulse sequence for a ^{13}C-observe version of the experiment is shown in Figure 17.

Following cross-polarization, a high-power rf field is applied to decouple the protons during data acquisition. Immediately after cross-polarization, the ^{13}C signal is acquired, while a *recoupling* field satisfying the condition $\omega_1 = n \times \omega_r$ (n is an integer) is applied to the ^{15}N spins. In the R^3 experiment, the centerband and each individual spinning sideband of any ^{13}C spin coupled to the ^{15}N splits into either a doublet or quartet (Levitt et al., 1988; Oas et al., 1988). The recoupling of the heteronuclear dipolar interaction by R^3 is sensitive to the magnitudes and orientations of both the ^{13}C and ^{15}N chemical-shift tensors. However, the ^{13}C spectrum itself is insensitive to the carbon CSA if the ^{13}C signal is observed synchronously with the sample rotation (Levitt et al., 1988).

Recoupled ^{13}C lineshapes are affected by mismatching the rf field strength away from the R^3 condition. Simulations show clear lineshape changes occurring for rf field strength mismatches as small as 5% (Levitt et al., 1988). Thus, the R^3 experiment requires good control of rf power levels. Simulations also show that R^3 lineshapes are sensitive to the uniformity of the directionality of the irradiation field. Deviations as small as 10° influence the recoupled carbon lineshapes, and it has been suggested that specially wound NMR coils be used to help eliminate this effect (Levitt et al., 1988). Although the heteronuclear dipolar coupling can be obtained by the R^3 method, no R^3 results beyond the initial three publications have been reported. By contrast, the REDOR experiment is much more easy to implement and, unlike R^3, is insensitive to the CSAs of the coupled spins.

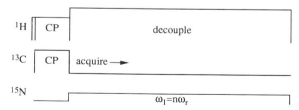

Figure 17. A ^{13}C – ^{15}N R^3 pulse sequence. A low-power ^{15}N rf field satisfying the condition $\omega_1 = n\omega_r$ is applied throughout data acquisition.

3.3 ^{13}C–^{13}C Distances by MAS NMR

In Section 3.2, we described the measurement of ^{13}C–I (I = ^{15}N, ^{31}P, ^{19}F, etc.) REDOR and TEDOR NMR distances. In heteronuclear experiments such as REDOR, dipolar couplings can be cleanly separated from other spin-system parameters, and be independently manipulated. Because of the large difference in Larmor frequencies, hard rf pulses may be applied selectively to one member or the other of a heteronuclear spin pair. As described by Eq. (3.14), the normalized REDOR difference signal depends only on the carbon–nitrogen dipolar coupling and is independent of the breadth or orientation of the CSA of either spin.

A natural complement to the ^{13}C–I REDOR work is the measurement of dipolar couplings between homonuclear ^{13}C–^{13}C pairs. Carbon is ubiquitous in organic and biological systems and the accurate determination of carbon–carbon distances provides another source of important structural information. Although distance measurements between homonuclear spin pairs are conceptually similar to those between heteronuclear spins, some important differences affect the design of homonuclear experiments.

Because homonuclear spin pairs are often strongly coupled, it is harder to manipulate homonuclear dipolar interactions by sample spinning and/or pulse trains. In particular, it is impossible to apply hard rf pulses to only one of two spins of a homonuclear spin pair. Assuming the two members of the spin pair are well separated in frequency, low-power rf pulses (Abragam, 1983) or DANTE pulse trains (Bodenhausen *et al.*, 1976; Morris and Freeman, 1978) can be used for selective inversion. However, the length of these selective pulses makes the inclusion of many such pulses in an extended pulse train impractical. Also, in general, the dipolar coupling and CSA cannot be cleanly separated in homonuclear cases, with signals depending on the breadth and relative orientations of the CSA tensors as well as on the magnitude of the dipolar coupling. Although sensitivity to CSA magnitude and orientation provides another potential source of structural information, it clearly complicates the measurement of internuclear distances.

Despite the difficulties posed by the interplay of dipolar couplings and CSA, a number of techniques have been developed to measure homonuclear dipolar couplings. In this section, we will review three of these techniques: dipolar recovery at the magic angle (DRAMA) (Tycko and Dabbagh, 1990, 1991; Tycko and Smith, 1993), simple excitation for the dephasing of rotational amplitudes (SEDRA) (Gullion and Vega, 1992), and rotational resonance (R^2)(Andrew *et al.*, 1966; Colombo *et al.*, 1988; Kubo and McDowell, 1988; Levitt *et al.*, 1990; Maas and Veeman, 1988; Raleigh *et al.*, 1987, 1988a,b, 1989; Robyr *et al.*, 1989). In the DRAMA experiment, a rotor-synchronous train of $\pi/2$ and π pulses scales the full dipolar interaction during part of each rotor cycle. In SEDRA, a train of rotor-synchronous π pulses, one per rotor period, prevents the dipolar flip-flop term from being averaged to zero. Rotational resonance also operates through the dipolar flip-flop term, but works by matching the sample spinning speed to differences in isotropic chemical shift between coupled spins. Of these three methods, rotational resonance is the most developed in terms of its application to structural problems. Therefore, we begin by briefly discussing the DRAMA and SEDRA techniques, and then we concentrate on R^2 and its application to biological solids.

3.3.1 Dipolar Recovery at the Magic Angle

In REDOR, the heteronuclear dipolar interaction is recovered by applying trains of rotor-synchronous rf pulses. These pulses manipulate the spin part of the dipolar interaction and have the effect of interfering with the spatial averaging process. In effect, the pulses produce a new, time-independent dipolar interaction. The homonuclear dipolar interaction can be similarly recovered by applying rotor-synchronized rf pulses.

One experiment of this type, dipolar recovery at the magic angle (DRAMA), uses a train of rotor-synchronous $\pi/2$ pulses (Tycko and Dabbagh, 1990; Tycko and Smith, 1993). The reintroduction of the like-spin dipolar interaction by $\pi/2$ pulses can be illustrated by the pulse sequence in Figure 18a. A single rotor cycle with two $\pi/2$ pulses, one at $T_r/4$ and the other at $3T_r/4$, is shown. For simplicity of discussion, we ignore the chemical shift. If the data are sampled at the end of each rotor cycle, the evolution of the spin system can be described by an average dipolar Hamiltonian. The two $\pi/2$ pulses applied at $T_r/4$ and $3T_r/4$ produce an average dipolar interaction of the form $\overline{D}(3I_{z1}I_{z2} - 3I_{y1}I_{y2})$ (Tycko and Dabbagh, 1990), where $\overline{D}$ is a lengthy expression containing the familiar dipolar constants, the polar and azimuthal angles describing the orientation of the dipolar vector with respect to the magnetic field, the spinning speed, and the times τ and τ'. Importantly, this factor is time independent, being generally nonzero. Thus, by changing the form of the spin part of the dipolar Hamiltonian, this simple pulse scheme prevents the interaction from averaging to zero and net dipolar dephasing occurs.

When the chemical shift is included, measuring the pure dipolar interaction becomes more difficult, since any dephasing from the chemical shift must be either eliminated or measured. The chemical shift anisotropy and isotropic chemical shift can be refocused by application of a π pulse at the end of every odd-numbered rotor cycle as illustrated in the four-rotor-cycle DRAMA building block shown in Figure 18b. The evolution time in the

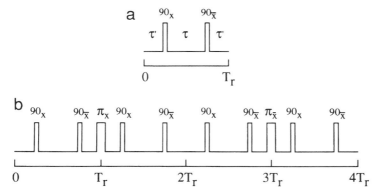

Figure 18. DRAMA pulse sequences. (a) Basic one-rotor-cycle building block for recovering the homonuclear dipolar interactions. As described in the text, scaling of the recovered dipolar interaction depends on the times τ and τ'. (b) General four-rotor-cycle building block that refocuses anisotropic and isotropic chemical shifts while recovering homonuclear dipolar interactions.

DRAMA experiment can be increased by simply adding multiple four-rotor-cycle blocks and, as in REDOR, the amount of dipolar dephasing accumulates with increasing evolution time.

A full two-dimensional version of the DRAMA experiment of Figure 18 has been demonstrated for a sample of ($^{13}CH_3)_2C(OH)SO_4Na$ (Tycko and Dabbagh, 1990). In this experiment, signal was collected as a function of dipolar evolution time, which was incremented in blocks of four rotor cycles. The Fourier transform of the data with respect to this evolution time produced a powder spectrum from which the homonuclear dipolar coupling was calculated. Because measuring dipolar couplings by this approach requires detailed lineshape analysis, it is limited to relatively large couplings, corresponding to short $^{13}C-^{13}C$ distances (<3.5 Å).

Longer distances can be probed by using the DRAMA experiment as a double-quantum filter (Tycko and Dabbagh, 1991; Tycko and Smith, 1993). Figure 19a shows an unfiltered spectrum of a methyl–^{13}C labeled undecapeptide (Tycko and Smith, 1993). (The peptide also contains a ^{13}C-enriched carbonyl carbon not relevant to this discussion.) In addition to an intense methyl–carbon peak, this spectrum shows a large number of natural-abundance resonances. The spectrum can be greatly simplified by performing a DRAMA-based double-quantum filtering experiment. In addition to the strong methyl peak, the filtered spectrum, shown in Figure 19b, has several other signals. From the known structure of this undecapeptide, these addi-

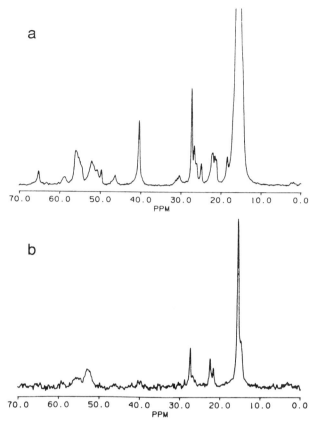

Figure 19. The ^{13}C MAS NMR spectra of Boc–L–Ala–[Aib–Ala]$_2$–OMe, ^{13}C labeled at the methyl carbon of Ala-5 and the carbonyl carbon of Ala-1. (a) Unfiltered spectrum centered near the methyl resonance. The intense peak at 15.5 ppm is the labeled methyl carbon. The carbonyl carbon appears at 175 ppm and is not shown in the spectrum. (b) Double-quantum filtered spectrum. Only those resonances within approximately 5 Å of the labeled methyl appear in the spectrum. Experimental: H_0 = 9.4 T; $H_1(^{13}C)$ = 50 kHz; $H_1(^{1}H)_{DCP}$ = 120 kHz; v_r = 2.5 kHz. (Reproduced with permission from Tycko and Smith, 1993. Copyright © 1993 American Institute of Physics.)

tional signals can all be assigned to carbon nuclei that are within 5 Å of the ^{13}C-labeled site. Thus, this DRAMA-based filtering experiment provides an excellent way to select carbon spins that are within approximately 5 Å of a labeled site.

3.3.2 Dipolar Recovery by Synchronously Applied π Pulse Trains

Recent experiments have shown that the ^{13}C–^{13}C dipolar interaction can also be recovered under MAS conditions by simply applying a train of rotor-synchronous π pulses. This class of experiments includes simple dephasing of rotational amplitudes (SEDRA) (Gullion and Vega, 1992; Weintraub and Vega, 1993), rf-driven dipolar recovery (rfdr) (Bennett et al., 1992), broadband dipolar recoupling (BDR) (Sodickson et al., 1993), and dipolar correlation spectroscopy (DICSY) (Ok et al., 1992). At first glance, the reintroduction of the homonuclear dipolar coupling by applying one π pulse per rotor period may seem counterintuitive. We will provide insight into this class of experiments by examining the SEDRA technique.

If two ^{13}C spins have identical chemical shifts, irradiation of the spins with on-resonance, δ-function, π pulses leaves the dipolar Hamiltonian unchanged. Over a complete rotor cycle, H_D therefore averages to zero. However, if the two ^{13}C spins have different chemical shifts (especially different isotropic chemical shifts), applying a train of π pulses leads to a net dipolar evolution of the spin system. This evolution arises not from the $I_{z1}I_{z2}$ part of the dipolar interaction, but from the $I_{+1}I_{-2} + I_{-1}I_{+2}$ flip-flop term. This can be shown by considering the combined effect of the dipolar interaction and isotropic chemical shifts. In the rotating frame defined such that $\delta_1 + \delta_2 = 0$, the Hamiltonian is

$$H(t) = D(t)[I_{z1}I_{z2} - 1/4(I_{+1}I_{-2} + I_{-1}I_{+2})] + \Delta/2(I_{z1} - I_{z2}), \quad (3.15)$$

where $\Delta = \delta_1 - \delta_2$ and $D(t)$ contains fundamental constants and the spatial coordinates of the dipolar interaction. Note that $D(t + nT_r) = D(t)$ and that $D(t)$ averages to zero over one rotor cycle. For an experiment in which a π pulse is applied at the end of each rotor cycle, the average Hamiltonian in the interaction representation defined by Δ becomes

$$\bar{H} = -\frac{1}{8T_r} \int_0^{2T_r} D(t)(I_{+1}I_{-2}e^{i2\Delta t} + I_{-1}I_{+2}e^{-i2\Delta t}) \, dt. \quad (3.16)$$

The π pulse eliminates the last term in Eq. (3.15) and the $I_{z1}I_{z2}$ term disappears because $D(t)$ has an average of zero. However, because the flip-flop term now has a time-independent component, the dipolar interaction is not averaged to zero.

Several limiting cases can be explored. If $\Delta \ll v_r$, the exponential terms in Eq. (3.16) are approximately one and the average dipolar interaction becomes zero. Thus, as described, reintroducing the dipolar coupling with a train of π pulses requires that there be a chemical shift difference between the coupled spins. If, however, $\Delta \gg v_r$, the exponential terms in Eq. (3.16) oscillate rapidly and the effective average dipolar interaction is again zero. Generally, a train of π pulses can effectively reintroduce the dipolar interaction if the condition $0 < \Delta < 2v_r$ is satisfied.

Figure 20 shows the pulse sequence for SEDRA, which, like REDOR, is performed as a difference experiment. The sequence in Figure 20a, consisting of a single π pulse at the middle of the dipolar evolution period, serves as a control experiment. This control experiment accounts for T_2 decay during the evolution period and produces the SEDRA *full* spectrum. For the sequence in Figure 20b, a train of rotor-synchronous π pulses is used to recover the dipolar interaction, producing the SEDRA *reduced* spectrum. In this sequence, one π pulse is applied per rotor cycle and *xy*-8 phase cycling (Gullion *et al.*, 1990a,b) helps to minimize pulse imperfections.

Experimental results are shown in Figure 21. Figure 21a shows results for a sample of [1,3-^{13}C$_2$, ^{15}N]alanine. The dipolar evolution period consists of 34 rotor cycles at a spinning speed of 3.2 kHz. At the bottom of Figure 21a results of the SEDRA control experiment for the alanine sample are shown. This spectrum was phased without any frequency-dependent phase

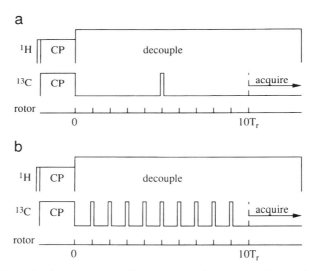

Figure 20. SEDRA pulse sequences. Following cross-polarization, ^{13}C magnetization evolves in the transverse plane for N_c rotor cycles ($N_c = 10$ is shown here). A strong proton decoupling field is applied during the evolution and data acquisition periods. (a) In the control experiment, which accounts for T_2 decay, a single π pulse is applied at the middle of the evolution period to refocus isotropic chemical shifts. (b) The SEDRA experiment recovers the dipolar interaction with a train of rotor-synchronous π pulses.

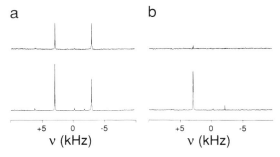

Figure 21. Experimental SEDRA results. (a) SEDRA ^{13}C NMR spectra of [1,3-^{13}C$_2$, ^{15}N]alanine (99 at.% ^{13}C and ^{15}N), diluted 1 part to 9 in natural-abundance alanine. (b) SEDRA ^{13}C NMR spectra of [1,3-^{13}C$_2$]malonic acid (99 at. % ^{13}C), diluted 1 part to 9 in natural-abundance malonic acid. The respective full-amplitude (bottom) and reduced-amplitude (top) spectra are shown for a 34-rotor-cycle SEDRA experiment. Significant dephasing occurs only when there is a significant difference in the isotropic chemical shifts of the two coupled spins. Experimental: H_0 = 3.5 T; $H_1(^{13}C)$ = 40 kHz; $H_1(^1H)_{DCP}$ = 110 kHz; v_r = 3.2 kHz.

terms, illustrating the effectiveness of the *xy*-8 phase scheme in compensating for pulse imperfections. Figure 21a (top) is the SEDRA *difference* spectrum obtained by subtracting the SEDRA *reduced* spectrum from the SEDRA *full* spectrum. Clearly, the dipolar interaction has been reintroduced in the experiment. Note that only those carbon spins coupled to other carbon spins appear in the *difference* spectrum, suggesting that SEDRA can also be used as a prefilter for other experiments.

The importance of having a nonzero chemical-shift difference between the coupled spins is illustrated by the spectra in Figure 21b. These spectra were obtained on a sample of [1,3-^{13}C$_2$]malonic acid under identical experimental conditions as those used for the alanine sample. In malonic acid, there is a small isotropic chemical-shift difference between the two carboxyl resonances due to crystal packing. This difference gives rise to a partially resolved doublet in the SEDRA control spectrum shown in Figure 21b (bottom). Figure 21b (top) shows the SEDRA *difference* spectrum. As predicted by Eq. (3.16), the *difference* signal is small and probably due more to the large chemical shift anisotropy of the two carboxyl carbons in malonic acid than to the small difference in their isotropic chemical shifts.

3.3.3 Rotational-Resonance NMR

3.3.3.1 *A Simple Description of* R^2

A consequence of the spatial averaging produced by MAS is that, like the heteronuclear dipolar interaction, the homonuclear dipolar interaction generally has little effect on the observed ^{13}C resonances during MAS. Special conditions do exist, however, under which like-spin dipolar interactions can be effectively reintroduced under MAS conditions without the assistance of rf irradiation. Under these *rotational-resonance* (R^2) conditions (Andrew et al., 1966; Colombo et al., 1988; Kubo and McDowell, 1988; Levitt et al., 1990; Maas and Veeman, 1988; Raleigh et al., 1987, 1988a,b, 1989; Robyr et al., 1989), which simply require a judicious choice of spinning speeds, the effects of homonuclear dipolar interactions can be observed and their strengths measured.

To gain insight into the mechanism responsible for recovery of ^{13}C–^{13}C dipolar couplings under R^2, we consider a single, isolated ^{13}C–^{13}C spin pair. We label the two ^{13}C nuclei 1 and 2 and identify them by their respective chemical shifts, δ_1 and δ_2. For the present, we assume these shifts are distinct, the dipolar coupling is weak, and we ignore the CSA (Raleigh et al., 1987). For a *static* sample, the spin part of the flip-flop term of the dipolar interaction in the interaction frame, $(I_{+1}I_{-2} + I_{-1}I_{+2})$, is time dependent with frequency $\Delta = \delta_1 - \delta_2$. The time average of this term over the period Δ^{-1} is zero. Under these conditions, this term makes no contribution to the NMR spectrum. The $I_{z1}I_{z2}$ term is time independent and produces the familiar dipolar doublet in the NMR spectrum.

This situation changes dramatically in a solid rotating at the magic angle. Under MAS, the spin part of the flip-flop term remains time dependent with frequency Δ. However, as described previously, spinning introduces a time dependence in the angular component of the spatial part of the dipolar interaction. Combining these two time-dependent terms, the flip-flop term now has time-dependent contributions of the form $\cos(\Delta - \omega_r)t$ and $\cos(\Delta - 2\omega_r)t$ (Maas and Veeman, 1988). Under most spinning conditions, the flip-flop term is averaged to zero and makes no significant contribution to the MAS NMR spectrum. However, if the spinning frequency is chosen to satisfy the rotational-resonance condition $\Delta = n\omega_r$, where n is an integer, the flip-flop term becomes time *independent* and contributes to the spectrum. Under R^2 conditions, this time-independent flip-flop term reintroduces the ^{13}C–^{13}C dipolar interaction.

This description of R^2 is somewhat simplistic; the situation becomes more complicated when chemical shift anisotropies are included, because the resonance frequencies δ_1 and δ_2 become time dependent as well. Furthermore, this treatment suggests that the spinning speed must be chosen to satisfy exactly the rotational-resonance condition. In fact, the R^2 condition *is* sensitive to spinning speed and careful regulation of this speed is required. However, because each ^{13}C resonance has finite width (e.g., broadening due to incomplete ^{1}H decoupling), small variations in spinning speed can be tolerated.

For arbitrary spinning speeds, the MAS NMR lineshape of each reso-

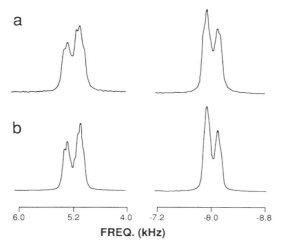

Figure 22. The cross polarization magic-angle spinning (CPMAS) ^{13}C NMR spectra of Zn[1,2-^{13}C$_2$]acetate obtained at the $n = 3$ rotational-resonance condition. (a) Experimental spectrum with $v_r = 4.384$ kHz, showing the asymmetric lineshapes for carboxyl (left) and methyl (right) carbons, which develop at rotational resonance. (b) Simulated lineshapes using the CSAs of the carbonyl and methyl carbons, ^{13}C–^{13}C dipolar and J couplings, and the Euler angles relating the two shielding tensors to the principal axis system of the dipolar tensor. Experimental: $H_0 = 7.45$ T; $v_r = 4.384$ kHz. (Reproduced with permission from Raleigh et al., 1988b. Copyright © 1988 Elsevier Science Publishers BV.)

nance of a dipolar-coupled ^{13}C–^{13}C spin pair is only slightly broadened compared with the corresponding signals from isolated spins. However, under R^2 conditions the observed lineshape for this spin pair can change dramatically. This is illustrated in Figure 22 for the $n = 3$ rotational-resonance spectrum of Zn[1,2-^{13}C$_2$]acetate (Raleigh et al., 1988b). The experimental methyl and carboxyl resonances shown in Figure 22a are split into doublets with characteristic and different lineshapes. Simulations of these R^2 spectra are shown in Figure 22b. To reproduce the entire structure of the experimental spectra, these simulations include CSAs, relative orientations of the two CSA tensors with respect to the dipolar tensor principal axis system, the isotropic chemical-shift difference, the ^{13}C–^{13}C-scalar coupling, and the ^{13}C–^{13}C dipolar coupling. The observed lineshapes are quite sensitive to minor changes in the spinning speed. Under the $n = 3$ rotational-resonance condition for zinc acetate, spinning speed variations of ± 5 Hz from the rotational-resonance condition produce detectable changes in lineshape.

3.3.3.2 The Spin-Exchange Experiment

As illustrated earlier, the flip-flop term plays a crucial role in the dipolar interaction when the rotational-resonance condition is satisfied, and spin exchange occurs between spins satisfying this condition. In principle, internuclear distances can be obtained by detailed analysis of R^2 lineshapes, such as those of Figure 22. However, such lineshape analysis is difficult and distances obtained in this manner are of limited accuracy. Instead, most distance measurements are obtained from rotational-resonance spin-exchange experiments.

The easiest way to monitor spin exchange under rotational-resonance conditions is by creating a population imbalance by inverting one of the resonances. A typical R^2 spin-exchange experiment is shown in Figure 23. The protons are used to enhance the ^{13}C signal by cross-polarization and then are decoupled from the ^{13}C spins by a strong rf field during the remainder of the experiment. Following cross polarization, a hard $\pi/2$ pulse aligns the ^{13}C spins along the z axis. Immediately following this z restoration pulse, one of the ^{13}C resonances is inverted by applying either a long, weak, on-resonance pulse or a DANTE sequence (Raleigh *et al.*, 1988b; Peersen *et al.*, 1992). The spin-inversion pulse is followed by a variable mixing time. During this mixing time, magnetization exchange occurs between the two coupled ^{13}C nuclei, causing a change in the spin-system population. This exchange of magnetization is observed by applying a $\pi/2$ inspection pulse and acquiring the free-induction decay.

An example of the spin-exchange experiment performed on a sample of Zn[1,2-$^{13}C_2$]acetate is shown in Figure 24 (Raleigh *et al.*, 1988b). These R^2 data were collected using the pulse sequence in Figure 23 for the $n = 3$ rotational-resonance condition. The methyl resonance was initially inverted and the data show the exchange of magnetization between the methyl and carboxyl carbons for various mixing times, τ_m. It is clear from the positive-to-negative oscillations of the signals that magnetization is exchanged between

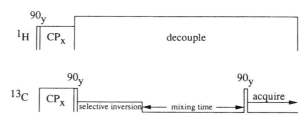

Figure 23. Pulse sequence for the rotational-resonance, spin-exchange experiment. After cross-polarization, ^{13}C magnetization is returned to the z axis by a hard $\pi/2$ rf pulse. Following the z restoration pulse, one of the resonances is inverted by either a weak rf pulse or a DANTE sequence. After a variable period of spin exchange, τ_m, magnetization is recorded following a $\pi/2$ read pulse. High-power proton rf removes the ^1H–^{13}C coupling during the spin-exchange period and proton-decoupled spectra are collected. The decoupling power is lowered (typically to 40 kHz) during the long mixing time to prevent sample heating and/or probe arcing.

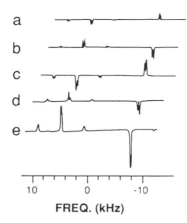

Figure 24. Spin-exchange ^{13}C NMR spectra of Zn[1,2-^{13}C$_2$]acetate obtained at the $n = 3$ rotational-resonance condition. Figure (e) was collected immediately after selective inversion of the methyl signal ($\tau_m = 0$ spectrum). Spectra in figures (d) through (a) were collected following mixing periods of 1.0, 3.0, 5.5, and 7.0 ms, respectively, and illustrate magnetization exchange occurring between the methyl and carboxyl carbons. Experimental conditions are given in Figures 22 and 23. (Reproduced with permission from Raleigh et al., 1988b. Copyright © 1988 Elsevier Science Publishers BV.)

the two resonances. The data in Figure 24 also suggest an inherent decay of the spin order. This decay, characterized by a time constant known as the zero-quantum relaxation time, T_2^{ZQ} (Kubo and McDowell, 1988; Levitt et al., 1990), is due primarily to incomplete proton decoupling and homonuclear dipolar interactions to nearby natural abundance ^{13}C nuclei not satisfying the rotational-resonance condition (Levitt et al., 1990). Unlike the more common transverse relaxation time T_2^*, magnet inhomogeneity does not contribute to T_2^{ZQ}.

Figure 25a is a graphical representation of the data shown in Figure 24. In this figure, the magnetization *difference* of the methyl and carboxyl resonances is plotted as a function of mixing time. (The difference signal was normalized to one for zero mixing time.) Both the oscillatory nature of the magnetization exchange and the decay of the difference signal are clearly evident. The solid lines are fits to the data calculated using the same parameters as used in the legend to Figure 22. When the CSA parameters and the relative orientation between the shift tensors are known, the ^{13}C–^{13}C internuclear separation can be obtained from this type of data analysis. The sensitivity of the spin-exchange experiment to spinning speed is illustrated in Figure 25b, which shows that magnetization exchange is severely reduced if the spinning speed is moved 100 Hz off the rotational-resonance condition.

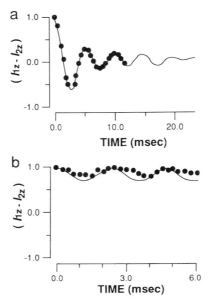

Figure 25. Rotational-resonance spin-exchange data for Zn[1,2-^{13}C$_2$]acetate on and off the $n = 3$ rotational-resonance condition. The discrete points are the experimental data and the solid lines are calculated curves. (a) On rotational resonance, $v_r = 4.384$ kHz. (b) 100-Hz off rotational resonance, $v_r = 4.260$ kHz. (Reproduced with permission from Raleigh et al., 1988b. Copyright © 1988 Elsevier Science Publishers BV.)

3.3.3.3 The $n = 1$ Experiment

As illustrated in Figure 25, internuclear distances can be extracted from detailed simulations of magnetization exchange curves. These exchange curves depend on several parameters, including the homonuclear dipolar coupling between the labeled sites, their CSAs, and T_2^{ZQ}. For $n > 1$ rotational resonance, magnetization exchange also depends on the relative orientation of the two CSA tensors. To measure accurately dipolar couplings using R^2, these parameters must be measured or estimated and included in the simulations.

The contribution of CSA to magnetization exchange depends on the magnitude of the anisotropy. Small anisotropies, such as those for methyl and methylene carbons, have little effect on the observed magnetization exchange curves; large anisotropies make important contributions to the exchange of magnetization (Thompson et al., 1992). For large anisotropies, CSA parameters can be obtained from a Herzfeld-Berger analysis of the spinning-sideband intensities in slow-spinning MAS experiments (Herzfeld and Berger, 1980). Such analyses can be difficult and time-consuming, especially in samples with substantial natural-abundance carbon signals. Accurate measurement of distances from the magnetization exchange curves also depends strongly on an accurate determination of T_2^{ZQ}. Unfortunately, no direct experimental measurement of T_2^{ZQ} is possible. Reasonable estimates of T_2^{ZQ} can often be obtained from the linewidths of the two resonances (Kubo and McDowell, 1988; Thompson et al., 1992). Caution must be used in estimating T_2^{ZQ} in this way, however, especially if there is a spread of carbon resonance frequencies due to chemical-shift dispersion or sample heterogeneity.

To increase the utility and applicability of R^2 spin-exchange techniques, experimental strategies are being developed to eliminate the need for measuring CSA and T_2^{ZQ} values in every sample. One approach has been to create a set of standard exchange curves for particular labeled spin pairs in particular molecular systems (e.g., peptides and proteins) (Peersen et al., 1992). The inherent assumption in this approach is that T_2^{ZQ} and the CSA are similar for similar chemical and physical environments. This approach was used for measuring distances between methyl and carbonyl carbons in a series of double ^{13}C-enriched small peptides (Peersen et al., 1992). Magnetization exchange curves were measured for a series of known distances ranging from 3.7 to 6.8 Å. Representative spectra from this work are shown in Figure 26.

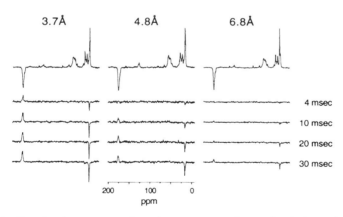

Figure 26. Rotational-resonance spin-exchange spectra of crystals of the undecapeptide Boc–L–Ala–{Aib–Ala}$_2$–Glu(OB,L)–Ala–{Aib–Ala}$_2$–OMe, Aib ≡ aminoisobutyric acid, obtained at the $n = 1$ rotational-resonance condition. The top spectrum in each column was collected at $\tau_m = 0$, the remaining spectra were obtained by subtracting this $\tau_m = 0$ spectrum from the spectra collected after the indicated τ_m mixing times. Intensities from these difference spectra were used to produce the spin-exchange curves of Figure 27. Magnetization exchange between methyl and carbonyl resonances clearly occurs more rapidly for shorter distances. (Reprinted with permission from Peerson et al., 1992. Copyright © 1992 American Chemical Society.)

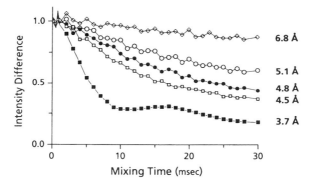

Figure 27. Rotational-resonance spin-exchange data for five double-^{13}C-enriched undecapeptides obtained at the $n = 1$ R^2 condition. The undecapeptides are identical except for the positions of the ^{13}C labels. The labeled sites are between the backbone carbonyl and alanine methyl groups in each of the peptides. The lines are meant to illustrate the trend of the data and are *not* theoretical simulations based on the molecular structures. Only the 3.7-Å data show a distinct dipolar oscillation. (Reprinted with permission from Peerson *et al.*, 1992. Copyright © 1992 American Chemical Society.)

The top spectrum of each column in this figure was collected immediately after the inversion of the carbonyl resonance. The remaining spectra were obtained by subtracting the spectrum at the top of the column from the spectrum collected after each labeled mixing time. The data show that magnetization exchange between the methyl and carbonyl resonances occurs at a faster rate for shorter distances. They also show that resonances not satisfying the rotational-resonance condition do not contribute to the difference spectra. In addition, these $n = 1$ spectra show little of the lineshape fine structure seen for the $n = 3$ spectra of Figure 22. Figure 27 plots the difference between the methyl and carbonyl intensities as a function of mixing time for different methyl–carbonyl distances. Only the 3.4-Å data show oscillations due to magnetization exchange.

3.3.3.4 ^{13}C–^{13}C Distances in Bacteriorhodopsin by Rotational-Resonance NMR

Bacteriorhodopsin (bR), a membrane-bound protein, plays an important role as a model system for understanding the light-driven mechanism of proton pumping. It is stable, well studied, and, at 25 kDa, is a relatively small protein. Bacteriorhodopsin contains retinal, connected to the ϵ-amino group of Lysine-216 by a Schiff base linkage, as the light-absorbing group (Stryer, 1988). The conformation of retinal changes significantly on binding to bR. In fact, chemical-shift data (Harbison *et al.*, 1985a,b) suggest that the 6-s-

trans form of retinal is the favored form in bR (see Figure 28). Isomerization about the 6–7 bond of retinal would lead to a lengthening of the polyene chain, thereby contributing to the observed red shift of the pigment on binding.

Although chemical-shift analysis can provide structural information, the interpretation of such data is necessarily model dependent. A more direct approach to gaining structural information is to measure internuclear distances. By combining selective ^{13}C isotopic labeling with rotational-resonance NMR, it was demonstrated that the retinal conformation is indeed 6-s-*trans* (Creuzet et al., 1991). We briefly describe those experiments here as an illustration of the rotational resonance technique.

Two model compounds, 6-s-*cis*[8,18–^{13}C$_2$] and 6-s-*trans*[8,18–^{13}C$_2$] retinoic acid, were prepared and each was crystallized in the presence of the corresponding natural-abundance compound in a ratio of 1:5. This dilution helped to isolate the ^{13}C$_8$–^{13}C$_{18}$ spin pairs from one another. The CSA of each labeled site was determined by Herzfeld-Berger analysis of spinning-sideband patterns under slow sample spinning (2.5 kHz). The orientation of each CSA tensor was determined from the crystal structure of retinoic acid and from known tensor orientations in similar chemical species. From these analyses, the most shielded position for C-8 is perpendicular to the polyene plane and the intermediate element is parallel to the double bond; for the C-18 position, the most shielded position is parallel to the C–C bond.

As illustrated by the structures drawn in Figure 28, the C$_8$–C$_{18}$ internuclear distance differs significantly in the two model compounds. The *cis* molecule forms a triclinic crystal with a C$_8$–C$_{18}$ separation of 3.1 Å; the *trans* molecule forms a monoclinic crystal with a C$_8$–C$_{18}$ separation of 4.2 Å. The plane of the ring is in the polyene plane for the *trans* crystal form, and is approximately 35° out of the polyene plane for the *cis* form. Rotational-resonance $n = 1$ experiments were performed on each of these samples, and

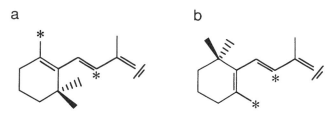

Figure 28. Geometries of the ring-polyene linkage for the *cis* and *trans* conformations of retinoic acid. The asterisk marks specific ^{13}C label sites. (a) 6-s-*cis*[8,18–^{13}C$_2$]retinoic acid. (b) 6-s-*trans*[8,18–^{13}C$_2$]retinoic acid. (Adapted with permission from Creuzet et al., 1991. Copyright © 1991 by the AAAS.)

the R^2 data were analyzed using the CSA parameters described earlier. The value of T_2^{ZQ} was obtained from the linewidths of the coupled spins. The measured C_8-C_{18} distances were 3.0 and 4.1 Å for the *cis* and *trans* forms, respectively, which is in excellent agreement with the X-ray data.

Similar experiments were then performed on [8,18-$^{13}C_2$]retinal–bR (Creuzet *et al.*, 1991). In this sample, there is a substantial natural-abundance ^{13}C background signal due to the protein. Also, resonances in this sample are significantly broader than in the model compounds. The increased linewidth in this sample is due, in part, to heterogeneity of chemical environments. Unfortunately, it is difficult to account quantitatively for this heterogeneity, leading to greater uncertainty in determining T_2^{ZQ} from linewidths.

Figure 29 shows spectra collected immediately after applying the inversion pulse ($\tau_m = 0$) to the methyl (C-18) resonance. Figure 29a is the R^2 spectrum of the labeled sample immediately following inversion of the C-18 resonance. Clearly, there are many signals near the methyl resonance. Since the inversion pulse is not perfectly selective, resonances near the labeled C-18 carbon are inverted as well. Therefore, the total linewidth of the in-

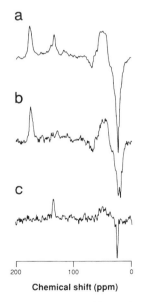

Figure 29. CPMAS ^{13}C NMR spectra at $-30°C$ of bR and retinal–bR following selective inversion of the methyl carbons. (a) [8,18-$^{13}C_2$]retinal–bR; (b) natural-abundance bR; (c) difference spectrum. Intensities from difference spectra, such as this one, were used to produce the spin-exchange curve of Figure 30. (Reproduced with permission from Creuzet *et al.*, 1991. Copyright 1991 by the AAAS.)

verted peak cannot be used to measure T_2^{ZQ}. Although not as cluttered, the region surrounding the C-8 resonance also displays significant natural-abundance signals. To analyze these R^2 data, all natural-abundance signals must be eliminated. In analogy to the procedure described earlier for analyzing REDOR data, the R^2 spectrum of an unlabeled sample (Figure 29b) is collected under identical conditions. Subtraction of the spectrum of the labeled spectrum from the natural-abundance sample yields signals from the enriched C-8 and C-18 sites only. This difference spectrum is shown in Figure 29c. Lines in this difference are much sharper than those in the spectrum of the labeled sample. All of the spectra obtained with variable mixing times τ_m were treated similarly. The magnetization exchange curves derived from the spectra of Figure 29 are plotted in Figure 30. Also plotted are simulations calculated using CSAs and distances obtained from the two model compounds and T_2^{ZQ} obtained from linewidths of the labeled bR sample. The simulation derived assuming a *trans* configuration provides a much better fit of the experimental data.

In addition to showing that the conformation of retinal bound to bR is 6-s-*trans*, rotational-resonance NMR has also been used to determine the conformation of the retinal–protein linkage in bR (Thompson *et al.*, 1992). The dark-adapted state of bR can exist in two different conformations, labeled bR_{555} and bR_{568}. At equilibrium, the ratio of bR_{555} to bR_{568} is 60:40. As illustrated in Figure 31, the internuclear separation of the 14-C position of retinal and the ϵ-C position of Lys_{216} is sensitive to the conformation of the Schiff linkage. To facilitate R^2 experiments, ^{13}C labels were selectively incorporated at these two sites in bR. The bR_{555} and bR_{568} protein conformations can be identified spectroscopically by the isotropic chemical shift

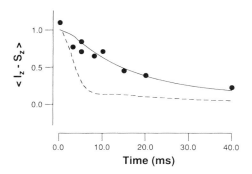

Figure 30. Rotational-resonance spin-exchange data at $-30°C$ for [8,18-$^{13}C_2$]retinal–bR. Calculated curves are based on the known distances and geometries for the 6-s-*cis* (dashed) and 6-s-*trans* (solid) configurations combined with values of T_2^{ZQ} derived from the measured linewidths for [8,18-$^{13}C_2$]retinal–bR. (Reproduced with permission from Creuzet *et al.*, 1991. Copyright © 1991 by the AAAS.)

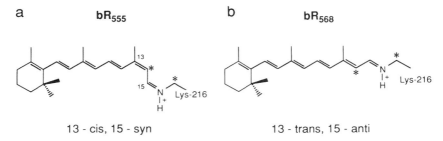

Figure 31. Proposed structures of two dark-adapted bR conformations, bR$_{555}$ and bR$_{568}$. The internuclear distance between C$_{14}$ of retinal and C$_\epsilon$ of Lys$_{216}$ in bR depends on the configuration of the C=N bond. (Adapted with permission from Thompson et al., 1992. Copyright © 1992 American Chemical Society.)

of these ^{13}C spin labels. In bR$_{555}$, the signal from the Lys$_{216}$ spin label is at 48 ppm and the retinal spin label is at 110.5 ppm, whereas in bR$_{568}$, the signal from the Lys$_{216}$ spin label is at 53 ppm and the retinal spin label is at 122.5 ppm. Thus, Δv_{iso}, the isotropic chemical-shift difference between the two spin labels, is different for the two conformations. (At a field strength of 9.4 T, $\Delta \theta_{iso}$ = 6250 Hz in bR$_{555}$ and $\Delta \theta_{iso}$ = 6950 Hz in bR$_{568}$.)

Using the pulse sequence of Figure 23 with mixing times from 0 to 25 ms, a series of n = 1 R^2 spin-exchange experiments were performed to determine the internuclear separation of the two spin labels. Because the four signals from the labeled sites are chemical-shift resolved and Δv_{iso} is different for bR$_{555}$ and bR$_{568}$ it was possible, in separate experiments, to determine the ^{13}C–^{13}C distance in each of the bR conformers. Prior to analysis of the magnetization exchange spectra, the 14-^{13}C tensor was determined by Herzfeld-Berger analysis of sideband intensities in a slow spinning experiment. The small ϵ–^{13}C tensor, although difficult to obtain by this method, has only a minor effect on the distance measurement. From the linewidths of the two coupled resonances, T_2^{ZQ} was estimated to be between 1.1 and 1.7 ms. From simulations of the R^2 data, the distances between the labeled spin pairs were determined to be between 2.8 and 3.1 Å for bR$_{555}$, and 4.0 to 4.6 Å for bR$_{568}$. These distances are consistent with conformations of 13-*cis*, 15-*syn*, and 13-*trans*, 15-*anti* for bR$_{555}$ and bR$_{568}$, respectively.

3.4 Concluding Remarks

The accurate determination of internuclear distances can provide important structural information in a wide variety of solids. These distances are easily derived from the $1/r^3$ distance dependence of the dipolar couplings between the spin pairs, if these couplings can be measured. Rapid MAS, while providing the line-narrowing required to produce high-resolution solid-state NMR spectra, generally averages these dipolar interactions to zero. In this chapter, we reviewed a number of techniques for recovering and measuring dipolar couplings under rapid MAS conditions. The popularity of the two most used techniques, REDOR (heteronuclear interactions) and R^2 (homonuclear interactions), comes from their ease of implementation and their proven ability to obtain accurate internuclear distances in complex molecular systems. The development of solid-state NMR techniques for structure determination remains an active area of research and we are confident that NMR will continue to play a prominent role in the characterization of biological and nonbiological solids.

References

Abragam, A. (1983). "The Principles of Nuclear Magnetism." Oxford Univ. Press, New York.
Ando, I., Yamanobe, T., and Asakura, T. (1990). *Prog. NMR Spectrosc.* **22**, 349.
Andrew, E. R., Bradbury, A., and Eades, R. G. (1958). *Nature (London)* **183**, 1802.
Andrew, E. R., Clough, S., Farnell, L. F., Gledhill, T. D., and Roberts, I. (1966). *Phys. Lett.* **21**, 505.
Banner, D. W., D'Arcy, A., James, W., Gentz, R., Schoenfeld, H.-J., Broger, C., Loetscher, H., and Lesslauer, W. (1993). *Cell (Cambridge, Mass.)* **73**, 431.
Bax, A., Szeverenyi, N. M., and Maciel, G. E. (1983). *J. Magn. Reson.* **55**, 494.
Bennett, A. E., Ok, J. H., Griffin, R. G., and Vega, S. (1992). *J. Chem. Phys.* **96**, 8624.
Bodenhausen, G., Freeman, R., and Morris, G. A. (1976). *J. Magn. Reson.* **79**, 168.
Chen, L., Durley, R. C. E., Matthers, F. S., and Davidson, V. L. (1994). *Science* **264**, 86.
Christensen, A. M., and Schaefer, J. (1993). *Biochemistry* **32**, 2868.
Clore, G. M., and Gronenborn, A. M. (1991). *Prog. NMR Spectrosc.* **23**, 43.
Colombo, M. G., Meier, B. H., and Ernst, R. R. (1988). *Chem. Phys. Lett.* **146**, 189.
Copie, V., Kolbert, A. C., Drewry, D. H., Bartlett, P. A., Oas, T. G., and Griffin, R. G. (1990). *Biochemistry* **29**, 9176.
Creuzet, F., McDermott, A., Gebhard, R., van der Hoef, K., Spijker-Assink, M. B., Herzfeld, J., Lugtenburg, J., Levitt, M. H., and Griffin, R. G. (1991). *Science* **251**, 783.
Fyfe, C. A., Mueller, K. T., Grondey, H., and Wongmoon, K. C. (1992). *Chem. Phys. Lett.* **199**, 198.
Gan, Z. (1992). *J. Am. Chem. Soc.* **114**, 8307.
Garbow, J. R., and Gullion, T. (1992). *Chem. Phys. Lett.* **192**, 71.
Garbow, J. R., and McWherter, C. A. (1993). *J. Am. Chem. Soc.* **115**, 238.
Griffiths, J. M., and Griffin, R. G. (1993). *Anal. Chim. Acta* **283**, 1081.
Gullion, T. (1989). *J. Magn. Reson.* **85**, 614.
Gullion, T. (1993). *J. Magn. Reson.* **101**, 320.
Gullion, T., and Conradi, M. S. (1990). *J. Magn. Reson.* **86**, 39.
Gullion, T., and Schaefer, J. (1989a). *J. Magn. Reson.* **81**, 196.
Gullion, T., and Schaefer, J. (1989b). *Adv. Magn. Reson.* **13**, 57–83.
Gullion, T., and Schaefer, J. (1991). *J. Magn. Reson.* **92**, 439.
Gullion, T., and Vega, S. (1992). *Chem. Phys. Lett.* **194**, 423.
Gullion, T., Poliks, M. D., and Schaefer, J. (1988). *J. Magn. Reson.* **80**, 553.
Gullion, T., Schaefer, J., Baker, D. B., Lizak, M., and Conradi, M. S. (1990a). *31st, Exp. NMR Conf.* Asilomar, CA.

Gullion, T., Baker, D. B., and Conradi, M. S. (1990b). *J. Magn. Reson.* **89**, 479.
Gullion, T., McKay, R. A., and Schmidt, A. (1991). *J. Magn. Reson.* **94**, 362.
Harbison, G. S., Smith, S. O., Pardoen, J. A., Courtin, J. M. L., Lugtenburg, J., Herzfeld, J., Mathias, R. A., and Griffin, R. G. (1985a). *Biochemistry* **24**, 6955.
Harbison, G. S., Mulden, P. P. J., Pardoen, J. A., Lugtenburg, J., Herzfeld, J., and Griffin, R. G. (1985b). *J. Am. Chem. Soc.* **107**, 4809.
Henry, E. R., and Szabo, A. (1985). *J. Chem. Phys.* **82**, 4753.
Herzfeld, J., and Berger, A. E. (1980). *J. Chem. Phys.* **73**, 6021.
Hing, A. W., and Schaefer, J. (1993). *Biochemistry* **32**, 7593.
Hing, A. W., Vega, S., and Schaefer, J. (1992). *J. Magn. Reson.* **96**, 205.
Hing, A. W., Vega, S., and Schaefer, J. (1993). *J. Magn. Reson.* **103**, 151.
Holl, S. M., McKay, R. A., Gullion, T., and Schaefer, J. (1990). *J. Magn. Reson.* **89**, 620.
Holl, S. M., Marshall, G. R., Beusen, D. D., Kociolek, K., Redlinski, A. S., Leplawy, M. T., McKay, R. A., Vega, S., and Schaefer, J. (1992). *J. Am. Chem. Soc.* **114**, 4830.
Hu, J. Z., Alderman, D. W., Ye, C., Pugmire, R. J., and Grant, D. M. (1993). *J. Magn. Reson. A* **105**, 82.
Kubo, A., and McDowell, C. A. (1988). *J. Chem. Soc., Faraday Trans. 1* **84**, 3713.
Lakshmi, K. V., Auger, M., Raap, J., Lugtenburg, J., Griffin, R. G., and Herzfeld, J. (1993). *J. Am. Chem. Soc.* **115**, 8515.
Levitt, M. H. (1989). *J. Magn. Reson.* **82**, 427.
Levitt, M. H., Oas, T. G., and Griffin, R. G. (1988). *Isr. J. Chem.* **28**, 271.
Levitt, M. H., Raleigh, D. P., Creuzet, F., and Griffin, R. G. (1990). *J. Chem. Phys.* **92**, 6347.
Lima, C. D., Wang, J. C., and Mondragon, A. (1994). *Nature (London)* **367**, 138.
Lowe, I. J. (1959). *Phys. Rev. Lett.* **2**, 285.
Maas, W. E. J. R., and Veeman, W. S. (1988). *Chem. Phys. Lett.* **149**, 170.
Maricq, M. M., and Waugh, J. S. (1979). *J. Chem. Phys.* **70**, 3300.
Marshall, G. R., Beusen, D. D., Kociolek, K., Redlinski, A. S., Leplawy, M. T., Pan, Y., and Schaefer, J. (1990). *J. Am. Chem. Soc.* **112**, 963.
McDermott, A. E., Creuzet, F., Griffin, R. G., Zawadzke, L. E., Ye, Q., and Walsh, C. T. (1990). *Biochemistry* **29**, 5767.
McDowell, L. M., Holl, S. M., Qian, S., Ellen, L., and Schaefer, J. (1993). *Biochemistry* **32**, 4560.
Mishra, R. K., Chiu, S., Chiu, P., and Mishra, C. P. (1983). *Methods Find. Exp. Clin. Pharmacol.* **5**, 203.
Morris, G. A., and Freeman, R. (1978). *J. Magn. Reson.* **29**, 433.
Munowitz, M. G., and Griffin, R. G. (1982). *J. Chem. Phys.* **76**, 2848.
Murray, D. K., Chang, J., and Haw, J. F. (1993). *J. Am. Chem. Soc.* **115**, 4732.
Oas, T. G., Griffin, R. G., and Levitt, M. H. (1988). *J. Chem. Phys.* **89**, 692.
Ok, J. H., Spencer, R. G. S., Bennett, A. E., and Griffin, R. G. (1992). *Chem. Phys. Lett.* **197**, 389.
Olejniczak, E. T., Vega, S., and Griffin, R. G. (1984). *J. Chem. Phys.* **81**, 4804.
Pake, G. E. (1948). *J. Chem. Phys.* **16**, 327.
Pavlovskaya, G., Hansen, M., Jones, A. A., and Inglefield, P. T. (1993). *Macromolecules* **26**, 6310.
Peersen, O. B., and Smith, S. O. (1993). *Concepts Magn. Reson.* **5**, 303.
Peersen, O. B., Yoshimura, S., Hojo, H., Aimoto, S., and Smith, S. O. (1992). *J. Am. Chem. Soc.* **114**, 4332.
Raleigh, D. P., Harbison, G. S., Neiss, T. G., Roberts, J. E., and Griffin, R. G. (1987). *Chem. Phys. Lett.* **138**, 285.
Raleigh, D. P., Kolbert, A. C., Oas, T. G., Levitt, M. H., and Griffin, R. G. (1988a). *J. Chem. Soc., Faraday Trans. 1* **84**, 3681.

Raleigh, D. P., Levitt, M. H., and Griffin, R. G. (1988b). *Chem. Phys. Lett.* **146**, 71.
Raleigh, D. P., Creuzet, F., Das Gupta, S. K., Levitt, M. H., and Griffin, R. G. (1989). *J. Am. Chem. Soc.* **111**, 4502.
Reed, L. L., and Johnson, P. L. (1973). *J. Am Chem. Soc.* **95**, 7523.
Robyr, P., Meier, B. H., and Ernst, R. R. (1989). *Chem. Phys. Lett.* **162**, 417.
Schaefer, J., and Stejskal, E. O. (1976). *J. Am. Chem. Soc.* **98**, 1031.
Schaefer, J., McKay, R. A., Stejskal, E. O., and Dixon, W. T. (1983). *J. Magn. Reson.* **52**, 123.
Schmidt, A., and Vega, S. (1992). *Isr. J. Chem.* **32**, 215.
Schmidt, A., McKay, R. A., and Schaefer, J. (1992). *J. Magn. Reson.* **96**, 644.
Schmidt, A., Kowalewski, T., and Schaefer, J. (1993). *Macromolecules* **26**, 1729.
Sodickson, D. K., Levitt, M. H., Vega, S., and Griffin, R. G. (1993). *J. Chem. Phys.* **98**, 6742.
Spencer, R. G. S., Halverson, K. J., Auger, M., McDermott, A. E., Griffin, R. G., and Lansbury, P. T. (1991). *Biochemistry* **30**, 10382.
Stallings, W. C., Abdel-Meguid, S. S., Lim, L. W., Shieh, H. S., Dayringer, H. E., Leimgruber, N. K., Stegman, R. A., Anderson, K. S., Sikorski, J. A., Padgette, S. R., and Kishore, G. M. (1991). *Proc. Natl. Acad. Sci. U.S.A.* **88**, 5046.
Steinrucken, H. C., and Amrhein, N. (1980). *Biochem. Biophys. Res. Commun.* **94**, 1207.
Stryer, L. (1988). "Biochemistry." Freeman, New York.
Terao, T., Miura, H., and Saika, A. (1986). *J. Chem. Phys.* **85**, 3816.
Thompson, L. K., McDermott, A. E., Raap, J., van der Wielen, C. M., Lugtenburg, J., Herzfeld, J., and Griffin, R. G. (1992). *Biochemistry* **31**, 7931.
Tonelli, A. (1989). "NMR Spectroscopy and Polymer Microstructures: The Conformational Connection." VCH Publishers, New York.
Torchia, D., and Szabo, A. (1982). *J. Magn. Reson.* **49**, 107.
Tycko, R. (1994). *J. Am. Chem. Soc.* **116**, 2217.
Tycko, R., and Dabbagh, G. (1990). *Chem. Phys. Lett.* **173**, 461.
Tycko, R., and Dabbagh, G. (1991). *J. Am. Chem. Soc.* **113**, 9444.
Tycko, R., and Smith, S. O. (1993). *J. Chem. Phys.* **98**, 932.
Tycko, R., Dabbagh, G., and Mirau, P. A. (1989). *J. Magn. Reson.* **85**, 265.
Wagner, G. (1990). *Prog. NMR Spectrosc.* **22**, 101.
Weintraub, O., and Vega, S. (1993). *J. Magn. Reson. A* **105**, 245.
Wyckoff, R. W. G. (1966). "Crystal Structures," Vol. 5, p. 650. Wiley (Interscience), New York.
Yarim-Agaev, Y., Tutunjian, P. N., and Waugh, J. S. (1982). *J. Magn. Reson.* **47**, 51.
Zanotti, G. (1992). *In* "Fundamentals of Crystallography" (C. Giacovazzo, ed.). pp. 535–597. Oxford Univ. Press, New York.

CHAPTER 4

^{13}C NMR Studies of the Interactions of Fatty Acids with Phospholipid Bilayers, Plasma Lipoproteins, and Proteins

James A. Hamilton
Department of Biophysics, Boston University School of Medicine,
Boston, Massachusetts 02118

4.1 Introduction

Our laboratory has been concerned with the application of nuclear magnetic resonance (NMR) spectroscopy, primarily that of the ^{13}C nucleus, to biological questions regarding lipids. We have addressed both long-standing and new issues about lipid metabolism and transport, lipid–membrane interactions, and lipid–protein interactions. New approaches using NMR spectroscopy, which overcame limitations of other methodologies, were formulated and tested on model physiological systems of increasing complexity. Our strategy also has been to begin with the simplest NMR experiments and to proceed to more complex experiments if they become useful. Because the physiological questions motivated the NMR experiments, the following illustrations of our research are prefaced by the questions that interested us and the rationale for these new NMR approaches.

This chapter focuses on unesterified fatty acids, which constitute a minor fraction of lipids in the body but play central roles in metabolism and in the regulation of cellular processes. With ^{13}C enrichment of the carboxyl carbon, it was possible to detect fatty acids in model membranes and to study properties such as ionization and the molecular environment by high-resolution and magic-angle spinning ^{13}C NMR spectroscopy. High-resolution NMR of ^{13}C-carboxyl-enriched fatty acids was used to characterize binding to serum albumin and to intracellular fatty acid binding proteins; specifically, to elucidate structural features of binding sites and to determine their relative binding affinities and locations within the protein. In a more complex system modeling an important biological process, ^{13}C NMR monitored transfer of fatty acids between proteins and phospholipid bilayers, which provided quantitative data about kinetics and equilibrium partitioning. When feasible to perform on biological systems, ^{13}C NMR spectroscopy has important advantages over classical techniques, such as the ability to study a complex mixture without separation of the components and consequent potential perturbation of the equilibrium, and the ability to provide chemical, structural, and dynamic information simultaneously.

4.2 Properties of Fatty Acids

Unesterified fatty acids (FAs) are a key intermediate in lipid metabolism. They also constitute an important metabolic fuel in certain cells (Neely and Morgan, 1974). In addition to these long-recognized roles, FAs have diverse biological activities. For example, they serve as second messengers in signal transduction (Hannigan and Williams, 1991) and activate K^+ (Ordway *et al.*, 1989) and Ca^{++} channels (Huang *et al.*, 1992). Structurally, FAs consist simply of a hydrocarbon chain terminating in a carboxyl group. In human physiology, FAs are predominantly long-chain species (>12 carbons) of variable length and saturation. Typical dietary fats include palmitic (16:0), oleic (18:1), stearic (18:0), and linoleic (18:2) acid. Certain less abundant FAs are important in diseases. For example, the very-long-chain FA hexacosanoic acid (26:0) accumulates in the tissues of patients with the rare but sometimes fatal inherited diseases adrenoleukodystrophy and Zellweger syndrome (Moser, 1992). Medium-chain FAs (octanoic, 8:0 and decanoic, 10:0) are not prevalent in normal diets but are fed intravenously to patients with impaired lipid uptake and metabolism (Kuo and Huang, 1965).

Long-chain FAs are very insoluble in an aqueous medium (John and McBain, 1948; Vorum *et al.*, 1992) and form acid-soap lamellar aggregates at pH 7.4 rather than micelles (Gebecki and Hicks, 1973; Cistola *et al.*, 1986). *In vivo*, FAs bind to extracellular and intracellular binding proteins and to membranes. In general, the affinities of membranes and such proteins are sufficiently high to keep the aqueous concentration of FAs below the solubility limit, preventing precipitation. In human plasma most of the FAs are bound to albumin, a 65-kD protein that transports FAs to, and removes FAs from, tissues. A small, but sometimes significant, portion of FAs is bound to plasma lipoproteins. A family of intracellular proteins, fatty acid binding proteins (FABPs), bind FAs with high affinity *in vitro*, but their precise physiological role remains to be determined (Veerkamp *et al.*, 1991). Cell membranes bind FAs avidly and serve as the sites for some metabolic events and probably for many of the biological activities of FAs.

The interactions of FAs with binding proteins and membranes are therefore of general interest. The illumination of molecular details of such interactions by classical methods other than NMR has proved difficult, in part because of the simple structure of FAs. The lack of chromophoric or reporter groups, which could serve as intrinsic probes of their interactions, has led some experimentalists to add nitroxide or fluorescent groups to FAs, but such covalent modifications alter the properties of FAs. The FA studies described in the following sections are based on the use of specific ^{13}C enrichment, which causes minimal perturbation of the FA molecule and averts the need to use non-native FAs.

4.3 Interactions of Fatty Acids with Serum Albumin

Albumin is the most abundant protein in serum and binds a number of water-insoluble compounds such as FA, bilirubin, and bile acids. It is unusual in its ability to bind several moles of the same ligand (FA) with high affinity, a property elucidated many years ago by equilibrium binding studies (Goodman, 1958; Spector, 1975). However, equilibrium binding studies do not yield *site-specific* information, and numerous classical chemical and biophysical studies spanning many years failed to provide a detailed molecular description of FA binding, although an elegant speculative model has been formulated by Brown and Shockley (1982). Moreover, some of the structural information about FA binding was derived from electron spin resonance studies of nitroxide-labeled FAs (Kuznetsov *et al.*, 1975; Morrisett *et al.*, 1975), NMR spectroscopy of fluorinated FAs (Muller and Mead, 1973), and fluorescence studies of highly conjugated FAs (Berde *et al.*, 1979; Sklar *et al.*, 1977), and such information may not be relevant to binding of native FAs.

A number of specific unresolved questions were potentially suitable for ^{13}C NMR studies. For example, the number of high-affinity sites and their locations within the multidomain protein were not clearly established. The dynamics of FAs in their binding sites and the kinetics of exchange between binding sites were unknown quantities. Other questions included what types of amino acids interact with FAs, whether FAs fill binding sites sequentially or simultaneously, and how the interactions in a specific binding site are affected by normal physiological variables such as pH and the amount of FA bound to the protein. Our laboratory has conducted detailed studies of the binding of seven different ^{13}C-labeled FAs (8:0, 10:0, 12:0, 14:0, 16:0, 18:0, and 18:1) to human serum albumin (HSA) and bovine serum albumin (BSA). Illustration of the key results, primarily for BSA, are presented later. Although some specific features of binding as elucidated by ^{13}C NMR are different for the two proteins, the general results are similar, as expected from previous studies of these two closely homologous proteins (Brown and Shockley, 1982).

4.3.1 Mole Ratio Studies

^{13}C NMR spectroscopy can provide unique and fundamental data for biological systems, but the low sensitivity of this method may limit its applications. In the present case the relevant question was how would the signals of a few lipid molecules with a molecular weight of ~200 to 300 be detected in a complex with a protein of molecular weight 65,000 Daltons? Our answer (which is simple to an NMR spectroscopist) was to use specific ^{13}C enrichment in native FAs to make it feasible to detect signals from the bound FAs at the low molar ratios of FA to albumin considered to be important in physiology. The general protocol, except as noted, was to add one molecular species of singly ^{13}C-enriched FA to defatted albumin in aqueous solution.

What then does ^{13}C NMR spectroscopy reveal about binding of FAs to albumin? The binding of oleic acid (18:1 *cis*), the most common FA bound to albumin in plasma, to BSA is illustrated in Figure 1. Albumin alone exhibits a spectrum (Figure 1a) with generally broad lines and poor resolution, as expected for a protein of this size with an organized structure (~70% α-helix; Reed *et al.*, 1975). Following the addition of ^{13}C carboxyl-enriched oleic acid to BSA (5 moles/mole of protein) the spectrum (Figure 1b) contains at least five additional narrow peaks encompassing a small spectral region (180.6 to 183.8 ppm) shown in the inset of Figure 1b. The only interfering signal from the protein is the small peak at 180.9 ppm from aspartate and glutamate carboxyl groups. If oleic acid were unbound it would give a very broad signal at ~180 ppm (Cistola *et al.*, 1988). Therefore, the data in Figure 1b show that there are at least five structurally distinct binding environments. Furthermore, the oleic acid must exchange slowly between different environments, as discussed later.

The molar ratio of FAs to albumin in plasma isolated from individuals is generally in the range of 0.5 to 1.5 (Court *et al.*, 1971; Spector, 1975). In the microenvironment of the vessels and capillaries, however, the mole ratio is variable and may be much higher. It was therefore of interest to obtain spectra for a wide range of mole ratios, and these spectra for oleic acid/BSA are shown in Figure 2. It is apparent that at low ratios there are at least two FA signals; at 2 to 3 moles there are three signals of about equal intensity. Because the spin-lattice relaxation time T_1 and nuclear Overhauser enhancement (NOE) of the different carboxyl signals are the same, the peak intensity reflects the occupancy of binding sites (Parks *et al.*, 1983) and hence the *relative* affinity of the different sites. We can therefore conclude that there are (at least) three high-affinity binding sites for oleic acid on BSA. Previous analysis of equilibrium binding data suggested two or three high-affinity sites, de-

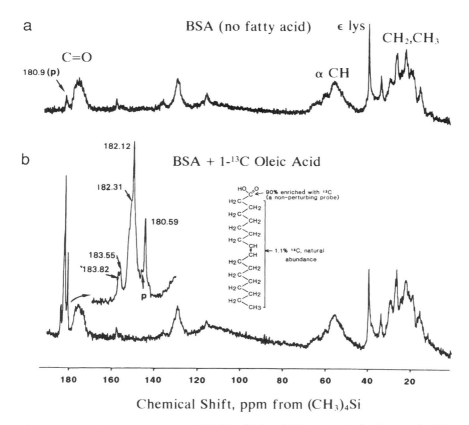

Figure 1. (a) ^{13}C NMR spectrum at 50.3 MHz of defatted BSA, 120 mg/ml, pH 7.4, and 35°C. Regions containing overlapping peaks from various groups are indicated. The ε-lysine peak at ~40 ppm is noted. A small peak (p) from glutamate carboxyls is seen at 180.9 ppm. (b) ^{13}C NMR spectrum of a complex of BSA and oleic acid with 90% ^{13}C enrichment of the carboxyl carbon. The mole ratio of oleic acid to albumin is 5:1. The carbonyl region of the spectrum (inset) is attributable mainly to the ^{13}C-enriched carboxyl carbon of oleic acid. There are five identifiable FA signals indicated by their chemical shifts and a weak signal from glutamic acid in BSA (p). Also shown is the structural formula of oleic acid (cis-9,10-octadecanoic acid), the major FA bound to serum albumin *in vivo*. Both spectra were recorded after 18,000 spectral accumulations with a repetition time of 2.0 s. (Reproduced with permission from Hamilton, 1992. Copyright © The American Physiological Society.)

pending on the method of data analysis (Goodman, 1958; Spector, 1975). The simplest interpretation is that each NMR signal represents a binding site, and the issue as to the number of high-affinity sites is answered by the more direct measurements of ^{13}C NMR.

At higher mole ratios the "picture" of binding as seen by NMR is not as

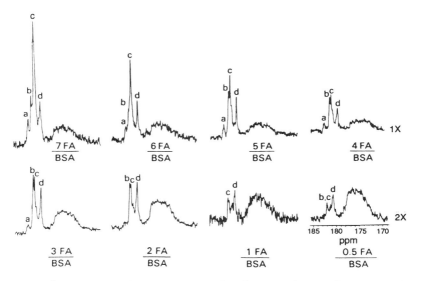

Figure 2. The 50.3-MHz ^{13}C NMR spectra (carbonyl region) of BSA with 0.5 to 7 moles of oleic acid/BSA at pH 7.4. The spectra (bottom row) for 0.5, 1.0, 2.0, and 3.0 moles of oleic acid/mole of BSA were obtained following 20,000, 4,000, 24,000 and 31,327 accumulations, respectively (2.0-s repetition times). Intensities are normalized to the protein carbonyl envelope; all spectra in the bottom row are printed with about a twofold higher vertical gain than spectra in the top row. Each spectrum in the top row (4.0, 5.0, 6.0, and 7.0 moles of oleic acid/mole of BSA) was obtained following 4000 accumulations and a 2.0-s repetition time. The four major oleic acid carboxyl peaks are labeled arbitrarily as a–d, beginning with the most downfield peak. The broad signal is primarily from protein backbone carbonyl groups. (Reproduced with permission from Parks *et al.*, 1983.)

clear. The signal intensity of the 180- to 184-ppm region increases, indicating binding of FA to the protein, but the new signals mainly overlap those for the high-affinity sites. These additional molecules of oleic acid are unlikely to pack into the same sites on the protein (Brown and Shockley, 1982), but their different environments are not distinguishable by chemical shift. At mole ratios equal to or above a value of 3, a small peak is detected at ~183.5 ppm, which represents a very distinct lower affinity site. Although the chemical shift is close to that for micellar oleic acid (Cistola *et al.*, 1988), oleic acid does not form miscelles below pH ~10 (Cistola *et al.*, 1988); therefore, this peak must also represent bound FA.

Other long-chain FAs, such as palmitic acid (16:0) and stearic acid (18:0), gave results very similar to those for oleic acid illustrated in Figure 2 (Cistola *et al.*, 1987a). By increasing the mole ratio of FA/BSA until the signal intensity of the FA carboxyl region came to a plateau, the maximum binding capacity of BSA was determined as ~10 moles/mole BSA. In the case of saturated FAs, the unbound FAs did not give an observable NMR signal and, when separated from the bound FAs, was determined by differential scanning calorimetry and X-ray scattering to be a crystalline acid-soap (Cistola et al., 1987a). Other methods can provide estimates of maximal binding capacity, and Scatchard binding data have suggested a much higher binding capacity (Spector, 1975). However, the NMR data are probably more reliable because the Scatchard analysis of low-affinity sites is difficult to perform.

4.3.2 pH Studies

A unique contribution of ^{13}C NMR in studies of FA–albumin interactions was to monitor the ionization behavior of FAs bound to albumin. pH titrations of albumin have provided estimates of the apparent pK's and the number of ionizable groups on the protein, but the behavior of individual amino acids or of the small number of bound FAs was not observed (Tanford et al., 1955). By detecting signals for FAs bound in different sites, NMR is uniquely suited to monitor the site-specific ionization behavior of FAs bound to albumin. Figure 3 presents NMR titration curves obtained from spectra

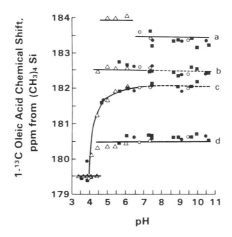

Figure 3. Plot of oleic acid carboxyl chemical shift as a function of pH for mixtures of 90% [1-^{13}C]oleic acid BSA: ●, 6 moles of 90% [1-^{13}C]oleic acid/mole of BSA, adding BSA to neat oleic acid and titrating to high pH; ○, same sample titrated from pH 10.5 to low pH; ■, 4 moles of 90% [1-^{13}C]oleic acid/mole of BSA to potassium oleate and titrating to low pH; △, 5 moles of 90% [1-^{13}C]oleic acid/mole of BSA, sample prepared by adding BSA to neat oleic acid and titrating from pH 6 to pH 3. The four curves represent data from resonances a–d, as designated in Figure 2. Note that peak a in spectra of the latter sample was slightly downfield (~0.5 ppm) in the other samples at higher pH values. Dashed lines indicate chemical-shift values that could not be unequivocally assigned to peak b or c. (Reproduced with permission from Parks et al., 1983.)

of 4, 5, and 6 moles of oleic acid/mole BSA. Three of the peaks (a, b, and d) are insensitive to pH. A titration curve is seen for peak c, which represents medium-affinity sites that have a coincident chemical shift to that for one of the three primary sites (182.0 ppm). The total chemical shift change for peak c (2.5 ppm) was about half that for complete titration of aqueous (short-chain) FAs (4.7 ppm; Cistola *et al.*, 1982). The incomplete titration probably results from unfolding of the protein at pH ~4.0 and release of FAs. The apparent pK_a of oleic acid in site c was 4.0; when the protein ionization in this range is taken into account, this apparent pK_a is very close to that of monomeric FAs in water (Cistola *et al.*, 1982). The pH-dependent ^{13}C spectra for all long-chain FAs studied (12 to 18 carbons) in complexes with albumin showed that the FAs are ionized at pH 7.4.

FA anions in high-affinity sites are likely stabilized by salt bridges with basic amino acid residues, which would depress the pK_a of the FA carboxylate. Interactions of FAs with basic amino acid residues of albumin have been suggested by numerous studies, but mostly from indirect measurements (Brown and Shockley, 1982). To investigate which positively charged side-chain residues of BSA interact with bound FA, ^{13}C NMR titrations were performed with a low mole ratio (3:1) complex of myristic acid with BSA. The spectrum at pH 7.4 for this 14-carbon FA differed somewhat from that for oleic acid at the same mole ratio. It exhibited a peak (designated b′) at a chemical shift between that of peaks b and c for oleic acid (Figures 1, 2, and 3), and a reduced intensity of peak d. The three FA carboxyl signals were monitored in the pH range of 7.4 to 11.3 (Figure 4): one carboxyl peak (d) showed no change, one (peak b) showed a small change, and the third (b′) a large change. The upfield shift of b′ correlated with a downfield shift of the ϵ-lysine signal in the range of ionization of the ϵ-ammonium group of lysine on BSA (Tanford *et al.*, 1955). These results suggest a close association of a lysine residue and the FA carboxyl in binding site b′. The remaining two sites showed little or no change over the entire pH range investigated (3.6 to 11.6) and are likely stabilized by an amino acid with a higher pK_a (tyrosine or arginine).

In the binding sites of albumin the FA carboxylate is expected to interact strongly with water molecules and one or more amino acids. The identification of specific amino acids and assessment of the relative importance of different interactions could be made by X-ray crystallography. To date, the X-ray structure of albumin with bound FA has not been published. Only the structure of FA-free HSA with and without several small organic ligands has been determined to 2.8 Å (He and Carter, 1992). HSA contains three homologous domains, each with two subdomains, as predicted by a variety of earlier experiments (Brown and Shockley, 1982). Binding of a small aromatic carboxylic acid occurs in subdomain IIA, where the carboxylate interacts with Arg 222 and 257 and Lys 199, and in subdomain IIIA, where the

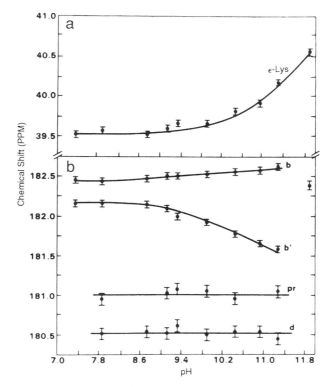

Figure 4. The ^{13}C NMR chemical-shift titration curves at high pH for 3 moles of [1-^{13}C]myristic acid/mole of BSA at 34°C. The NMR sample was titrated with 1 N KOH from pH 7.4 to 11.9. (a) Chemical shift of ϵ-carbon (lysine) resonance as a function of pH. (b) Carboxyl chemical shift of FA carboxyl peaks b, b', and d (refer to Figure 3) and BSA glutamate carboxyl peak pr as a function of pH. Peaks b and d have the same chemical shifts as the correspondingly designated peaks in Figures 1 and 2. Peak b' occurs at a chemical shift between those of peaks b and c in Figure 1. (Reproduced with permission from Cistola et al., 1987b. Copyright © 1987 The American Society for Biochemistry and Molecular Biology.)

carboxylate interacts primarily with Arg 410 but is also close to Tyr 411 (He and Carter, 1992). These binding sites have been implicated in the binding of medium-chain FAs (discussed later). It will be important to provide a better three-dimensional picture of the FA binding site by multidimensional NMR methods for comparison of the solution structure to the crystal structure.

4.3.3 Locations of High-Affinity Binding Sites

The preceding NMR results reveal several structurally distinct binding sites for FA on albumin and implicate certain types of amino acids in polar interactions with the acid anion. However, these data do not show where the binding sites are located within the multidomain structure. Previous approaches also had not mapped the locations of the binding sites with certainty. These studies generally used non-native FA analogues or modified specific protein residues in the protein, which may perturb the native structure, resulting in loss of native sites or in the production of new, non-native sites. When the results of such studies are in disagreement with predictions of various models, it is impossible to tell whether the disagreement reflects a limitation of the model or the technique.

To assign binding sites represented by NMR peaks to locations on albumin, Hamilton *et al.* (1991) compared the binding of oleic acid to native BSA and to fragments of BSA (Figure 5). To date, molecular biological techniques have not produced albumin or fragments thereof in quantities sufficient for NMR spectroscopy. Fortunately, large proteolytic fragments of BSA encompassing interesting regions of the molecule can be obtained by proteolysis; furthermore, these fragments bind long-chain FAs with high affinity, according to equilibrium binding assays, and appear to retain a native structure (Reed *et al.*, 1975). A schematic representation of the primary structure of BSA and the domain structure as described by Brown and Shockley (1982) is shown on the right-hand side of Figure 5. The fragments used for NMR studies (PA, PB, and TA) were generated by cleavage at the sites indicated by arrows. On the left-hand side of Figure 5, ^{13}C NMR spectra (carboxyl region) of oleic acid bound to three different fragments are compared with the spectrum of 3 moles of oleic acid/mole of native BSA. Figure 5a shows two peaks for oleic acid bound to a fragment (TA) encompassing the third domain of BSA (residues 377–582). The same two peaks are observed (Figure 5b) for a complex of oleic acid with the COOH-terminal one-half of the molecule (residues 307–582). Both spectra (Figures 5a and 5b) were obtained for 1 mole of oleic acid/mole of protein fragment and indicate (at least) two binding sites.

Addition of a second mole of oleic acid results in an increase in intensity of resonances already seen and a broadening of the downfield side of the resonance envelope (shown for PA in Figure 5c), but no additional identifiable peaks. Addition of 1 mole of oleic acid to a fragment encompassing the NH_2-terminal one-half of the molecule (residues 1–306) produces a single sharp peak at a much different chemical shift than the other fragments (Figure 5d). The three resonances seen for the complexes of oleic acid with the different fragments are indistinguishable in chemical shift and linewidth from the three major resonances for 3 moles of oleic acid bound to intact BSA (Figure 5e).

These comparative NMR data suggest that the structure of the three primary binding sites is preserved in the fragments and the binding sites can be

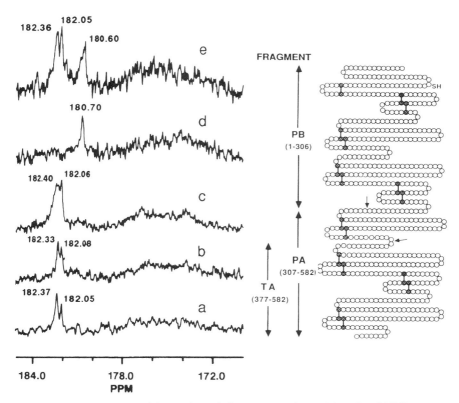

Figure 5. Carbonyl region of the ¹H-decoupled Fourier transform 50.3-MHz ^{13}C NMR spectra at 35°C. (a) The TA fragment (residues 377–582) with 1 mole of oleic acid (12,000 spectral accumulations). (b) The PA fragment with 1 mole of oleic acid (7500 spectral accumulations). (c) The PA fragment (residues 307–582) with 2 moles of oleic acid (10,000 spectral accumulations). (d) The PB fragment (residues 1–306) with 1 mole of oleic acid (6000 spectral accumulations). (e) Native bovine albumin with 3 moles of oleic acid (3000 spectral accumulations). Spectrum (e) is similar to previously published spectra (Parks et al., 1983). All spectra were obtained with 10 to 30 mg of protein fragment or 50 mg of bovine albumin in 0.05 M phosphate buffer at pH 7.4 or 8.6 and a repetition time of 2.0 s. All spectra were processed with 3-Hz line-broadening except for spectrum (e). (Reproduced with permission from Hamilton et al., 1991.)

reliably localized to domains within the protein. The two downfield signals reflect FAs bound in domain III, and the upfield signal, FAs bound to the NH_2-terminal one-half of BSA (most likely domain I, which is unmodified by the proteolytic cleavage).

Another fascinating result was obtained in the study of binding of oleic acid to BSA fragments. It was known that the fragments comprising the two halves of the molecule [residues 1–306 (PB) and 307–582 (PA)] form a complementary complex that retains some properties of albumin not observed for either of the fragments alone (Reed et al., 1975). Hamilton et al. (1991) recorded the spectrum of oleic acid bound to a stoichiometric mixture of PA and PB to determine if any secondary sites for oleic acid were regenerated in the complex. When the spectra of 3 moles of oleic acid bound to PA and 2 moles of oleic acid bound to PB were added digitally, the resultant spectrum was quite different from that for 5 moles of oleic acid bound to intact albumin (see Figure 2 in Hamilton et al., 1991). Differences in chemical shift and relative peak intensities suggested that the secondary sites on the individual fragments are different from the secondary sites on intact albumin. The spectrum of 5 moles of oleic acid bound to a stoichiometric complex of PA and PB bore a striking similarity with respect to chemical shifts, linewidths, and relative intensities to that for 5 moles of oleic acid bound to intact albumin. Thus, secondary binding sites of albumin were restored to a native form when the fragments were mixed to form a complementary complex. Since the individual fragments retain the tertiary structure for the primary binding sites for oleic acid, it appears likely that the secondary sites lie in domain II, which reforms its tertiary structure in a noncovalent complex.

4.3.4 Molecular Motions and Environments of Different Fatty Acid Carbons

The microenvironment of any FA carbon in a FA–albumin complex may be studied by the approach described earlier provided that ^{13}C enrichment can be introduced into a specific carbon atom. The environments of different carbons of myristic acid (14:0) were studied by the use of 90% [1-^{13}C]-, [3-^{13}C]-, and [14-^{13}C]myristic acids (Hamilton et al., 1984). As expected, the carboxyl carbon exhibited multiple resonances; moreover, the methylene C_3 and terminal methyl C_{14} also showed multiple resonances. Figure 6 shows spectra of the methyl-labeled myristic acid as a function of the mole ratio bound to BSA. At the lowest ratio examined (1.4 moles FA/mole BSA), three methyl resonances were observed, demonstrating at least three different high-affinity binding sites, in accordance with the results for carboxyl-labeled FA. From the significant chemical shift dispersion, it is clear that amino acids are in close contact with different parts of the FA chain and that the local amino acid environments vary somewhat in the binding sites.

Another reason for using a FA molecule with ^{13}C enrichment at sites other than the carboxyl carbon was to obtain information about the molecular motions of FA bound to BSA. Molecular motions are difficult to assess from relaxation measurements of carboxyl resonances because the ^{13}C relaxation mechanism of the carboxyl carbon is not known. However, ^{13}C relaxation of protonated carbons occurs via $^{13}C-^{1}H$ dipolar interactions of

4. Interactions of FAs with Phospholipids, Lipoproteins, and Proteins / 129

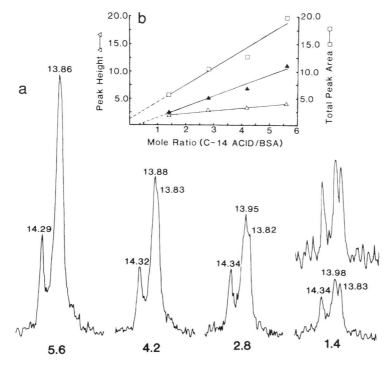

Figure 6. (a) Methyl region of the 50-MHz ^{13}C NMR spectra at 34°C of mixtures of 90% [14-^{13}C]myristic acid and BSA at four mole ratios (1.4, 2.8, 4.2, and 5.6). (Inset) Convolution difference spectrum from the same time domain data as the bottom spectrum. (b) Plots of peak heights of the main resonance at ~13.9 ppm (▲) and of the resonance at 14.3 ppm (△) and the integrated intensity of the entire resonance envelope (□) as a function of molar ratio. Plots were linear with $r > 0.98$. In the spectrum taken at a molar ratio of 0.7 (not shown), two methyl resonances (at 14.27 and 13.95 ppm) were detected. (Reproduced with permission from Hamilton et al., 1984.)

magnetic fields comparable with that used in this study (Norton et al., 1977). If the myristic acid molecule were rigidly bound to BSA, the overall tumbling of the protein molecule would be the dominant motion determining the relaxation of protonated carbons. The rotational reorientation of the BSA molecule is anisotropic, with correlation times in the range of 25 to 100 ns (Moser et al., 1966; Wilbur et al., 1977). When the shortest correlation time (25 ns) and the theoretical analysis for isotropic reorientation are used, the predicted linewidth of a methylene or methyl resonance is >100 Hz and the predicted NOE is the minimum value of 1.15 (Doddrell et al., 1972). The observed linewidths for the myristic acid C_3 and C_{14} resonances (which include a significant contribution from chemical shift inhomogeneity) were much smaller, and the observed NOEs much larger, than the predicted values. Therefore, both the C_3 and C_{14} carbons must have internal motions that are rapid compared with protein tumbling. Comparison of T_1 and NOE values of the C_3 and C_{14} resonances suggest that the motions of the C_3 carbon are relatively restricted compared with those of the methyl carbon; also the methyl peaks appear to be narrower than the C_3 peaks. Thus, the polar end of the FA chain may be more rigidly attached to the binding site than the nonpolar end, which is consistent with the hypothesis that strong and specific interactions between FA and BSA play a significant role in FA binding (Parks *et al.*, 1983). In addition, T_1, NOE, and linewidth values did not differ markedly for different peaks within a resonance envelope, so that the molecular motions are similar in different binding environments.

Molecular motions of FA bound to BSA have also been assessed by electron spin resonance (ESR) with the use of spin-labeled FAs. ESR studies produced conflicting ideas concerning (1) whether the (derivatized) FA is rigidly or loosely attached to BSA and (2) which portion of the FA molecule is most rigidly bound (see, e.g., Kuznetsov *et al.*, 1975; Langercrantz and Setaka, 1975; Morrisett *et al.*, 1975). For example, stearic acid molecules containing nitroxide probes near the carboxyl group appear to be immobilized in the albumin complex. However, nitroxide probes may significantly alter both the FA binding environment and FA molecular motions, whereas the ^{13}C probes are nonperturbing. The NMR results show significant freedom of motion of bound FA and suggest a much more dynamic structure of the lipid–protein complex than does ESR.

4.3.5 Exchange of Fatty Acids between Binding Sites

Another important aspect of the dynamics of FA binding to albumin is the exchange of FA molecules between different binding sites. Detection of different binding sites for a long-chain FA by ^{13}C NMR as described earlier means not only that the sites are structurally distinct but that FA exchange between sites is slow (the half-time for exchange must be longer than approximately 66 ms, as discussed later). In conducting NMR studies with FAs of different chain lengths, we found a striking contrast between the spectra of FAs with 8 and 10 carbons (octanoic and decanoic acid) and those with longer chains. Octanoic or decanoic acid in the presence of BSA (Hamilton, 1989) or HSA (Kenyon and Hamilton, 1994) gave a single carboxyl peak under conditions where long-chain FAs gave rise to multiple peaks (e.g., low mole ratios and temperatures greater than 30°C). Thus, the NMR spectrum for 3 moles of octanoic acid/mole BSA at 38°C (Figure 7a) showed a single narrow FA carboxyl peak, in contrast to the multiple peaks for the 12-carbon dodecanoic acid (Figure 7d) or for oleic acid (Figures 1 and 2). This result

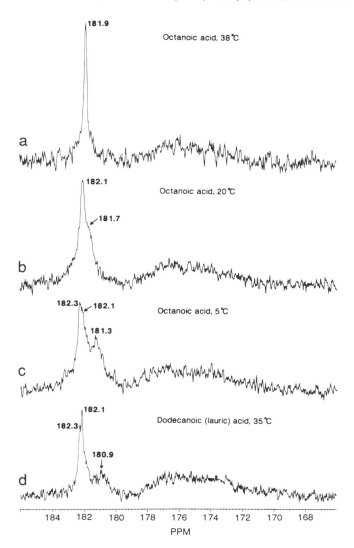

Figure 7. (a)–(c) Carboxyl region of the ^{13}C NMR spectrum of 3 moles of 90% [1-^{13}C]octanoic acid per mole of BSA, pH 7.4, at different temperatures. (d) Spectrum of 3 moles of 90% [1-^{13}C]dodecanoic acid per mole of albumin. Note that the low-temperature spectrum of octanoic acid complex resembles the spectrum for the dodecanoic acid complex obtained at a higher temperature. (Reproduced with permission from Hamilton, 1992. Copyright © 1992 The American Physiological Society.)

indicated that, in contrast to long-chain FAs, medium-chain FAs either do not occupy highly distinct sites or they exchange between binding sites fast enough to yield an "exchange-averaged" resonance.

To explore the latter possibility, NMR spectra were obtained at lower temperatures where exchange (if occurring) might be slowed to the point where individual sites could be observed. Figures 7a, 7b, and 7c show that the spectrum of octanoic acid/BSA separates into multiple peaks with decreasing temperature and exhibits (at least) three signals at 5°C [spectrum (c)]. The latter spectrum is somewhat similar to that of the 12-carbon dodecanoic (lauric) acid with BSA at 35°C (Figure 7d). Thus, octanoic acid, like FAs of greater chain length, occupies multiple, structurally distinct binding sites, but rapid exchange of octanoic acid among sites obscures them at higher temperatures (Hamilton, 1989).[1]

The mean lifetime of a FA in a binding site can be determined from spectra that show temperature-dependent changes in chemical shift and/or linewidth by the use of standard NMR equations (Roberts and Jardetzky, 1985). The mean lifetime at 33°C calculated from our data was 3 ms for octanoic and 50 ms for decanoic acid (Hamilton, 1989). When the spectrum shows separate, narrow peaks at higher temperatures, as in the case of dodecanoic acid (Figure 7d) and oleic acid (Figures 1 and 2), and no changes at lower temperatures, the slow exchange limit has been reached. In this case only the lower limit of the mean lifetime can be calculated. Dodecanoic acid and fatty acids with longer chains have a mean lifetime in a high-affinity binding site of >66 ms. Independent measurements by fluorescence have shown dissociation of long-chain FAs from albumin on the time scale of seconds (Daniels *et al.*, 1985), or faster (Kamp and Hamilton, 1992).

Hydrophobic interactions, which increase with increasing FA chain length, are thus critical in determining the rate of release of a FA from a binding site, a key aspect in the delivery of FAs to tissues by albumin (as discussed later).

4.4 Binding of Medium-Chain Fatty Acids to Human Serum Albumin

Because of the special relevance of medium-chain FAs in human nutrition, we conducted detailed ^{13}C NMR studies of the interactions of octanoic

---

1. Note that, as seen in Figure 7, the signal at 180.8 ppm corresponding to a strong binding site for oleic acid (peak d in Figures 1 and 2) and other long-chain FAs was of diminished relative intensity for FAs with fewer than 16 carbons. This binding site, which is located in domain I (Figure 5), is either altered in structure in the presence of medium-chain FAs or is not as strong a binding site.

and decanoic acid with HSA (Kenyon and Hamilton, 1994). The carboxyl spectrum of each medium-chain FA showed a single FA signal at 35°C at various mole ratios of FA/HSA (spectra not shown); at mole ratios of >2, multiple signals were observed at low temperature, as in the case of BSA (Figure 8). Studies at low temperature demonstrated sequential filling of high-affinity binding sites as opposed to simultaneous filling for longer chain FAs (see Figure 2). Thus, a single peak was seen at a FA/HSA mole ratio of 1.0, a second peak at a ratio of 2.0, and a third at a ratio of 3.0 (in contrast to the results for oleic acid in Figure 2). Interestingly, as shown in Figures 8a and b, the chemical shift corresponding to the first mole of octanoic acid (181.9 ppm) differs significantly from that for the first mole of decanoic acid (182.3 ppm), even though the reference chemical shift (aqueous FA at 183.4 ppm)

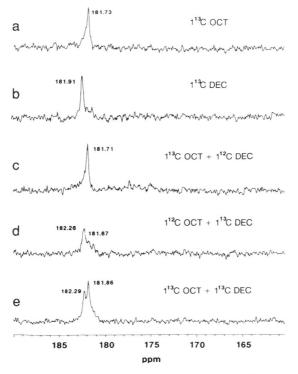

Figure 8. Carboxyl region of the ^{13}C NMR spectrum at 50.3 MHz of HSA (92 mg/ml in 0.07 M KCl; pH 7.4) with medium-chain FAs. (a) 1 mole of ^{13}C octanoic acid (OCT). (b) 1 mole ^{13}C decanoic acid (DEC). (c) 1 mole of ^{13}C OCT and 1 mole of unenriched DEC. (d) 1 mole of ^{13}C DEC and 1 mole of unenriched OCT. (e) 1 mole of ^{13}C OCT and 1 mole of ^{13}C DEC. All spectra were obtained at 6°C, with a repetition rate of 2.0 s and 6000 spectral accumulations. (W. Guo, M. Kenyon, and J. A. Hamilton, unpublished results, 1994.)

is the same for both FAs. According to this result, octanoic and decanoic acid either bind to different sites or to the same site in a significantly different manner. To distinguish these two possibilities, competition studies were carried out in which one FA contained ^{13}C enrichment and the second FA was unenriched. In the presence of 1 mole of unenriched decanoic acid, the chemical shift observed for 1 mole of [1–^{13}C]octanoic acid bound to HSA is unaltered (Figure 8c). Similarly, the chemical shift for [1–^{13}C]decanoic acid (1 mole/mole HSA) is unaffected by 1 mole of (unenriched) octanoic acid (Figure 8d), although the peak is somewhat broader. In a sample containing 1 mole of each ^{13}C-enriched FA/mole HSA, the spectrum showed signals for both FAs at the expected chemical shift (Figure 8e).

The highest affinity binding sites for octanoic and decanoic acid on HSA thus appear to be structurally distinct. This result is surprising because of the close structural similarity of the two acids and because previous studies had not differentiated binding sites for these acids (Brown and Shockley, 1982). The binding sites characterized by X-ray crystallography for the small aromatic carboxylic acid in subdomains IIA and IIIA are putative sites for medium-chain FAs. Although the aromatic acid binds equally in both sites, it is possible that these sites discriminate between octanoic acid and decanoic acid. Regardless of where these two FAs bind, the NMR results suggest interesting X-ray crystallographic experiments.

4.5 Binding of Fatty Acids to Human Serum Lipoproteins

Albumin binds most of the FAs in human plasma under normal conditions, but FAs can also bind to the less abundant lipoprotein particles, whose primary function is to transport cholesterol and triacylglycerol. The major classes of lipoproteins are chylomicrons, very-low-density (VLDL), low-density (LDL), and high-density (HDL) lipoproteins. Each density class has a different lipid-to-protein ratio, lipid composition, and protein component(s). All lipoproteins have an emulsion structure, consisting of an amphipathic surface comprised primarily of protein(s), phospholipids, and cholesterol covering a core of triacylglycerol and cholesteryl ester. The triacylglycerols in chylomicrons and VLDL are hydrolyzed by lipoprotein lipase in the plasma, especially in the capillary endothelium, to provide FAs for tissues. These two classes of lipoproteins therefore can contain high amounts of FAs, at least transiently. HDL is primarily a vehicle for transport of cholesterol from tissues to the liver. LDL transports cholesterol to tissues and can enter cells by a receptor-mediated mechanism, which is influenced by the presence of long-chain FAs (Bihain *et al.*, 1989).

Very little is known about the physical properties of FAs in lipoproteins, and such knowledge is crucial in understanding the biological activities of FAs. In spectroscopic studies the technical problem is to distinguish the signals of a minor constituent (FA) that closely resembles most of the other molecules (acylated lipids) except for the headgroup. The technical barrier may be overcome by attaching a label to the FA (e.g., a fluorescent or nitroxide group) or, more satisfactorily, by using isotopically enriched FA and NMR spectroscopy. Studies of selectively deuterated FAs incorporated into VLDL, LDL, and HDL have provided information about the dynamics of the monolayer surface of the particles (Chana *et al.*, 1990a,b; Parmar *et al.*, 1985). A correspondence of lineshapes of palmitic acid and phosphatidylcholine with selective deuteration in the acyl chain suggested that the FA mobility was similar to that of the phospholipid. Palmitic acid showed a complex lineshape with a broad and a narrow component, which was interpreted to reflect two pools of different mobility in slow exchange.

Our laboratory has studied binding of [$1-^{13}$C]oleic acid to model chylomicron particles (triacylglycerol-rich emulsions; Spooner *et al.*, 1989), and to VLDL, LDL, and HDL. A single resonance for bound oleic acid was observed at all oleic acid ratios and all pH values investigated, in contrast to the findings for albumin. The carboxyl signal was narrow under all conditions, with no evidence of a broad component. Thus, the ^{13}C NMR results show that oleic acid occupies a single type of binding site in the lipoprotein, or it exchanges rapidly among different sites. In the model chylomicron particles, it was possible to demonstrate at least two pools of oleic acid, one in the surface (the majority) and the other in the core, which are in fast exchange. Only the un-ionized FA is soluble in the triacylglycerol core, so that the partitioning favors the phospholipid surface (Spooner *et al.*, 1989).

The ionization behavior of oleic acid in VLDL, LDL, and HDL is shown in Figure 9. The protein content of VLDL is very low, and the titration curve for oleic acid bound to VLDL falls very close to that for oleic acid in a phospholipid vesicle, which serves as a model for the phospholipid-rich surface of the lipoprotein, as well as for a membrane bilayer (discussed later). This curve suggests little partitioning of oleic acid into the triacylglycerol-rich core, which would shift the titration curve to higher pH and give a higher apparent pK_a for the FA (Spooner *et al.*, 1989). The titration curve for oleic acid in LDL is shifted to higher ppm and lower pH and shows a greater deviation from the titration curve for oleic acid in phospholipid vesicles at high pH than at low pH. The latter effect may reflect transient binding of ionized oleic acid to the protein moiety (apo B), which occupies about 20% of the surface of LDL (Walsh and Atkinson, 1986). However, the overall curve closely resembles that for oleic acid in phospholipid vesicles with 30 mol% cholesterol (see Figure 11a), a proportion similar to that in the surface mono-

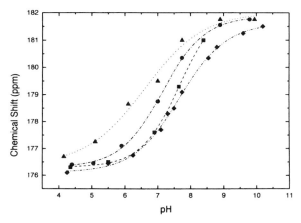

Figure 9. NMR titrations (plots of carboxyl chemical shifts versus pH) at 50.3 MHz and ~35°C of [1-^{13}C]oleic acid in: ◆, egg PC vesicles; ■, VLDL; ●, LDL; and ▲, HDL. The dotted lines represent curve fitting of the data; the PC titration curve follows predicted Henderson-Hasselbach behavior (see Figure 10). (J. A. Hamilton, unpublished results, 1994.)

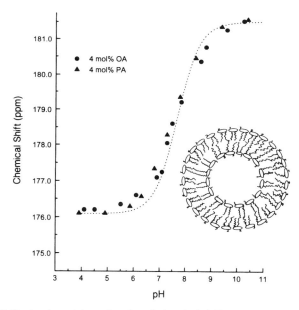

Figure 10. NMR titrations (plots of carboxyl chemical shift versus pH) at 50.3 MHz and ~35°C of [1-^{13}C]oleic acid (OA) or palmitic acid (PA) in egg PC vesicles in 0.075 KCl. FAs were incorporated into vesicles either by cosonication of FA and PC or by addition of K$^+$ FA$^-$ to PC vesicles. Titrations were performed by addition of small aliquots of 0.1 N HCl or KOH. The data for oleic acid and palmitic acid are not significantly different. The dotted line represents the predicted Henderson-Hasselbach titration curve for palmitic acid as calculated in Cistola *et al.*, 1987. Also shown is a representation of the structure of an egg PC vesicle. (S. P. Bhamidipate and J. A. Hamilton, unpublished results, 1994.)

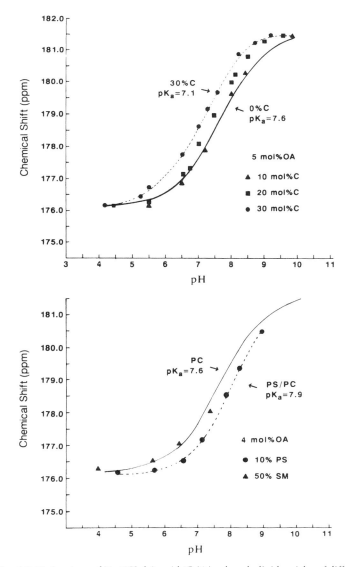

Figure 11. NMR titrations of [1–^{13}C]oleic acid (OA) in phospholipid vesicles of differing composition carried out as in Figure 10. The top curves are for 5 mol% OA in egg PC vesicles without cholesterol and with 10, 20, or 30 mol% cholesterol. The dotted line connects the data for 30 mol% cholesterol (apparent pK_a = 7.1). The bottom curves are for 4 mol% OA in PC vesicles with 10 mol% phosphatidylserine (PS) or 50 mol% sphingomyelin (SM). The solid line represents the curve for OA in egg PC vesicles. Dotted lines represent curve fitting of the data. Spectra were recorded after 500 to 1000 accumulations and a repetition rate of 2.0 s. (J. A. Hamilton, unpublished results, 1993.)

layer of LDL. The apparent pK_a (the pH corresponding to one-half the maximum chemical-shift difference) estimated from the LDL titration (7.0) is close to that for the model vesicle system (7.1). The titration curve for oleic acid in HDL is further shifted to lower pH, giving an apparent pK_a of 6.75. Since the HDL particle is 50% protein by weight, it is likely that the protein environment influences the ionization behavior of FAs. The curve could suggest partitioning of FAs between protein and phospholipid at all pH's and a suppression of the apparent pK_a by interactions of FAs with basic amino acids, as described for albumin.

4.6 Binding of Fatty Acids to Cytosolic Fatty Acid Binding Proteins

A small protein (~14 kDa), which binds FA with high affinity, the cytosolic fatty acid binding protein (FABP), is found inside cells of many tissues. Whereas serum albumin of a given species has a highly invariant primary structure (Carlson *et al.*, 1992), FABP exists in at least five structurally distinct types (Veerkamp *et al.*, 1991). These are named after the tissue from which they were first isolated. This designation does not mean however that the specified FABP is found only in that tissue or that it is present in all cell types of the tissue. Although present in high abundance in the cytosol, FABPs perform yet unknown functions.

The first FABP to be studied by ^{13}C NMR spectroscopy was rat liver FABP (Cistola *et al.*, 1988). Binding of [1–^{13}C]oleic acid yielded a single carboxyl resonance at all pH's and mole ratios examined, in contrast to the case for serum albumin. The single carboxyl peak titrated with an apparent pK_a of 4.8, suggesting that the carboxyl group was not strongly bound to basic amino acid residues and was water accessible. Although different sites were not visualized by ^{13}C NMR, liver FABP appeared to have a maximal binding capacity of 2 or 3 moles of FA/mole of protein from peak intensity measurements. In contrast, rat intestinal FABP binds only 1 mole of FA and the carboxyl signal does not change with pH (Cistola *et al.*, 1989). It was suggested that the carboxyl FA was ionized and in close contact with an arginine residue, a prediction that was verified by the crystal structure of the protein with bound palmitic acid (Sacchettini *et al.*, 1989).

The ability to express FABPs in *E. coli* in large quantities has permitted detailed comparison of binding properties of highly pure proteins (Cistola *et al.*, 1990). Thus, although the structure and binding properties of FABPs are simpler than albumin, the heterogeneity of types of FABPs and the availability of genetically produced FABPs will certainly stimulate further ^{13}C NMR binding studies.

4.7 Binding of Fatty Acids to Model Membranes

FAs move in and out of cells continuously through the plasma membrane and thus reside there at least temporarily. Furthermore, because FAs are very water insoluble but bind avidly to lipid membranes, a substantial portion of FAs in cells is probably bound to the plasma membrane and certain intracellular membranes. In cells containing FABP, a fraction of the FA will be bound to this protein (Veerkamp et al., 1991). Most of the biological activities of FAs [such as metabolic conversion to acyl CoA and remodeling of lipids (Spector, 1992) and activation of ion channels (Huang et al., 1992; Ordway et al., 1989)] probably take place in membranes. FAs are formed directly in membranes by the action of phospholipase A_2.

Studies of the physical properties of FAs in biomembranes by NMR represent a challenging task, which has not yet been undertaken to our knowledge. However, we and others have examined properties of FAs in model membranes (phospholipid bilayers). ^{13}C NMR spectroscopy is uniquely suited to measure one especially important property of FA in membranes—their ionization behavior—as described originally by Kantor and Prestegard (1978) and Ptak et al. (1980). The extent of ionization of FAs in a membrane will influence the net charge, determine the amount of ionized FAs available for (potential) protein interactions, and possibly affect the fusion of cell membranes (Kantor and Prestegard, 1978) and the rate and mechanism of FA movement across the membrane bilayer.

The ionization of the FA carboxyl group in a specific environment can be determined readily by measuring the carboxyl carbon chemical shift as a function of bulk pH, as illustrated earlier for FAs bound to albumin (Figure 3). In membranes FAs are normally present in small proportions to the phospholipid. In model systems where the FA is the minor component, the NMR signal for a FA is enhanced by ^{13}C enrichment. Figure 10 shows titration curves for oleic acid (18:1) and palmitic acid (16:0) (4 mol% with respect to phospholipid) in phospholipid (specifically, phosphatidylcholine) bilayer vesicles. Small unilamellar vesicles (schematized in Figure 10) produced by sonication are used to ensure a homogeneous distribution of FA and, most important, to give narrow NMR signals suitable for high-resolution NMR spectroscopy. The titration curve for oleic acid or palmitic acid in vesicles is well represented by the classical Henderson-Hasselbach equation (dotted line in Figure 10), as observed for other FAs by Ptak et al. (1980). The apparent pK_a of oleic acid in the model membrane is 7.6, much higher than the pK_a of the water-soluble, shorter chain FAs (pK_a ~4.8; see Cistola et al., 1982). We have examined concentrations of 0.5 to 10 mol% oleic acid in PC vesicles and found the apparent pK_a to be invariant over this range of composition. Further, the titration curves for other long-chain FAs [myristic acid (C14:0) and stearic acid (C18:0)] obtained under identical conditions as for oleic acid were superimposible on the oleic acid titration curve. For FAs with chain lengths shorter than 14 carbons, the NMR titration curves are skewed to higher ppm because of partitioning of the FA into the water.

The general conclusion can be made that, on binding to the phospholipid bilayer, the FA resides in an environment that effectively raises its pK by about 3 pH units. Therefore, at pH values near physiological pH, there will be an almost equal population of un-ionized and ionized species in the lipid interface.

Note that only a single resonance is observed for FAs bound to vesicles, irrespective of (1) the particular FA, (2) the exact phospholipid composition of the vesicle, and (3) the pH. The carboxyl lineshape is symmetrical, and the signal is very narrow (<10 Hz) at low and high pH values and somewhat broader near the apparent pK_a.[2] Therefore, only one environment is reported by the NMR signal. However, it was recently demonstrated by fluorescence measurements that native FAs equilibrate to both leaflets of a phospholipid bilayer, so that at least two environments are present in small vesicles (Kamp and Hamilton, 1992). The rapid flip-flop between the two leaflets via the un-ionized form of the FA can explain the apparent equivalency of FA environments in vesicles.

Although phosphatidylcholine is the predominant lipid in most mammalian membranes, the composition of biomembranes is complex and variable. We therefore examined the ionization behavior of FAs in vesicles of differing composition (Figure 11). The top plots show the ionization of oleic acid in phosphatidylcholine (PC) vesicles with 0, 10, 20, and 30 mol% unesterified cholesterol. With 20 mol% cholesterol, a small shift in the curve to lower pH occurs; with 30 mol% cholesterol, the shift is more significant, and the apparent pK_a is 7.1. The bottom plots show the ionization of oleic acid in vesicles comprised of PC with 10 mol% phosphatidylserine (PS) or 50 mol% sphingomyelin (SM). While the partial titration curve for PC/SM vesicles suggests no change in apparent pK_a compared to PC vesicles, the presence of only 10% PS in PC shifts the curve to higher pH, with an apparent pK_a of 7.9. The differences in apparent pK_a can be explained by changes in the interfacial charge (Ptak et al., 1980). These authors observed an increase in the apparent pK_a of myristic acid in PC vesicles when 10 mol% of the negatively charged phosphatidic acid was added and a decrease in pK_a when 20 mol% stearyl amine was added. In our model membranes the presence of the negatively charged serine group of PS is expected to increase the apparent pK_a of the FA. Cholesterol dilutes the surface charges of the PC, which should decrease the apparent pK_a.

---

2. Broadening near the apparent pK_a may reflect a slower exchange of H$^+$ at half-ionization.

Small unilamellar vesicles are suitable vehicles for monitoring FA ionization by high-resolution ^{13}C NMR spectroscopy, but the important question remains as to how applicable the results obtained with highly curved surfaces of such vesicles are to the more planar surfaces of cell membranes. Do the surfaces of small vesicles distort the interactions of FAs with phospholipids and their ionization behavior? To address such questions, we recently performed ^{13}C NMR titrations of oleic acid in PC multilayers by magic-angle spinning (MAS) NMR spectroscopy. Although MAS NMR has been used extensively in polymer chemistry and in organic chemistry for a number of years, its application to biological samples has been limited. Recently it was demonstrated that ^{13}C MAS NMR spectra of phospholipid multilayers can exhibit the resolution of sonicated vesicles (Oldfield *et al.*, 1987), and that the properties of minor components with ^{13}C enrichment can be studied quantitatively (Hamilton *et al.*, 1991). Figure 12a illustrates the poor resolution of ^{13}C signal multilayers in the absence of MAS (notably, the incorporated ^{13}C-carboxyl-enriched oleic acid was not detected) and the greatly improved resolution with MAS (Figure 12b). Reliable chemical shifts were obtained for the oleic acid carboxyl resonance. The minimum and maximum chemical shifts were the same as that found for vesicles (see earlier), and the NMR titration curve (Figure 12, inset) gave an apparent pK_a of 7.2. The difference between the pK_a for oleic acid in multilayers and small vesicles is likely a function of the ionic strength of the sample rather than structural effects of the phospholipid matrix. Vesicle titrations were typically performed in 0.075 M KCl, whereas the multilayer titration was performed in a 0.50 M buffer because of the much higher lipid concentration. The data of Ptak *et al.* (1980) for single-walled vesicles predict that this high ionic strength will reduce the pK_a to ~7.2. On the other hand, if the differences in apparent pK_a of the FA in small vesicles compared to multilayers are in fact due to curvature differences, they are not large.

The NMR titrations of long-chain FAs in model membranes show apparent pK_a values in the range of physiological pH's. Because the pH titration curve is steepest near the pK_a, small changes in pH in the vicinity of a membrane could result in significant changes in the proportion of the un-ionized and ionized forms of FAs. Membrane composition also can be expected to influence the apparent pK_a of FAs in the membrane; though the changes in pK_a may be small, the ratio of ionized and un-ionized species will change significantly. Results of all model membranes reveal a sizable pool of un-ionized FAs in the interface. These NMR findings led to the prediction that FA could move readily across a phospholipid bilayer via the un-ionized form, which was recently verified by a fluorescence technique (Kamp and Hamilton, 1992, 1993). The misconception that FAs exist entirely in the ionized form in membranes led to the erroneous concept that FAs can cross a membrane *only* via a protein transporter (Potter *et al.*, 1989).

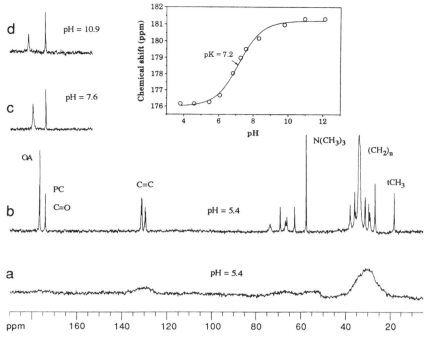

Figure 12. Unsonicated egg PC with 2 mol% [1-^{13}C]oleic acid in phosphate buffer (500 mM) at various pH's and a buffer-to-lipid ratio of 4:1 w/w. Spectra were obtained at 75 MHz in a solid-state MAS probe at 35°C. Samples (~400 μl) were placed in a 7-mm ZrO$_2$ rotor. Each titration point represents an individual sample in buffer at the specified pH. (a) Sample at pH 5.4 without spinning, showing a typical low-resolution spectrum of phospholipid multilayers. Note that the carbonyl region shows no detail. Spectra (b), (c), and (d) were obtained for samples at the indicated pH values with sample spinning of 3 kHz at the magic angle. A ^{1}H decoupling power of ~10 W was used, and 4000 spectral accumulations were Fourier transformed. The inset shows the dependence of the chemical shift of the oleic acid carbonyl on the pH of the buffers used for lipid hydration. The measured pH of the hydrated lipid–buffer mixture was close to that of the buffer (±0.05). (W. Guo and J. A. Hamilton, unpublished results, 1994.)

The NMR titration curves also provide information about the localization of the FA molecule in the phospholipid bilayer. Long-chain FAs are too insoluble in water to determine a titration curve for the monomeric FAs. However, data for octanoic acid (C8:0), the longest chain FA for which an NMR titration curve for the monomeric form has been reported, provide a useful reference. Compared to the NMR titration curve for aqueous, monomer octanoic acid (Cistola *et al.*, 1987a), the titration curve for long-chain FAs in vesicles is shifted upfield by about 2 ppm; the difference between the minimum and maximum chemical shifts is only slightly larger than that for the monomer titration. The upfield shift suggests diminished solvation and H-bonding with the FA carboxyl group in vesicles compared to water, consistent with localization of the carboxyl group at the aqueous interface, where the FA carboxyl group will only be partially hydrated. The small difference in the chemical shift range could be indicative of a modest change in the exact position of the carboxylate as a function of ionization, possibly a deeper immersion in the lipid bilayer of the un-ionized form, as suggested by ESR studies (Miyazaki et al., 1992). The correspondence of titration curves for different long-chain FAs (C14:0, C16:0, C18:0, C18:1) indicates the same position of the FA carboxyl group in the interface. Interestingly, we also found the titration curve for a very-long-chain FA in PC vesicles (hexacosanoic acid; C26:0) to be the same as for the long-chain FAs (J. Ho and J. A. Hamilton, unpublished results). Since the carboxyl group must therefore be localized in the same position as other FAs, the acyl chain of C26:0 will cause significant perturbations in the middle of the phospholipid bilayer, which has chain lengths of primarily 16 and 18 carbons. Accumulation of hexacosanoic acid in membranes is associated with a group of severe inherited diseases in which this FA is not metabolized, as noted in the introduction.

Future ^{13}C NMR studies of FAs in biomembranes will utilize the data from simple (protein-free) lipid bilayers as well as some of the same strategies. Incorporation of ^{13}C-enriched FAs will be necessary since typical levels of FA in biomembranes are lower than 5 mol% with respect to phospholipid. Furthermore, since membranes do not typically give high-resolution spectra, the technique of MAS NMR should prove useful. Questions to be posed include the following: Will the FA show a single signal representative of binding to the lipid bilayer? Will additional signals or an exchange-averaged signal representing binding to proteins be seen? Will the ionization behavior be similar to a simple bilayer? Will saturated and unsaturated FAs show identical behavior, as in vesicles? NMR approaches will also yield unique information about the mobility of FAs in membranes.

4.8 Exchange of Fatty Acids between Albumin and Model Membranes

As crucial sources of metabolic energy and intermediates for lipid remodeling in cells, FAs move into (and out of) cells rapidly and continuously. To the physiologist, the question of how FAs are transferred between serum albumin and cell membranes is likely to be more important than the details of binding interactions with albumin and lipid bilayers. How does a protein with such a high affinity for FAs release this ligand to cells? How fast does this occur? What factors govern the partitioning of FAs between serum albumin and cells? Some investigators have concluded that receptors for FA–albumin complexes or FA-binding membrane proteins are required to sequester FAs from the plasma (Veerkamp *et al.*, 1991). A simpler hypothesis is that FA equilibration occurs via the small amount of FAs dissolved into the aqueous phase and that diffusion from albumin is not facilitated by other proteins. One of the problems in defining mechanisms and discriminating among different hypotheses lies in the simple structure of FAs, which makes it difficult to track FA movement. As in binding studies, physical chemists have often relied on modified FAs to conduct their studies. Furthermore, experimental strategies have generally relied on separation of donor and acceptor species to determine where the FA is bound.

To the biophysicist it is essential to simplify complex systems into model systems with a limited number of controllable variables. Therefore, to study the partitioning of FAs between serum albumin and cells, we chose to "model" the cell with a phospholipid vesicle (Figure 10 inset), which has the essential structural element of plasma membranes, the phospholipid bilayer, but no proteins or other membrane constituents. The interior of the vesicle consists of water (or buffer) and is devoid of metabolic machinery to metabolize FAs and soluble proteins to bind FAs. From inspection of ^{13}C chemical-shift data for binding of FAs to a phospholipid bilayer (Figure 10) and to albumin (Figure 3), it is apparent that the chemical shifts are well differentiated at pH values in the physiological range. Therefore, it should be possible to add FAs to a mixture of vesicles and albumin, and determine the partitioning of FAs by the chemical shift(s) of the FA peak(s). The NMR measurements could thus be performed without separating the donor and acceptor species and potentially disturbing the equilibrium.

The strategy for studying partitioning of FAs is illustrated in Figure 13, which shows results of a transfer experiment starting with 6 mol% [1-^{13}C]oleic acid in vesicles, to which aqueous FA-free BSA was added. The carboxyl region of the ^{13}C spectrum of oleic acid in vesicles (Figure 13a) shows a single carboxyl peak at the expected chemical shift at pH 7.4 (Figure 10). To this "donor" system, FA-free BSA was added and the pH adjusted to 7.0. The spectrum of the mixture (Figure 13b) shows signals in the spectral region (between 180 and 184 ppm) where resonances are observed for FA-albumin complexes alone (Figures 1 and 2), a peak at the chemical shift for FA bound to PC vesicles at pH 7.0 in the absence of BSA (Figure 10), and carboxyl peaks from PC on the outer and inner leaflets of the vesicles. Thus, the presence of PC vesicles did not perturb the FA binding properties of BSA with respect to the local magnetic environments of the FA carboxyl, and the

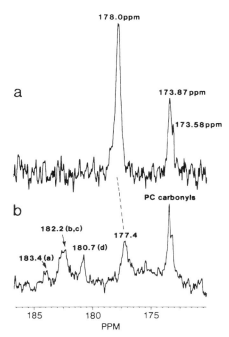

Figure 13. The carboxyl and carbonyl region of the ^{13}C NMR spectrum at 50.3 MHz of (a) egg PC vesicles (50 mg/ml) with 6 mol% of 90% [1-^{13}C]oleic acid at pH 7.4 after 1000 spectral accumulations and a pulse interval of 1.8 s; (b) a 50:50 (vol/vol) mixture of PC/oleic acid vesicles (final concentration, 50 mg/ml) and 10% (wt/vol) BSA [final concentration, 5% (wt/vol)] at pH 7.0 after 4000 spectral accumulations and a pulse interval of 1.8 s. In (b), the number given above each peak is the chemical shift in parts per million. Peaks from oleic acid bound to BSA are designated by letters in parentheses, as in Figure 2. The change in chemical shift of the peak from oleic acid in PC (dashed line) is a titration shift. (Reproduced with permission from Hamilton and Cistola, 1986.)

binding of FAs to vesicles and BSA could be clearly distinguished by chemical shift. There were no signals for unbound oleic acid, as expected from the high affinities of both BSA and PC for FAs. The distribution of FAs between BSA and vesicles could be determined directly from the relative intensities of peaks a through d compared to the single peak for oleic acid at 177.4 ppm. In the case illustrated in Figure 13, ~33% of the oleic acid was bound to PC vesicles and 66% to BSA.

The ^{13}C NMR spectrum of the preceding mixed system also provides some information about the kinetics of exchange. Net transfer of FAs was complete in the time required for sample equilibration and signal averaging (~1 h), since spectra taken during longer time intervals were identical. The spectrum shows that the exchange rate of oleic acid was slow on the NMR timescale ($k <$ 30 per second), since narrow (<10-Hz) resonances were observed in the mixtures at the same chemical shift as the isolated system; i.e., there was no evidence of exchange either in the peak lineshape or chemical shift. These results are consistent with a half-time of seconds for dissociation of long-chain FAs from albumin (Daniels *et al.*, 1985; Svenson *et al.*, 1974). Newer measurements by a fluorescent technique show that long-chain FAs such as oleic or palmitic acid exchange between albumin and phospholipid vesicles faster than 1 s^{-1} (Kamp and Hamilton, 1992). Exchange rates for FAs of different chain lengths were measured directly by NMR, as discussed later.

Results for differing phospholipid vesicles (such as mixed phosphatidylcholine/phosphatidylserine vesicles) showed no large differences in the partitioning of oleic acid, predicting that the lipid structure of biomembranes will not drastically alter the partitioning between albumin and the membrane. Another important physiological variable that might influence partitioning is pH. Studies with whole cell systems have led to the suggestion that acidic pH gradients promote release of FAs from albumin and increased binding to the cell (Spector, 1975). However, in such complex systems it is difficult, if not impossible, to elucidate a mechanism. To explore whether simple partitioning between albumin and phospholipid bilayers is pH dependent, we used the NMR procedure described earlier. Again, factors such as the metabolic machinery of the cell and membrane proteins, which could potentially contribute to pH-dependent changes in FA binding to cells, were eliminated in our simple model system.

Figure 14 shows the pH dependence in the acid range (pH 5.5 to 7.3) of the transfer of oleic acid between BSA and PC vesicles, as monitored by changes in the carboxyl spectrum. In this experiment, BSA with 4 moles of oleic acid/mole protein [spectrum (a)] was added to PC vesicles and the pH of the mixture adjusted for each spectrum; all other experimental conditions were unchanged. At pH 7.3 [spectrum (b)] most of the oleic acid is bound to BSA, whereas a minor fraction transferred to PC, as evidenced by the small peak at 178.2 ppm. A major redistribution of oleic acid occurs as a result of decreasing the pH to 6.6, where the distribution between PC and BSA becomes nearly equal. At pH 5.5, nearly all (>90%) of the oleic acid is associated with the vesicles, as indicated by the intense, narrow resonance at the predicted chemical shift for FA in PC vesicles (176.15 ppm; see Figure 10) and by the weak resonance at 181.8 ppm reflecting oleic bound to BSA [spectrum (d)]. After increasing the sample pH to 7.5, the mixture yields a spectrum [spectrum (e)] very similar to the initial spectrum at pH 7.3 [spectrum (b)], indicating reversibility of FA transfer. Thus, the partitioning of oleic acid between a phospholipid bilayer and albumin is pH dependent, and acidic pH

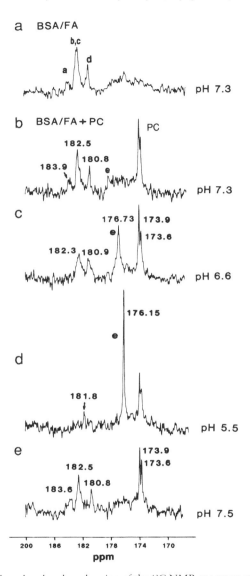

Figure 14. The carboxyl and carbonyl region of the ^{13}C NMR spectrum at 50.3 MHz of (a) 4 moles of 90% [1-^{13}C]oleic acid per mole of BSA at pH 7.3. (b) A 50:50 (vol/vol) mixture of the BSA–FA complex from (a) and egg PC vesicles containing no oleic acid at pH 7.3. (c)–(e) The same mixture at (c) pH 6.6; (d) pH 5.5; and (e) pH 7.5. All spectra were recorded after 2000 accumulations with a pulse interval of 1.8 s. The letter designations in (a) are the same as those in Figure 2 and represent oleic acid bound to BSA; e stands for oleic acid bound to PC vesicles. The number given above each peak is the chemical shift in parts per million. (Reproduced with permission from Hamilton and Cistola, 1986.)

could provide a driving force for uptake of FAs into a cell by increasing the partitioning to the cell membrane. Keep in mind that the model system is a static one, and that utilization of FA inside cells would cause equilibration of FAs from albumin at any pH.

To explore further the mechanism of FA transfer, the equilibration of the medium-chain FA octanoic acid was studied at different pH's by ^{13}C NMR (Hamilton, 1989). In contrast to oleic acid, octanoic acid showed only one ^{13}C carboxyl signal in mixtures of PC vesicles and BSA. At acidic pH's the chemical shifts fell between the value for octanoic acid with BSA or PC alone (Figure 15). (Note that a measurement at pH ~7.4 would not be definitive, since the chemical shift for octanoic acid bound to vesicles and to BSA fortuitously coincide.) The observation of a single peak suggests that exchange of octanoic acid between BSA and vesicles was fast, and from standard NMR equations $k = 1230$ s^{-1} ($\tau = 0.81$ ms) was calculated for pH 6.5. The equilibrium distribution can be obtained from the exchange-averaged chemical shift. Fast exchange of octanoic acid between albumin and vesicles was expected, since the exchange of octanoic acid between binding sites on BSA (Figure 7) is fast [$k = 354$ s^{-1} ($\tau = 2.8$ ms) at the same temperature, 33°C (Hamilton, 1989)]. The NMR data suggest a mechanism in which the FA desorbs from the albumin molecule for exchange to another binding site or to a membrane and supports conclusions from other methods that FA exchange between albumin and phospholipid bilayers is a unimolecular process

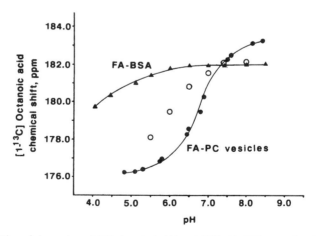

Figure 15. Plot of the carboxyl ^{13}C chemical shifts of 90% [1-^{13}C]octanoic acid in egg PC vesicles (●), bound to BSA (5 mol/mol, ▲) and in a mixture of 5 moles of [1-^{13}C]octanoic acid per mole of BSA and egg PC (1:1 weight ratio of BSA to PC, ○). (Reproduced with permission from Hamilton, 1989).

whose rate-limiting step is the dissociation of FA bound to albumin (Daniels et al., 1985). The physiological implication is that octanoic acid would be transferred to membranes from 10^3 to 10^4 times faster than common dietary long-chain FAs such as oleic, palmitic, and stearic acid. Although octanoic acid normally comprises a small portion of the FA bound to albumin in plasma, increased levels may be found in patients on therapeutic diets using medium-chain triacylglycerols, and the finding of very rapid desorption from albumin may help explain their ready uptake into cells.

Based on the results of the octanoic acid/vesicle/BSA system, it was of interest to perform parallel experiments using FAs with slightly longer chains. The 10-carbon FA decanoic acid showed results generally similar to those for octanoic acid: a single peak with BSA alone or in BSA/PC mixtures. In the latter mixture, the chemical shift at pH <7.0 was intermediate between the separated systems, and the decanoic acid peak was broader at a corresponding pH, suggesting intermediate exchange. The 12-carbon FA lauric acid showed separate, narrow resonances characteristic of slow exchange on the NMR timescale ($k \ll 200$ s^{-1}) when bound to BSA or in PC/BSA mixtures (Hamilton, 1989). These FA exchange rates between bilayers and albumin followed the trend for exchange between binding sites on BSA. One of the important effects of increasing chain length therefore is to increase the lifetime of FAs in a binding site on albumin and to decrease the rate of equilibration to model membranes.

At the other extreme of FA chain lengths are the very-long-chain fatty acids (VLCFAs), of which hexacosanoic acid (C26:0) is an example. These FAs are normally present in small amounts in human plasma and cells, but in genetic disorders where their metabolism (β-oxidation) is impaired, their accumulation leads to toxic physiological and neurological effects (see also Section 5.5.10). Because of the particular importance of hexacosanoic acid as a marker for such diseases of FA oxidation, we studied its transfer between model membranes and albumin. As described earlier, C26:0 bound to PC vesicles showed an ionization behavior characteristic of long-chain FAs. We attempted also to study binding of C26:0 to albumin by ^{13}C NMR but were unable to prepare complexes with this very hydrophobic acid. A strategy was adopted to load C26:0 in vesicles by cosonication of the two lipids and to monitor transfer to albumin, thereby studying both partitioning and binding. The spectrum (carboxyl region) of C26:0 in PC vesicles at pH 7.4 (Figure 16a) shows a single peak at the same chemical shift as a long-chain FA at the same pH. Aqueous FA-free BSA was added to a suspension of the vesicles, and spectra were obtained as a function of time. Unlike the case for any other FA investigated, equilibrium was not reached immediately. The ^{13}C spectra show a decreasing signal for FA bound to vesicles and an increasing signal for FA bound to BSA over a time period of hours (Figures 16b, c, and d). A

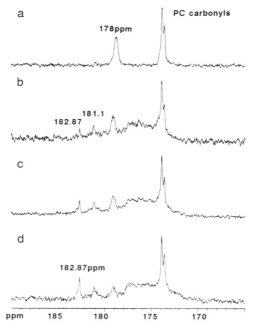

Figure 16. (a) The carboxyl region at 75 MHz of phospholipid PC vesicles with 2 mol% [1-^{13}C]hexacosanoic acid after 1800 spectral accumulations. The fatty acid carboxyl peak is at 178.0 ppm. (b) The carboxyl spectrum after the addition of aqueous FA-free BSA (1800 spectral accumulations). Two new peaks are seen at 181.13 ppm and 182.9 ppm. (c) The carboxyl region after 7200 spectral accumulations (the first 4 h after adding BSA). Note that the peak at 182.9 ppm has increased, the peak at 178 ppm has decreased, and the peak at 181.13 ppm has remained constant, in intensity. The latter peak is from amino acid carboxyl groups of BSA. (d) Between the fifth and seventh hour (3600 accumulations), the system has reached an equilibrium state. All spectra were obtained at 38°C and with a repetition time of 2.0 s. (J. Ho and J. A. Hamilton, unpublished results, 1994.)

half-time of ~3 h for transfer was estimated from the time-dependent decrease in the signal for the vesicle-bound FA. These spectra also show that BSA binds a maximum of only 1 mole of C26:0; further, the chemical shift of the single peak does not correspond to any of the chemical shifts for the three high-affinity sites for long-chain FAs. Thus, albumin appears to have a very limited capacity for VLCFAs, and the binding site may be different from the high-affinity sites for typical long-chain FAs. The NMR results suggest that the transfer of VLCFAs between membranes and albumin is extremely slow compared to normal dietary FAs and that albumin will not be as effective in removing VLCFAs that have accumulated in membranes.

In summary, the ^{13}C NMR approach described measures partitioning of FAs between two complex systems noninvasively and without separating the systems—important advantages over most conventional methods. ^{13}C NMR also permits the following:

1. Estimation of the kinetics of transfer.
2. Determination of the FA ionization state in the vesicles.
3. Observation of specific binding sites on albumin.
4. Assessment of vesicle integrity (through observation of the phospholipid carbonyl signals).

Fluorescence approaches that use native FAs and do not require sample separation have also been developed recently (Daniels *et al.*, 1985; Kamp and Hamilton, 1992, 1993). However, ^{13}C NMR is unique in providing structural information simultaneously, as mentioned in points 2, 3, and 4.

4.9 Exchange of Fatty Acids among Proteins, Model Membranes, and Lipoproteins

From inspection of the chemical shift data obtained for binding of FA to a single acceptor in water (i.e., a protein, a model membrane, or a plasma lipoprotein), it is evident that FA binding in various mixtures might be monitored by ^{13}C NMR. For example, the chemical shift of oleic acid bound to LDL or HDL is distinct from that for oleic acid bound to phospholipid vesicles at all pH values (Figure 9). The single signal for oleic acid bound to plasma lipoproteins (Figure 9) is also distinct from the multiple signals for oleic acid bound to BSA (Figure 3) at most pH values. Thus, in addition to the studies of equilibration of FA between albumin and model membranes, described in detail earlier, numerous combinations of FA acceptors can be studied by a similar protocol.

^{13}C NMR studies of transfer of long-chain FAs (palmitic, oleic, or stearic) between chylomicrons and BSA, between FABPs and phospholipid vesicles, and between HSA and low-density or high-density lipoproteins have been reported. Partitioning of FAs between albumin and plasma lipoproteins is pH dependent, with acidic conditions promoting increased partitioning to the plasma lipoproteins (Cistola and Small, 1991; Spooner *et al.*, 1988; Small, 1992). The surface monolayer of lipoproteins, which is comprised mainly of phospholipids, therefore acts similarly to a phospholipid bilayer and binds the un-ionized form of the FA more readily than does albumin. At a given pH, increasing the FA/albumin mole ratio results in increased partitioning to LDL and HDL (Cistola and Small, 1991).

A particularly informative experiment was the addition of ^{13}C-enriched FAs (palmitic acid) to whole human blood and plasma. ^{13}C NMR spectra distinguished binding to albumin and to the plasma lipoproteins (Figure 17).

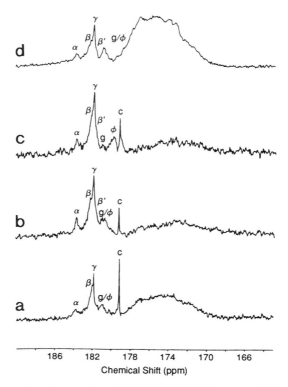

Figure 17. Carboxyl/carbonyl region of 90.6-MHz ^{13}C NMR spectra of (a)–(c) whole human plasma and (d) whole human blood with added ^{13}C-enriched palmitate, all at 37°C. Other conditions are: (a) moles of FA/mole of albumin, pH, 7.4; (b) 6 FA/SA, pH 7.4; (c) 6 FA/HSA, pH 6.8; (d) 6 FA/HSA, pH 7.4. The resonance labeled c represents carboxyl carbons of citrate, the anticoagulant used in this sample. Resonances for palmitate bound to albumin are designated α, β, γ, and β' as in Cistola (1985); resonance g is from protein glutamate and resonance ϕ is attributable to palmitate bound to lipoproteins. The plasma and blood samples were obtained from a healthy, nondiabetic human donor after a 14-h fast. The plasma FA-to-albumin mole ratio prior to the addition of ^{13}C-enriched palmitate was 1.0. Other laboratory values were as follows: albumin, 4.7 g/dl; total cholesterol, 164 mg/dl; HDL cholesterol, 39 mg/dl; and triglycerides, 65 mg/dl. (Reproduced the *Journal of Clinical Investigation*, 1991, Vol. 87, by copyright permission of the American Society for Clinical Investigation.)

The resolution in this complex physiological system is similar to that of isolated model systems. The signals for palmitic acid bound to HSA (Cistola, 1985) are quite similar to those labeled α, β, γ, and β' in Figure 17; the small fraction of palmitic acid bound to plasma lipoproteins (designated ϕ) is seen more clearly at pH 6.8 (Figure 17c). FAs in plasma are generally assumed to be almost completely bound to albumin, because of its high affinity for FAs and its much higher abundance than lipoproteins. These NMR studies suggest that under certain physiological conditions in which plasma FA levels may be significantly increased, such as obesity and diabetes, plasma lipoproteins may transport a measurable fraction of plasma FA. The study also addressed questions of the relevance of binding studies carried out in buffer solutions compared to serum or plasma. Brodersen *et al.* (1990) argued that binding equilibria are different in these two cases. However, ^{13}C NMR spectra, which permit visualization of the heterogeneous binding sites and their relative occupancy, do not reveal differences between binding in buffered solutions and blood or plasma at a given pH.

The feasibility of studying complex mixtures in which three different acceptors for FAs are present is illustrated in Figure 18. To model the process of lipoprotein lipolysis, FA equilibration in a mixture of phospholipid/triacylglycerol emulsions, phospholipid vesicles, and albumin was examined. The complete ^{13}C NMR spectrum (Figure 18) shows excellent resolution. The narrow signals that dominate the spectrum are from the acyl chain car-

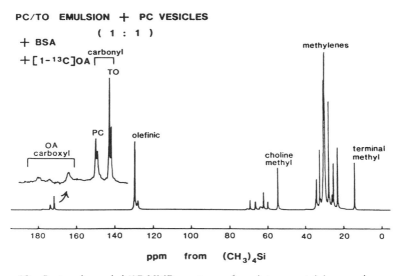

Figure 18. Proton-decoupled ^{13}C NMR spectrum of a mixture containing equal masses of emulsion triacylglycerol (triolein, TO) and phospholipid (PC) in the form of vesicles (10 wt% each), with BSA at 4.0 wt%. The albumin-to-oleic acid (OA) mole ratio is 1:5 with incorporated [1-^{13}C]OA present at ~0.4 wt% with respect to total lipid. The spectrum was recorded at 37°C over 16,384 time-domain points from 4000 accumulations using a repetition time of 2.0 s. The expanded region shows downfield peaks for PC and TO carbonyls and from the carboxyl of incorporated OA, which is distributed between albumin, emulsion, and vesicles. (Reproduced with permission from Spooner *et al.*, 1990. Copyright 1990 The American Society for Biochemistry and Molecular Biology.)

bons of phospholipids (PC) and triacylglycerol (triolein; TO). In the carboxyl/carbonyl region, the PC and TO carbonyls are readily distinguished, and the signals from ^{13}C oleic acid (OA) are well separated from the PC and TO carbonyl signals. The effects of increasing amounts of oleic acid and pH were investigated in this complex mixture. As in previous systems studied, the distribution of FA among the different acceptors was predictable from simple physicochemical principles. It should also be pointed out that in all systems, long-chain FAs showed slow exchange on the NMR chemical shift timescale, but equilibration was complete within the time required for data acquisition (minutes).

4.10 Concluding Remarks

The ^{13}C NMR approach described has furnished new and unique information about the structural and dynamic aspects of the binding of FA to albumin, providing insights into the structure of binding sites, the approximate locations of binding sites, and the molecular motions of FAs within the binding sites. Future studies, especially 2D and 3D NMR, could precisely locate and define the tertiary structure of binding sites on albumin. Although albumin is too large for complete structure determination by NMR, fragments encompassing the domains and the two halves of the molecule should prove feasible with current technology. Some of our NMR results, such as the distinction of binding sites on HSA for the first mole of octanoic and decanoic acid, were unexpected and suggest new experiments by other techniques. In this example, comparative crystal structures of HSA with octanoic and decanoic acid would be interesting. ^{13}C NMR spectroscopy also characterized FA binding to FABPs and to plasma lipoproteins. Other NMR approaches will provide additional characterization of FA interactions. The simple ^{13}C strategy for binding studies is also being applied to newly discovered proteins that bend FAs, such as the interphotoreceptor retinoid-binding protein, a large (140-kD) multidomain protein that binds both retinoids and FAs (Boylan *et al.*, 1993).

Monitoring FA exchange and partitioning in complex mixtures by ^{13}C NMR has provided insights into how FAs might behave in biological systems. Results from these model systems provide a basic framework for interpreting data from the more complex cellular systems. It is clear, for example, that FAs can move rapidly between high-affinity binding proteins and model membranes, without the aid of additional proteins or receptors. Data for uptake of FAs into cells must be carefully interpreted, and proof of the existence of a protein important in sequestering FAs into a cell will rest on isolation and characterization of such a protein.

Our results graphically illustrate that binding in systems with more than one acceptor of FA must always be described as a partitioning. In cell biology experiments, albumin is often added to cells or tissues without an understanding of how albumin might alter FA distribution in the cells. Depending on the exact experimental conditions, such as the albumin-to-cell membrane ratio and the FA content of the cell, addition of FA-free albumin could remove FAs from the cell and alter the metabolic and nutritional status. Similarly, FAs are often added to cells, sometimes in the presence of albumin, but the immediate fate of the FA is not known. Since the albumin-to-membrane lipid ratio is seldom determined, and the cellular content of FA is not readily measured, it is not surprising that there is a great variability in the reports of the threshold for toxic effects of FAs on cells (Corr *et al.*, 1984). Such variability could simply reflect differences in partitioning of FAs between the cell and albumin.

Thus, important future contributions of ^{13}C NMR would be to characterize building of FAs to cell membranes, to monitor partitioning of FAs between proteins and membranes, and to follow the movement of FAs from albumin to the plasma membrane and to the cytoplasm. ^{13}C NMR spectroscopy will also likely prove extremely useful for metabolic studies of long-chain FAs. The chemical shifts of acylated FAs—acyl coenzyme A, phospholipids, diacylglycerols, and triacylglycerols—are well separated from FAs (see, for example, Figure 18). Provided adequate sensitivity and resolution is achieved, the distribution of ^{13}C labels in different metabolites can be measured, as illustrated by a study of the metabolism of the medium-chain FA octanoic acid in the liver of an intact rat (Pahl-Wostl and Seelig, 1987). The advantage of NMR spectroscopy over many other methods is the ability to quantitate different *native* metabolites and pools of the same metabolite in a nonperturbing, noninvasive manner.

Acknowledgments

In addition to the numerous colleagues listed as coauthors in published work, I would also like to thank Dr. Shastri Bhamidipati, Dr. Wen Guo, and Jet Ho for providing unpublished results.

References

Berde, C. B., Hudson, B. S., Simoni, R. D., and Sklar, L. A. (1979). *J. Biol. Chem.* **254**, 391.
Bihain, B. E., Deckelbaum, R. J., Yen, F. T., Gleeson, A. M., Carpentier, Y. A., and Witte, L. D. (1989). *J. Biol. Chem.* **264**, 17316.

Boylan, J., Wiggert, B., Chader, G., Reed, R., and Hamilton, J. A. (1993). *Biophys. J.* **64,** A289.
Brodersen, R., Andersen, S., Vorum, H., Nielsen, S. U., and Pedersen, A. O. (1990). *Eur. J. Biochem.* **189,** 343.
Brown, J. R., and Shockley, P. (1982). *In* "Lipid-Protein Interactions" (P. C. Jost and O. H. Griffith, eds.), Vol. 1, pp. 25–68. Wiley, New York.
Carlson, J., Sakamoto, Y., Laurell, C.-B., Madison, J., Watkins, S., and Putnam, F. W. (1992). *Proc. Natl. Acad. Sci. U.S.A.* **89,** 8225.
Chana, R. S., Treleaven, W. D., Cushley, R. J., and Steinbrecher, U. P. (1990a). *Biochem. Cell Biol.* **68,** 180.
Chana, R. S., Treleaven, W. D., Parmar, Y. I., and Cushley, R. J. (1990b). *Biochem. Cell Biol.* **68,** 189.
Cistola, D. P. (1985). Ph.D. Dissertation, Boston University School of Medicine.
Cistola, D. P., and Small, D. M. (1991). *J. Clin. Invest.* **87,** 1431.
Cistola, D. P., Small, D. M., and Hamilton, J. A. (1982). *J. Lipid Res.* **23,** 795.
Cistola, D. P., Atkinson, D., Hamilton, J. A., and Small, D. M. (1986). *Biochemistry* **25,** 2804.
Cistola, D. P., Small, D. M., and Hamilton, J. A. (1987a). *J. Biol. Chem.* **262,** 10971.
Cistola, D. P., Small, D. M., and Hamilton, J. A. (1987b). *J. Biol. Chem.* **262,** 10980.
Cistola, D. P., Walsh, M. T., Corey, R. P., Hamilton, J. A., and Brecher, P. (1988). *Biochemistry* **27,** 711.
Cistola, D. P., Sacchettini, J. C., Banaszak, L. J., and Walsh, M. T. (1989). *J. Biol. Chem.* **264,** 2700.
Cistola, D. P., Sacchettini, J. C., and Gordon, J. I. (1990). *Mol. Cell. Biochem.* **98,** 101.
Corr, P. B., Gross, R. W., and Sobel, B. E. (1984). *Circ. Res.* **55,** 135.
Court, J. M., Dunlop, M. E., and Leonard, R. F. (1971). *J. Appl. Physiol.* **31,** 345.
Daniels, C., Noy, N., and Zakim, D. (1985). *Biochemistry* **24,** 3286.
Doddrell, D., Glushko, V., and Allerhand, A. (1972). *J. Chem. Phys.* **56,** 3683.
Gebecki, J. M., and Hicks, M. (1973). *Nature (London)* **243,** 242.
Goodman, D. S. (1958). *J. Am. Chem. Soc.* **80,** 3892.
Hamilton, J. A. (1989). *Proc. Natl. Acad. Sci. U.S.A.* **86,** 2663.
Hamilton, J. A. (1992). *News Physiol. Sci.* **7,** 264.
Hamilton, J. A., and Cistola, D. P. (1986). *Proc. Natl. Acad. Sci. U.S.A.* **83,** 82.
Hamilton, J. A., Cistola, D. P., Morrisett, J. D., Sparrow, J. T., and Small, D. M. (1984). *Proc. Natl. Acad. Sci. U.S.A.* **81,** 3718.
Hamilton, J. A., Bhamidipati, S. P., Era, S., and Reed, R. G. (1991). *Proc. Natl. Acad. Sci. U.S.A.* **88,** 2051.
Hannigan, G. E., and Williams, B. R. G. (1991). *Proc. Natl. Acad. Sci. U.S.A.* **86,** 2663.
He, X.-M., and Carter, D. C. (1992). *Nature (London)* **358,** 209.
Huang, J.M.-C., Xian, H., and Bacaner, M. (1992). *Proc. Natl. Acad. Sci. U.S.A.* **89,** 6452.
John, L. M., and McBain, J. W. (1948). *J. Am. Oil Chem. Soc.* **25,** 40.
Kamp, F., and Hamilton, J. A. (1992). *Proc. Natl. Acad. Sci. U.S.A.* **89,** 11367.
Kamp, F., and Hamilton, J. A. (1993). *Biochemistry* **32,** 11074.
Kantor, H. L., and Prestegard, J. H. (1978). *Biochemistry* **17,** 3592.
Kenyon, M., and Hamilton, J. A. (1994). *J. Lipid Res.* **84,** 458–467.
Kuo, P. T., and Huang, N. N. (1965). *J. Clin. Invest.* **44,** 1924.
Kuznetsov, A. N., Ebert, B., Lassmann, G., and Shapiro, A. B. (1975). *Biochim. Biophys. Acta* **379,** 139.
Langercrantz, C., and Setaka, M. (1975). *Acta Chem. Scand., Ser B* **B29,** 397.
Miyazaki, J., Hideg, K., and Marsh, D. (1992). *Biochim. Biophys. Acta* **1103,** 62.
Morrisett, J. D., Pownall, H. J., and Gotto, A. M., Jr. (1975). *J. Biol. Chem.* **250,** 2487.
Moser, H. W. (1992). *In* "New Developments in Fatty Acid Oxidation" (P. M. Coates and K. Tanaka, eds.), pp. 369–388. Wiley-Liss, New York.

Moser, P., Squire, P. G., and O'Konski, C. T. (1966). *J. Phys. Chem.* **70**, 744.
Muller, N., and Mead, R. J., Jr. (1973). *Biochemistry* **12**, 3831.
Neely, J. R., and Morgan, H. E. (1974). *Annu. Rev. Physiol.* **36**, 413.
Norton, R. S., Clouse, A. O., Addleman, R., and Allerhand, A. (1977). *J. Am. Chem. Soc.* **99**, 79.
Oldfield, E., Bowers, J. L., and Forbes, J. (1987). *Biochemistry* **26**, 6919.
Ordway, R. W., Walsh, J. V., Jr., and Singer, J. J. (1989). *Science* **244**, 1176.
Pahl-Wostl, C., and Seelig, J. (1987). *Biol. Chem. Hoppe-Seyler* **368**, 205.
Parks, J. S., Cistola, D. P., Small, D. M., and Hamilton, J. A. (1983). *J. Biol. Chem.* **258**, 9262.
Parmar, Y. I., Gorrissen, H., Wassall, S. R., and Cushley, R. J. (1985). *Biochemistry* **24**, 171.
Potter, B. J., Sorrentino, D., and Berk, P. D. (1989). *Annu. Rev. Nutr.* **9**, 253.
Ptak, M., Egret-Charlier, M., Sanson, A., and Bouloussa, O. (1980). *Biochim. Biophys. Acta* **600**, 387.
Reed, R. G., Feldhoff, R. C., Clute, O. L., and Peters, T., Jr. (1975). *Biochemistry* **14**, 4578.
Roberts, J. K. M., and Jardetzky, O. (1985). In "Modern Physical Methods in Biochemistry," (A. Neuberger and L. L. Van Deenen, eds.), Part A, pp. 1–67. Elsevier, Amsterdam.
Ruf, H. H., and Gratzl, M. (1976). *Biochim. Biophys. Acta* **446**, 134.
Sacchettini, J. C., Gordon, J. I., and Banaszak, L. J. (1989). *J. Mol. Biol.* **208**, 327.
Sklar, L. A., Hudson, B. S., and Simoni, R. D. (1977). *Biochemistry* **16**, 5100.
Small, D. M. (1992). In "Polyunsaturated Fatty Acids in Human Nutrition" (U. Bracco and R. J. Deckelbaum, eds.), pp. 25–39. Raven Press, New York.
Spector, A. (1975). *J. Lipid Res.* **16**, 165.
Spector, A. (1992). In "Polyunsaturated Fatty Acids in Human Nutrition" (U. Bracco and R. J. Deckelbaum, eds.), pp. 1–12. Raven Press, New York.
Spooner, P. S., Bennett, Clark, S., Gantz, D. L., Hamilton, J. A., and Small, D. M. (1988). *J. Biol. Chem.* **263**, 1444.
Spooner, P. S., Gantz, D. L., Hamilton, J. A., and Small, D. M. (1990). *J. Biol. Chem.* **265**, 12650.
Svenson, A., Holmer, E., and Anderson, L.-O. (1974). *Biochim. Biophys. Acta* **342**, 54.
Tanford, C., Swanson, S., and Shore, W. (1955). *J. Am. Chem. Soc.* **77**, 6414.
Veerkamp, J. H., Peeters, R. A., and Maatman, R. G. H. J. (1991). *Biochim. Biophys. Acta* **1081**, 1.
Vorum, H., Broderson, R., Kragh-Hansen, U., and Pedersen, A. O. (1992). *Biochim. Biophys. Acta* **1126**, 135.
Walsh, M. T., and Atkinson, D. A. (1986). In "Methods in Enzymology" (J. Segrest and J. Albers, eds.), Vol. **128**, p. 582. Academic Press, New York. Orlando, FL.
Wilbur, D. J., Norton, R. S., Clouse, A. O., Addleman, R., and Allerhand, A. (1977). *J. Am. Chem. Soc.* **98**, 8250.

CHAPTER 5

Application of ^{13}C NMR Spectroscopy to Metabolic Studies on Animals

Basil Künnecke
Biocenter of the University of Basel, CH-4056 Basel, Switzerland

5.1 Introduction

During the last two decades, ^{13}C magnetic resonance spectroscopy (MRS) on biological tissue has evolved to become an extremely powerful and versatile tool in biochemistry, which is outperforming other techniques in many aspects. One of the striking features of ^{13}C MRS and, in general, of magnetic resonance (MR) methods, is the nondestructive and noninvasive nature of the MR assessment. Repetitive measurements of the same sample and measurements in living organisms are hence well within the realm of MR techniques. Among MR methods, ^{13}C MRS of biological material is particularly appealing because it allows for the assessment of the metabolism of carbon, which is so elementary to life on earth. However, comparisons with other techniques including proton MRS, chromatography, spectrophotometry, and radiochemistry have led to ^{13}C MRS being considered a method with poor sensitivity. The intrinsically low sensitivity of MR techniques in general, the low gyromagnetic ratio of the ^{13}C nucleus, and the low natural abundance of the magnetically active ^{13}C nuclei among the total population of carbons have contributed to this unfavorable reputation and are indeed limiting factors for its application to biological material. The strength of ^{13}C MRS, however, does not lie in the detection of minute amounts of chemicals but in the wealth of information provided by a single spectrum.

Chemical shifts of ^{13}C are dispersed over a range of more than 200 ppm and are very characteristic for the physicochemical surroundings of specific

^{13}C nuclei. Consequently, separate resonances for almost every single carbon can be simultaneously detected and identified for a multitude of metabolites. ^{13}C MR spectra therefore give an overview of carbon-containing compounds even in complex mixtures of metabolites such as those encountered in biological tissues. The apparent disadvantage—that the ^{13}C nucleus has a natural abundance of only 1.1%—turns into a distinct advantage by allowing for the possibility of tagging a specific carbon position by selective enrichment with ^{13}C. ^{13}C labeling not only increases signal intensities but offers a novel approach for tracing cellular metabolism. The comprehensive view provided by ^{13}C MRS offers a convenient means to follow the fate of specific carbons throughout the metabolism without the need for the tedious isolation, purification, and carbon-by-carbon degradation procedures inherent in radio tracer experiments. The noninvasive nature of MRS and the fact that ^{13}C is a stable carbon isotope open an unprecedented avenue for labeling experiments *in vivo* where sequential measurements can reveal the time response of the metabolism and furnish intrinsic control measurements from the same organism. Furthermore, the concept of labeling can be extended to the enrichment of several adjacent carbon sites in the same molecule. In turn, the fate of all labels can be followed simultaneously but, more importantly, ^{13}C MRS additionally renders novel biochemical information from homonuclear spin couplings between adjacent ^{13}C labels, a property unique to MR methods.

This chapter focuses on the special features and versatility that distinguish ^{13}C MRS from other competing techniques used for assessing cellular biochemistry. The first section of this chapter outlines the concepts of ^{13}C-labeling in conjunction with ^{13}C MRS in metabolic research. The second section illustrates how mathematical models of cellular metabolism are used to extract biologically relevant parameters such as metabolic pathways and their activities from the information obtained by ^{13}C MRS. Considerations on the experimental design essential for ^{13}C MRS on animal tissues *in vivo* and *in vitro* are summarized in the third section. Finally, the fourth and largest section is intended to give an extensive, albeit incomplete, overview of ^{13}C MRS studies performed on intact animals, excised tissues from animals, and tissue extracts with special reference to the findings promoted by the peculiarities of ^{13}C MRS. The predominant use of rodents as laboratory animals has led to a strong bias of this review toward the metabolic interactions in mammals and this last section is therefore subdivided into paragraphs focusing on the metabolism of individual organs.

This review is, of course, not the first of its kind. A number of surveys have been published that encompass a wider scope on ^{13}C MRS in biology (Scott and Baxter, 1981; Cozzone *et al.*, 1985; Cerdan and Seelig, 1990; Badar-Goffer and Bachelard, 1991; Seelig and Burlina, 1992) or deal with specific aspects of cellular metabolism and its assessment by ^{13}C MRS (Cohen and Shulman, 1982; Bachelard et al., 1987; Nissim et al., 1988; Becker and Ackerman, 1990; Jans and Kinne, 1991; Bernard and Cozzone, 1992; Gallis and Canioni, 1992). The reader is referred to these latter reviews and to the original papers cited throughout the text for more detailed information.

5.2 ^{13}C MRS and the Concepts of ^{13}C Labeling

^{13}C MRS provides biochemical information in the form of spectra that are based on the resonance phenomena of ^{13}C nuclei. However, only 1.1% of the total carbon content of a sample consists of the magnetically active ^{13}C nuclei; the bulk is magnetically inactive ^{12}C, which escapes detection by MRS. Generally, the ^{13}C nuclei are uniformly distributed among the carbon positions within a molecule. Hence, natural abundance ^{13}C MR spectra of tissue samples potentially display distinct resonances for essentially every carbon. In the following we focus on the information that can be derived from the analysis of such ^{13}C MR resonances.

5.2.1   Natural Abundance Spectroscopy

The chemical shifts of ^{13}C MRS resonances are very characteristic of the various physicochemical surroundings of the corresponding nuclei and allow the identification of different compounds within a sample. Due to the large dispersion of the ^{13}C chemical shifts over more than 200 ppm, resonances from chemically closely related metabolites are most often well separated and clearly identifiable even under conditions of decreased resolution such as at low field strength and in vivo. Chemical shifts not only discriminate between different chemical surroundings but can give further information on the physicochemical environment of a metabolite. The chemical shifts of certain carbons are dependent on factors such as pH and ion concentrations as was shown for the Mg^{2+} and H^+ dependence of glycerol-3-phosphate, citrate, and carnosine resonances, respectively (Cohen, 1983, 1984; Arús et al., 1985; Chacko and Weiss, 1993).

Quantitative information on the concentration of metabolites can be inferred from the integral under the ^{13}C MR resonances. The area beneath a resonance is a measure of the amount of ^{13}C nuclei involved in generating the corresponding resonance and is proportional to the ^{13}C enrichment, to the number of coresonating carbons, and to the amount of the metabolite present. Signal saturation, the nuclear Overhauser effect (NOE), and the restricted MR visibility of metabolites further affect this proportionality (see Section 5.4.4.).

Longitudinal (T_1) and transverse (T_2) relaxation characteristics give clues to the physical environments of nuclei. Linewidths of MR resonances (measured at half-maximum height) are a measure of the transversal relaxation and carry information on magnetic field inhomogeneities and molecular motion. Fast molecular tumbling is an essential prerequisite for the detection of narrow MR resonances. However, high intracellular viscosity, a wide range of macromolecules, and molecules of low molecular weight bound to macromolecules are common features of biological samples and hamper fast isotropic tumbling. A reduction of molecular mobility significantly decreases T_2 with a concomitant increase in linewidth, which ultimately leads to the disappearance of the resonance in the background noise. As a result, some metabolites may be partially or even totally invisible by MR spectroscopy. The T_1 relaxation is also determined by molecular motion and structure. Information on T_1 is obtained by the circumstance that a long T_1 relaxation potentially leads to signal saturation with an ensuing reduction of the signal strength.

Spin couplings, a property measurable only by MR methods, provide further information on the magnetic surroundings of a particular ^{13}C nucleus. Spin couplings, as evidenced by the splitting of resonances into multiple lines, indicate the presence of other magnetically active nuclei such as ^1H, ^2H, ^{14}N, ^{15}N, ^{31}P, ^{19}F, etc., in the vicinity, that is, covalently bound to the ^{13}C nucleus under investigation. The size of the coupling strongly depends on the gyromagnetic ratio of the coupling partners and the distance (number of covalent bonds) between the coupled nuclei. The resulting multiline structure of a resonance is determined by the number of coupling partners and their nuclear spin. Nowadays, the very prominent ^1H–^{13}C couplings are most often suppressed by broadband proton decoupling for spectral simplification and signal enhancement (see Section 5.4.3.), and are hence not visible in standard ^{13}C MR spectra. Spin couplings with other nuclei are rarely detected in biological samples because the couplings are either too weak or the natural occurrence of the magnetically active nuclei is too low. However, unresolved spin couplings with coupling constants of the order of the ^{13}C MR linewidths may lead to significant and indicative increases in the linewidth of the corresponding coupled ^{13}C-resonances. In addition, deliberately introduced spin couplings caused by the artificial enrichment of magnetically active nuclei might provide novel information on the operation of biochemical pathways (see Section 5.2.3).

5.2.2 Specific ^{13}C Labeling of a One-Carbon Site

^{13}C labeling, or the selective enrichment of a carbon position with ^{13}C, has the obvious advantage of increasing the ^{13}C MR signal intensity of the corresponding resonance up to ninety-fold. However, the main objective of

^{13}C labeling in metabolic research is to tag a certain carbon position and follow its fate throughout metabolism. Ultimately, this procedure aims at discriminating between the contributions of different metabolic routes and at measuring their activities. Hence, the apparent disadvantage that the ^{13}C nucleus has a natural abundance of only 1.1% opens a new avenue for performing labeling experiments similar to ^{14}C radio tracer studies. Essentially the same information may be obtained from a single ^{13}C MR spectrum as is obtained from traditional ^{14}C tracer experiments, but without the need for extensive metabolite isolation, purification, and carbon-by-carbon degradation. A further advantage of ^{13}C MRS is its noninvasive nature, which allows the measurements to be performed *in vivo* and in a repetitive manner (e.g., in a time course) using the same biological sample. At this point it is worth emphasizing again that the ^{13}C nucleus is a stable carbon isotope and hence is not subject to the dangers related to radio tracers.

Let us first focus on selective ^{13}C enrichment at one carbon position only. In biochemical reactions, the artificially augmented occurrence of ^{13}C at a specific carbon position of the substrate(s) is transferred to the product(s). The increased intensity detected for particular resonances traces intermediary metabolites and carbon rearrangements involved in the associated metabolic pathway(s). Biologically relevant information on metabolic pathways and metabolic fluxes is derived by answering the questions as to where, how much, and how fast the label is accumulated. In a qualitative analysis of ^{13}C MR spectra, a signal increase on availability of a ^{13}C label is interpreted as an incorporation of label into the carbon position under observation. However, keep in mind that the signal intensity is intimately connected to both the degree of enrichment and the total amount of metabolite present. Hence, the degree of ^{13}C enrichment is quantitatively expressed as fractional enrichment, which is the ratio of ^{13}C-bearing molecules versus the total amount (^{13}C + ^{12}C) of the corresponding compound. This measure reveals only a minute part of the information potentially available with ^{13}C MRS, but it has a well-known counterpart, the specific (radio)activity in radio tracer experiments. As pointed out earlier, ^{13}C MRS can usually distinguish between the different carbons within a molecule. Hence fractional enrichments can be measured directly for every single-carbon position. These ^{13}C enrichments are referred to as *positional enrichments*. To simplify the interpretation of labeling data, fractional and positional enrichments are also expressed as *enrichments-in-excess*, which relate to the corresponding enrichments corrected for the 1.1% natural occurrence of ^{13}C.

5.2.3 Specific ^{13}C Labeling of Several-Carbon Sites

The concept of labeling can easily be extended to situations where several specific carbon positions are enriched simultaneously. In fact, this idea is not new and has been used in nuclear chemistry to enhance the specific activity of certain molecules. Similarly, bilabeled or multilabeled ^{13}C compounds have been used with the aim of intensifying certain ^{13}C MR resonances by directing the labels into the same carbon position in the end product of a biochemical pathway (Pahl-Wostl and Seelig, 1986; Cerdan et al., 1988). However, new fields of application for multilabeled precursors have emerged. On the one hand, tagging of different carbon positions in the same molecule with ^{13}C allows the observation of the fate of all labeled carbons simultaneously (Cohen, 1983, 1987b; Künnecke and Cerdan, 1989; Cerdan et al., 1990; Künnecke and Seelig, 1991). On the other hand, ^{13}C-labeling of two or several adjacent, covalently bound carbons results in homonuclear $^{13}C-^{13}C$ spin couplings, which can provide novel biochemical information (Cohen et al., 1979a; Malloy et al., 1988; Cerdan et al., 1990; Künnecke and Seelig, 1991; Künnecke et al., 1993).

Whereas spin coupling is a very common occurrence in phosphorus and proton MRS, ^{13}C homonuclear spin coupling is a rather unusual feature in ^{13}C MR spectra and deserves closer inspection. Figure 1a shows the proton-decoupled natural abundance ^{13}C MR spectrum of a –C–C–C– fragment. Small singlet resonances are observed for all three carbon positions. Upon selective enrichment of one or the other carbon position, the corresponding resonance will dramatically increase in intensity (Figures 1b and 1c). In contrast, simultaneous enrichment of two consecutive carbon positions in the same molecule not only increases the signal intensities for the two carbon positions but leads to $^{13}C-^{13}C$ spin coupling. This spin coupling is detected in the form of line splitting of the resonances into doublets (Figure 1d) and is characterized by the line separation within the doublets known as the apparent coupling constant $^1J_{12}$.

The concept of mutual interaction between two ^{13}C nuclei can be extended to a third covalently bound ^{13}C nucleus (Figure 1e). While the resonance of the third carbon will be split into a doublet with the coupling constant $^1J_{23}$, the resonance of the coupling partner (already coupled to another carbon) shows a superposition of both couplings, namely a doublet of doublets. By analogy, this series could be continued with further coupling partners yielding more and more complex spin-coupling patterns. The quintessence, however, is that ^{13}C MRS not only measures the positional enrichments but spin couplings detect the degree of labeling of adjacent, covalently bound carbons in the same molecule. This approach obviously increases the amount of information available to describe the metabolic events that have occurred. First, the multiplet structures of resonances from molecules labeled in contiguous positions are readily distinguishable from the singlet resonances of natural abundance contributions (due to the low natural abundance of ^{13}C, the intrinsic occurrence of homonuclear spin couplings has a

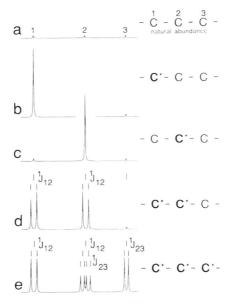

Figure 1. Simulated ^{13}C MR spectra obtained for different isotopomers of a C_3 unit. (a) The C3 unit is labeled at the natural abundance level (1.1% ^{13}C at every carbon position) and small singlet resonances are detected for all carbons. (b) Upon enrichment of C1 with ^{13}C a correspondingly strong singlet resonance is discernible. (c) Enrichment of C2 leads to a strong singlet resonance similar to (b), but at a different chemical shift. (d) Simultaneous enrichment of the adjacent carbon positions C1 and C2 increases the intensity of the corresponding two resonances. The magnetic interaction of the two carbons over the common chemical bond leads to homonuclear spin coupling. Spin coupling is manifested by the line splittings into two doublets characterized by the coupling constant $^1J_{12}$. (e) The concept of multiple labeling is expanded to three consecutive ^{13}C nuclei. Homonuclear spin couplings between C1 and C2 ($^1J_{12}$), and between C2 and C3 ($^1J_{23}$) are observable. As a consequence, the resonance of C2 is split into a doublet of doublets.

negligible probability, e.g., 0.012% for a C2 segment). Secondly, ^{13}C homonuclear spin couplings are only detected as long as two ^{13}C nuclei are chemically bound together, and rearrangements of the ^{13}C label by disintegration or formation of ^{13}C–^{13}C bonds result in characteristic changes of the homonuclear spin-coupling pattern observed by ^{13}C MRS. These concepts of multiple labeling and analyses of spin-coupling patterns can be easily extended to heteronuclear couplings between ^{13}C and artificially enriched magnetically active heteroatoms; e.g., ^{2}H–^{13}C and ^{15}N–^{13}C couplings are likely to give additional biochemical information on stereochemistry and nitrogen metabolism, respectively (Nieto et al., 1992).

5.2.4 Isotopomer Analysis

Every carbon of a molecule can be either a ^{12}C or a ^{13}C independent of the labeling state of the other carbons in the molecule. Hence, for a molecule with N carbons, a maximum of 2^N different isotope isomers, so-called *isotopomers*, exist. During metabolism, pathway-specific carbon rearrangements occur and, upon labeling with ^{13}C, a specific range of isotopomers is produced that store the biochemical history of the metabolite. In turn, the distribution of ^{13}C within and among the isotopomers is coded in the ^{13}C homonuclear spin-coupling patterns detected by ^{13}C MRS, thus providing a spectroscopic fingerprint of the biochemical history of the metabolite.

Isotopomer analysis is based on the premise that the label distribution within a molecule is solely dependent on the label distribution within its biochemical predecessor(s) and the rearrangements of carbon–carbon bonds in the particular biochemical pathway involved. Alternative pathways leading to the same metabolite are likely to produce different characteristic carbon rearrangements. By applying these ideas with rigor, it becomes clear that for a known labeling of the precursors, the distribution of isotopomers in the products reflects the relative metabolic activity of competing pathways and the chemical rearrangements involved in the transformation (see also Jeffrey *et al.*, 1991).

With these concepts of labeling in mind, the aim of ^{13}C MR spectral analyses is to deduce the isotopomer distribution from the information provided by the homonuclear spin-coupling patterns and thus to gain information on the biochemical transformations involved in the metabolism. Theoretically, each of the 2^N possible isotopomers could produce a different and therefore unequivocally distinguishable spectral contribution. However, two restrictions apply to this theoretical approach: (1) As a consequence of equal or nearly equal $^{13}C-^{13}C$ coupling constants for different couplings, spectral contributions of some isotopomers are identical and cannot be distinguished unambiguously; and (2) because long-range ^{13}C couplings over two and more bonds are very small and generally unresolved, a particular carbon resonance only provides the degree of labeling of the carbon in question and, due to spin coupling, the labeling of the directly bound carbons. No information on the labeling of the rest of the molecule is available and leaves the spectral pattern of some of the isotopomers indistinguishable. Therefore, ^{13}C MRS only discriminates between groups of isotopomers.

Higher enrichments give stronger signals and increase the probabilities for spin couplings and are thus favorable for a comprehensive and more accurate isotopomer analysis. However, this does not imply that 100% ^{13}C enrichment at all carbon sites of the substrate is the optimum labeling strategy. For a labeled precursor to be useful in isotopomer analyses, pathway-specific arrangements of $^{13}C-^{13}C$ and $^{13}C-^{12}C$ bonds have to occur during its metabolism. Carbon-12 is either provided intramolecularly by partially enriched substrates or from unlabeled material. The indirect influence of ^{12}C on the spin-coupling patterns by the absence of $^{12}C-^{13}C$ spin couplings thus allows isotopomer analysis not only to provide a quantitative measurement of the utilization of labeled metabolites but also to determine the flux of ^{12}C.

Besides the relative flux rates and the labeling pattern of the initial substrate, the degree of ^{13}C enrichment has a strong and in many cases nonlinear influence on the occurrence of a particular isotopomer. On the one hand, this implies that for slight variations of the ^{13}C enrichments, completely different spin-coupling patterns may be detected (Malloy et al., 1990b). As a consequence, the direct comparison of similar experiments is complicated and metabolic modeling may be necessary (see Section 5.3). On the other hand, however, this feature renders a method with high sensitivity for the detection of small changes of enrichments due to alterations in metabolism.

5.2.5 Higher-Order Effects of ^{13}C Labeling

While spin coupling provides additional biochemical information, the line splitting into multiplet patterns reduces the signal intensity of the individual spectral lines. As a result, the signal-to-noise ratio (SNR) deteriorates and longer data averaging is necessary to discern spectral contributions of lowly abundant isotopomers and of multiple labeled isotopomers with highly split spectral patterns. Spin coupling also increases spectral complexity and the probability of overlapping resonances. These effects associated with multiple ^{13}C labeling are particularly unfavorable for *in vivo* spectroscopy. However, in many cases deliberate introduction of $^{13}C-^{13}C$ spin couplings and the subsequent gain of information surpasses any disadvantages derived from multiple labeling and allows the elucidation of complex metabolic systems.

Labeling with ^{13}C nuclei has further implications on the appearance of ^{13}C MR spectra. In cases where homonuclear spin coupling occurs with a coupling constant of the order of the chemical-shift difference between the two coupling partners (in particular at low field strengths), higher-order effects become visible in the acquired spectra, e.g., multiplet patterns are no longer symmetrical around their centers and the multiplet resonances are spaced further apart than the corresponding natural abundance resonances (Figure 1d).

Substitution of a ^{12}C with a ^{13}C in the neighborhood of a second ^{13}C nucleus slightly affects the chemical surrounding of the latter by increasing the mass of the molecule by one unit. This isotope effect causes resonances from carbons with neighboring ^{13}C nuclei to shift to lower frequencies (high field shift) (Batiz-Hernandez and Bernheim, 1967). In particular, multiplet patterns are slightly displaced to the low-frequency side of the natural abundance resonances. The increased molecular mass can also alter the reactivity in chemical reactions, as is well known from the substitution of hydrogen with deuterium. However, since the increase in mass is only a small fraction of the total molecular mass, the kinetic isotopic effect of ^{13}C is negligible in most cases.

A further important effect due to neighboring ^{13}C nuclei is the accelerated T_1 relaxation. Dipolar coupling between the neighboring nuclei offers an additional mechanism for longitudinal relaxation (Moreland and Carroll, 1974; London *et al.*, 1975). This relaxation pathway is particularly efficient for spin-coupled carbons and may hence lead to different relaxation times for singlets and multiplets in the same resonance. For chemical groups where relaxation is genuinely slow due to the lack of directly bound protons, as for example with carboxylic groups, the difference in relaxation between monolabeled and multilabeled species can be significant. Thus, under conditions with partial signal saturation, an overestimate of the spectral contribution of the multilabeled molecular species would result. Likewise, dipolar coupling between two ^{13}C nuclei reduces the NOE, which, in turn, leads without appropriate mathematical correction to a relative overestimate of the central singlet resonance.

5.3 ^{13}C MRS, ^{13}C Labeling, and Models of Intermediary Metabolism

5.3.1 Metabolic Modeling and Labeling

As pointed out in the previous sections, isotopomer distributions store essential information both on the precursors and on the biochemical pathways by which a given metabolite is synthesized. Hence, the analysis of label distribution in the intermediates and end products of a pathway potentially provides a wealth of biochemically relevant information. However, because many of the biochemical pathways are complex and combine the simultaneous activity of different metabolic fluxes or even allow intermediates to be recycled, a direct backward conclusion from the labeling patterns to the actual processes involved in metabolism is most often not possible. This problem is not unique to ^{13}C labeling and ^{13}C MRS but has already been recognized for metabolic investigations with radio labels. As a consequence, mathematical models were developed with the purpose of describing the labeling data in terms of physiological parameters such as pathway activities (flux rates) and metabolite concentrations (pool sizes).

Since the early metabolic studies with radio tracers, the focus has primarily been the tricarboxylic acid (TCA) cycle with its central role in eucaryotic cellular metabolism. To account for the complex situation of continuous metabolite and label recycling that occur in the TCA cycle, several different approaches for data analysis have been proposed. Expressions for physiological parameters were derived from the sum of infinite convergent series (Strysower et al., 1952), differential equations (Garfinkel, 1966), and more recently input–output equations (Katz and Grunnet, 1979). These models, which were originally designed for the interpretation of ^{14}C tracer data, are, in principle, directly applicable to the analysis of ^{13}C MR spectra. Specific (radio)activity and positional activity are simply replaced by their ^{13}C MRS equivalents, fractional enrichment and positional enrichment. However, none of these models takes advantage of the additional ^{13}C MR-specific information provided by homonuclear spin couplings.

5.3.2 Metabolic Modeling and Isotopomer Analysis

Recently, models designed for the analysis of ^{14}C tracer data were specifically modified for the interpretation of ^{13}C MR data. Instead of defining the enrichment at every carbon position as a singular state, each carbon isotopomer is considered and hence the term *isotopomer analysis* was coined. With theoretically 2^N different isotopomers of a compound with N carbons, the number of different labeling states is dramatically increased compared with the corresponding N possible positional enrichments or the one fractional enrichment for the same molecule. With the availability of this additional information, complex questions can be addressed; in particular, those on the competition of alternative pathways, on metabolite cycling, and on metabolite compartmentation.

The purpose of metabolic modeling is to predict the isotopomer distribution resulting from an anticipated metabolic scheme in terms of physiological parameters such as flux rates and pool sizes of metabolites. Thus, the spectral contribution of MR-distinguishable groups of isotopomers can be calculated and compared with the corresponding multiplet structures in the experimental ^{13}C MR spectra. The resulting equation system can finally be solved for the physiological parameters by numerical approximation or analytical deduction.

For a unique solution, the equation system must be determined. The number of experimentally measured isotopomer contributions therefore has to be equal to or exceed the number of unknown independent physiological parameters. Thus, for complex models with numerous fluxes, a large body of experimental input is required and, in the view of the uncertainties in each measurement (e.g., due to noise), a considerable redundancy of experimental data is preferred. The strength of isotopomer analysis lies in the way in which it provides both a large amount of information and built-in redundancy. The multiplet patterns detected for different carbons within a molecule are not completely independent from each other due to the reciprocity of spin couplings. Hence, an inherent cross-check is available for the isotopomer quantification and the metabolic parameters calculated thereof.

5.3.3 Metabolic Modeling and the Tricarboxylic Acid Cycle

The theoretical outline of isotopomer analysis just given is now applied to the modeling of the TCA cycle and ancillary pathways. The choice of an appropriate model is of utmost importance for the isotopomer calculations because the underlying assumptions ultimately limit its scope. Generally, a model that can explain the experimental data with the smallest number of parameters is favored. The key assumptions for the reviewed models of TCA activity are either derived from the legacy of radio tracer experiments or are the result of adaptations that had to be made on the basis of the peculiarities of ^{13}C MRS:

1. The underlying metabolic scheme presumes a topology of the TCA cycle where acetyl-coenzyme A (CoA) condenses with oxaloacetate to citrate, which, in turn, is metabolized in several steps to α-ketoglutarate and further to succinate, fumarate, malate, and finally oxaloacetate (see Figure 2 for an overview).

2. All TCA cycle intermediates are generally present only in small concentrations, which most often preclude their detection by ^{13}C MRS. Therefore, isotopomer analysis has to rely on those metabolic products that accumulate in sufficiently high concentrations, e.g., glutamate, aspartate, alanine, lactate, or glucose. It is assumed that glutamate accurately reflects the labeling of the TCA intermediate α-ketoglutarate since the two metabolites are in a fast equilibrium via transamination. Similar assumptions have been made for lactate and alanine, and aspartate, which are thought to reflect the labeling of pyruvate and oxaloacetate, respectively. Furthermore, glucose emerging from gluconeogenesis has been used as an alternative probe for the labeling in oxaloacetate.

3. Due to the molecular symmetry of fumarate, label may be scrambled between the two halves of the molecule.

4. The reactions in the TCA cycle are commonly regarded as unidirectional with the exception of the sequence fumarate $\leftrightarrow$ malate $\leftrightarrow$ oxaloacetate, thus allowing metabolite cycling between oxaloacetate and fumarate.

5. It is evident that the models cannot account for label cycled through unanticipated pathways. However, most models also do not account for the all-present contribution of natural abundance ^{13}C, thus introducing

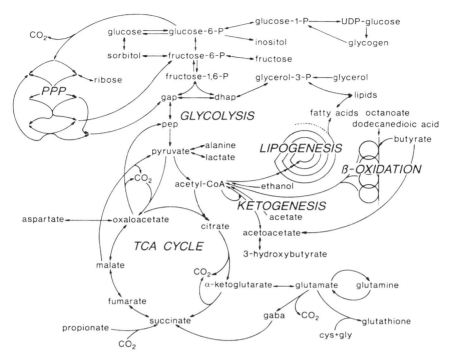

Figure 2. Schematic representation of metabolic pathways discussed in the text. The different pathways and their interconnections as depicted here are based on findings from both classical biochemistry and ^{13}C MRS assessments of animal tissues. This diagram shall by no means imply that the shown metabolic pathways do occur unequivocally in all tissues. For more details, see text in Section 5.5. Abbreviations: CoA, coenzyme A; cys, cysteine; gaba, 4-aminobutyric acid; gly, glycine; P, phosphate; PPP, pentose phosphate pathway; UDP, uridinediphosphate.

systematic errors that become more important under conditions of low enrichments.

6. For further simplification, the assumption is made (unless stated otherwise) that the metabolism is in metabolic steady state where neither flux rates nor pool sizes change during the experiment.

5.3.4 Isotopomer Analysis and Modeling with Differential Equations

Chance *et al.* (1983) proposed a metabolic model of the TCA activity based on a set of simultaneous differential equations. For every reaction step involved in the modeled metabolism, differential equations describe the time-dependent concentration change of all possible isotopomers. The resulting system of interdependent equations is then solved for the rate constants of the different reaction steps. This approach requires the knowledge of the time course of label incorporation into strategic metabolites of the TCA cycle and their absolute concentrations. Clearly, the former requirement is met by serial measurements with ^{13}C MRS, where pool sizes have to be determined by an alternative method. The accuracy of the calculation strongly depends on the achievable time resolution and SNR of the MR spectra. From a mathematical point of view, the enormous number of simultaneous differential equations resulting even from a simple model may impose further limitations and require judicious selection of data storage and computing algorithms.

Chance *et al.* (1983) successfully applied this method for the quantification of acetate and pyruvate metabolism in perfused rat hearts. Perchloric acid (PCA) extracts obtained from hearts perfused for designated times were analyzed with ^{13}C MRS for the fractional enrichment in individual carbons of glutamate and aspartate, while standard biochemical assays provided metabolite concentrations. Metabolic modeling then allowed the calculation of the flux rates through the TCA cycle, transamination reactions and of the consumption of the substrates (see Section 5.6.).

5.3.5 Isotopomer Analysis and Modeling with Infinite Convergent Series

Perhaps the most intuitive way of simulating the TCA cycle activity is a calculation based on infinite convergent series. The labeling patterns of the various TCA cycle intermediates are calculated after every full turn of the cycle with the assumption that the delivery of labeled substrates is constant. The results obtained after each turn can be interpreted as the discrete version of the mathematical approach with the differential equations described earlier. For an infinite number of TCA cycle turns, the calculated label distribution converges and reflects the condition of a labeling steady-state which is achieved after long, constant exposure to labeled substrates. For complex systems, the analytical derivation of infinite series turns into a tedious task and the analytical solution of the resulting equations is no longer trivial. Numerical summation of the label contributions for a finite number of TCA cycle turns provides an alternative approach and omits analytical solutions. This latter method can also be time-consuming since a sufficiently high number of iteration steps are required for calculating the label distribution resembling that at steady state. In contrast to the calculation based on differential equations, the steady-state approach requires only one ^{13}C MR spectrum for determining a set of physiological parameters.

Jans and Willem (1989, 1991) applied infinite convergent series for the interpretation of ^{13}C MR data obtained from cell suspensions of renal cells incubated with [^{13}C]alanine and [^{13}C]succinate. Perchloric acid extracts of cells that had attained metabolic and isotopic steady state were analyzed with ^{13}C MRS. From the labeling in glutamate, alanine, and glucose, the model provided measures for the relative contribution of pyruvate dehydrogenase (PDH) and pyruvate carboxylase (PC) to the flux through the TCA cycle. Furthermore, the relative flux rates for metabolites leaving the TCA cycle were estimated (see Section 5.9.).

5.3.6 Isotopomer Analysis and Modeling with Input–Output Equations

The most advanced method for calculating steady-state labeling patterns is, at present, the input–output concept originally proposed by Katz and Grunnet (1979). This approach relies on the premise that during metabolic and isotopic steady state neither the concentrations of metabolites nor their labeling changes. Expressions can be derived in a manner such that the sum of influxes of a particular isotopomer into a metabolite pool is counteracted by the sum of outfluxes of the same isotopomer. Thus, the probabilities of formation and disposal of particular isotopomers have to be equal. Based on these probability calculations, a single ^{13}C MR spectrum provides direct measurements of relative fluxes at the various branchpoints of the metabolism. Input–output equations give the same results as those obtained from infinite convergent series, but they are arrived at with less computational effort.

Metabolic modeling based on the input–output concept was used extensively for assessing the TCA cycle activity in perfused hearts (Malloy *et al.*, 1987, 1988, 1990a; Sherry *et al.*, 1988). Analysis of spin-coupling patterns detected for myocardial glutamate in PCA extracts provided estimates for the relative contribution of anaplerotic pathways (i.e., pathways that replenish TCA cycle intermediates) to the flux through the TCA cycle. Furthermore, substrate selection for the synthesis of acetyl-CoA entering the myocardial TCA cycle was quantified (see Section 5.6).

Recently, the input–output concept has been revised and a model applicable under non-steady-state conditions has emerged (Malloy *et al.*, 1990b; Sherry *et al.*, 1992). This "input" model is based on the premise that the labeling patterns of freshly formed metabolites are solely dependent on the specific influx of label. The means of disposal is of no direct concern because it affects only the labeling patterns of subsequent metabolites. Malloy and coworkers (1990b; Sherry *et al.*, 1992) have used this approach to quantify the relative substrate selection for fuel entering the myocardial TCA cycle under non-steady-state conditions (see Section 5.6).

The models described here were designed for metabolic networks without compartmentation, and metabolic information has been extracted from the resonances of aliphatic carbons of only a few ^{13}C MR-visible metabolites. Künnecke et al. (1993) have expanded the scope of the "input" model to the analysis of brain metabolism approaching isotopic steady state. Spin-coupling data were derived from the full spectral range including all resonances of ^{13}C MR-detectable metabolites. This approach has provided a comprehensive and quantitative picture of the compartmentalized TCA metabolism in rat brain (see Section 5.7). The analysis of spin-coupling patterns and the calculation of isotopomer distributions not only provided relative flux rates at the various branch points of the metabolism but also relative concentrations of metabolites. Appropriate summation of the different isotopomers revealed the positional enrichments at every carbon position. In turn, relative metabolite concentrations could be deduced from a single ^{13}C MR spectrum by integrating resonance intensities (Künnecke et al., 1993). This model of brain metabolism also fully appreciated the influence of label flow due to natural abundance ^{13}C contributions.

5.3.7 Simplified Models of Isotopomer Analysis

Many researchers have avoided comprehensive isotopomer analysis by devising simplified models that involve less computational effort or aim at explaining only very limited aspects of the metabolism. One of the most recent examples is the determination of competing fluxes from different substrates into acetyl-CoA entering the TCA cycle (Malloy et al., 1990b; Sherry et al., 1992). Relative flux rates were deduced from the analysis of spin-coupling patterns in glutamate carbons C3 and C4 or solely carbon C4, providing the advantage that the analysis is applicable to non-steady-state conditions.

Cohen (1983) presented a first-order analysis of the TCA cycle activity in liver where the flux of label is only considered for one turn of the TCA cycle. The assumption of such a truncated model was justified by the inability to detect significant amounts of isotopomers that resulted from metabolite recycling and by the finding that this model accounted for all spectral features within experimental errors.

Jans and Leibfritz (1989) addressed the influence of label recycling on the calculation of the relative flux through PDH and PC reactions in their own way. Relative flux rates were determined by the first-order model proposed by Cohen (1983) and the results, in turn, were corrected for label recycling. The corresponding correction factors were deduced from the theoretical distribution of label entering the cycle via PDH after several turns of the TCA cycle.

This list of simplified models could be continued with many more entries. The common concept, however, is the limited focus on a few strategic isotopomers, whereby the contribution of some relevant competing pathways may be deduced (e.g., Cohen *et al.*, 1981a; Cohen, 1987b; Shulman *et al.*, 1987)

5.3.8 Limitations of Models Based on Isotopomer Analysis

Despite the many advantages and the power of ^{13}C isotopomer analysis for the metabolic interpretation of ^{13}C labeling data, several drawbacks are inherent in this method. Isotopomer analysis as such only provides relative flux rates. This problem can generally be solved by independent measurements of at least one absolute flux rate using other techniques. A comprehensive isotopomer analysis for a large metabolic network also results in a very substantial computational effort. MR-distinguishable groups of isotopomers have to be identified and quantified, and an extensive, presumably nonlinear system of equations has to be solved.

A further drawback is the intrinsic noise in ^{13}C MR spectra, which interferes with the resonances, particularly with those from less abundant isotopomers and the highly split resonances of multiply enriched isotopomers. Since isotopomer analysis relies on an accurate determination of relative spectral contributions of multiplet patterns, this type of analysis is confined to metabolites present in sufficiently high concentrations. Malloy *et al.* (1990a) have assessed the error propagation from spectral noise to physiological parameters using a Monte Carlo simulation. Their findings suggest that when higher noise levels occur, more reliable results are obtained by using numerical analysis of all available data instead of calculating physiological parameters from analytical expressions. This can be explained by the fact that the numerical analysis takes advantage of the partial redundancy of information contained in the spectral patterns, whereas the analytical calculations deduce the physiological parameters from a minimum of the available data. In this context, it is important to note that the labeling pattern chosen for the substrate(s) has to create distinguishable label distributions for the competing pathways analyzed. Ambiguous labeling patterns and label flux through pathways with minor effects on the label distribution in the end products would lead to ill-defined relative flux rates through these pathways.

Low spectral resolution as encountered under *in vivo* situations is another factor that limits accurate quantification of multiplet patterns. A partial solution for this problem is achieved with editing methods in which the constituents of the multiplet patterns are separated by their different phase behaviors (Malloy *et al.*, 1990b). However, this procedure involves the time-consuming acquisition of several subspectra, which are then processed in an addition–subtraction scheme in order to extract the multiplet components.

5.4 Experimental Considerations

This section covers an area that is often only marginally addressed. Instead of providing a summary of the underlying MR theory, it takes a more practical approach and gives an overview on the ^{13}C MRS methods currently used for the assessment of metabolism in animal tissues. This section is thus intended as an introduction to the range and scope of different methods, and gives a short outline of their advantages and disadvantages. To obtain more comprehensive descriptions of certain methods, the reader is referred to the references in the text.

5.4.1 Experimental Systems and Sample Preparation

^{13}C MRS on intact animals *in situ* is certainly the experimental approach that complies the most with the exciting noninvasive nature of MR methods (see Le Moyec and Akoka, 1992). Despite its drawbacks, ^{13}C MRS *in situ* is still the only method for a comprehensive assessment of the carbon metabolism in the intact and alive animal with a minimum of interference derived from the measurement itself. Normally, ^{13}C MRS investigations *in situ* are focused on the metabolism in a particular organ or a specific region thereof. However, even with adequate means for discriminating between the spectral contributions of the organ of interest and other tissues (for a discussion of localization, see Section 5.4.2 and Chapter 6), the metabolic interaction with the rest of the body persists. Hence when using this approach, keep in mind that accumulated spectra represent an excerpt of the metabolic activity in the tissue under investigation and its interaction with the overall metabolism.

The overall metabolism in an intact animal is well regulated and is under tight internal control. As a consequence, direct manipulation of the metabolism in the target organ from the outside is very limited. Starvation/feeding, experimental diseases such as diabetes mellitus, and externally applied effectors such as the hormones glucagon and insulin have been used to influence the physiology of the animals. However, it is often difficult to know whether changes in the metabolism as detected by ^{13}C MRS are the result of direct or systemic responses to the imposed metabolic challenge. In this context and with special reference to ^{13}C-labeled substrates, further attention has to be paid to the modality of administration. While intravenous infusion or injection techniques allow for efficient control of the amount of label deposited into the blood stream, the normal sequence of digestion, resorption, and a first passage through the liver is bypassed (Cross *et al.*, 1984). Hence, intragastric or intraduodenal infusion may prove to be a valid alternative (Shalwitz *et al.*, 1989). Bolus administration may be less invasive than infusion techniques but has the disadvantage that the time course of label supply is yet another unknown (Künnecke and Seelig, 1991). Both dosage and route of administration of substrates can influence their further metabolism.

Experimental animals unwillingly undergo MRS investigations. Hence, restraints, sedation, or anesthesia have commonly been applied in order to keep the animals quiet. Stress and fear due to the animal handling, postoperative trauma (e.g., after cannulation of blood vessels), the depth of anesthesia, and the anesthetics themselves may have further direct or indirect impacts on the physiology of the investigated tissue and need judicious consideration. Thus, many researchers have adopted a procedure in which catheters, coils, etc., are implanted permanently by surgery well in advance of the MR experiments. The physiological changes due to a drop in body temperature and altered or even suppressed respiration rate often associated with anesthesia have been counteracted by warming pads together with the measurement of rectal temperature; mechanical respirators have been used to control respiration (Neurohr et al., 1984; Behar et al., 1986)

Perfused or superfused organs are very popular experimental systems (Cohen, 1987a,b,c). They represent functionally complete units that lack the metabolic interference with other tissues typically observed in the whole-body environment. This implies that essential functions normally maintained by the integral body (e.g., temperature, flow rate and composition of the perfusion medium, etc.) have to be controlled externally and their manipulation remains entirely in the hands of the experimenter. However, under appropriate conditions, such organ preparations are viable for several hours. Because no anaesthetics are required for sustaining the perfused organ in a stable state, the undesirable effects of anaesthesia can be kept to a minimum (organs being generally obtained from acutely anaesthetized animals). Perfused and superfused organs are normally maintained submersed in the perfusion solution in order to prevent drying and magnetic susceptibility changes between tissue and surroundings. This setup, however, has the disadvantage that significant contributions of signals can arise from extracellular metabolites dissolved in the perfusion buffer as well as from the buffer itself. Although inorganic buffers (e.g., phosphate buffer) lack any ^{13}C MR signals, separation between intracellular and extracellular contributions to the MR signals remains difficult. Shift reagents to which cell membranes are impermeable have been used for attempts to unravel this problem.

MR experiments on cultured or isolated cells, cell organelles, and subcellular preparations provide the advantage that spectra are obtained from selected cell species, thereby avoiding the metabolic heterogeneity encountered in intact organs. Metabolic interactions and responses to exterior effectors can be studied without interference from other tissues and in a state that is free of systemic effects. However, the relatively low sensitivity of ^{13}C MRS requires a high cell density in the sample tube. The lack of vascularization as found in intact tissue makes the adequate perfusion and the homogeneous supply of nutrients and oxygen to the cells a challenging task. Systems including cell suspensions, spheroids, cells embedded in beads of gel matrix, cells grown on microcarrier beads, and cells incubated in a hollow fiber bioreactor have been tested (for a review see Szwergold, 1992). Appropriately controlled perfusion conditions then allow the maintenance of cells in a metabolically stable state over a long period of time and the physiological surrounding of the cells can easily be manipulated (Seguin et al., 1992). The limited cell density (the maximum is approximately 70% v/v) imposes the same problems as already discussed for perfused organs: Extracellular metabolites may give rise to substantial spectral contributions.

Body fluids are of potential biochemical value since their composition may reflect the functional state of the body as a whole due to their intimate metabolic relationship with other tissues (Bell, 1989, 1992). MRS of body fluids therefore constitutes a convenient method for the nearly noninvasive analysis of metabolism. With little or no sample pretreatment, a wide variety of metabolites can be observed at a generally high spectral resolution and sensitivity. Repetitive sampling of body fluids such as blood and urine allows the dynamics of metabolism to be followed. Other body fluids such as bile, cerebrospinal fluid, or sweat may be harder to obtain since in small experimental animals the quantities available are very limited.

MRS on extracts obtained from tissues and cell preparations represents a powerful alternative to MR experiments in vivo. During the course of the extract preparation, the cellular metabolism is permanently arrested and the metabolites of interest are solubilized. Extracts are thus homogeneous and very stable, allowing the accumulation of MR spectra with an excellent spectral resolution and a high SNR. Most prominent are perchloric acid extractions for water-soluble and acid-soluble metabolites and, alternatively, chloroform/methanol extractions for lipophilic compounds. Resuspending the extracts in an appropriate buffer provides a constant pH and hence a more accurate alignment of pH-sensitive resonances in individual spectra. Chelating agents have been used to further increase the spectral resolution by removing traces of paramagnetic ions or bivalent ions such as calcium and magnesium that can cause line broadening. Despite all the obvious advantages of tissue extracts for the assessment of the cellular metabolism by ^{13}C MRS, extracts have the drawback that they only reflect the state of the metabolism at the time of extraction and the dynamics of metabolism cannot be followed in a single experiment. Furthermore, it has to be pointed out that even though many metabolites are quantitatively extracted with an appropriate extraction method, a wide range of metabolites is coextracted at a partial level. Furthermore, chemical decomposition of some metabolites may occur during the extraction procedure. Quantitative assessment of extracts thus needs a precise knowledge of the extraction profile.

5.4.2 Data Acquisition, Spatial Localization, and Spectral Discrimination of Tissues

The vast majority of ^{13}C MR spectra obtained from animal tissues were acquired using simple pulse-acquire sequences with broadband decoupling of the protons. The large chemical-shift range of the ^{13}C nucleus usually provides excellent spectral resolution without the need for spectral editing methods. Long acquisitions due to the limited sensitivity of ^{13}C MRS together with the time constraints imposed by the viability of living tissue, fast relaxation, and the inhomogeneous radio frequency (rf) field of surface coils have further restricted the application of multipulse sequences *in vivo*.

Focusing now first on the *in vivo* situation, the issue of spatial localization and spectral discrimination of the tissues of interest becomes pertinent. Surface coils are a convenient means for acquiring spectra of a selected part of tissue noninvasively. Usually, a single ^{13}C surface coil is used for excitation and reception. The advantages of this surface coil design are the simplicity of the setup and the large filling factor that can be achieved, which allows spectral acquisition with good sensitivity. The major drawback is that the sensitive volume of the coil is inhomogeneous and extends not much further than a sphere with the radius of the coil, thus restricting its application to tissues close to the external surface of the animal. Furthermore, all the tissue within the sensitive volume gives rise to MR-detectable signals. In particular, the resonances from the large fat deposits in the skin and adipose tissue of intact animals can conceal the smaller spectral contributions from the underlying tissue of interest.

To date, a large number of localization schemes are potentially available that enable the acquisition of MR signals from well-defined regions within the living body. However, localization techniques based on direct volume selection by magnetic gradients imposed on the main field, as known from proton and phosphorus MRS, have not found widespread application in ^{13}C MRS. The most constraining factors have been the large chemical-shift range for ^{13}C nuclei and the predominant use of surface coils, which make these localization techniques almost inapplicable in the case of ^{13}C MRS. However, ^{13}C MRS has recourse to alternative localization procedures as described later.

An easy method for increasing the spatial localization of surface coils is a method referred to as *surface nulling*. This method consists of a single 180° pulse applied at the center of the surface coil. This pulse results in an inversion of the magnetization of nuclei close to the surface coil (theoretically no detectable signal) and a 90° excitation at a distance of approximately 0.6 times the coil radius. Thus, partial localization is achieved by reducing the signals from tissue close to the surface coil and by simultaneously emphasizing the signals from deeper lying tissues. On the one hand, the coil's diameter strongly influences the depth of rf penetration and the total sensitive volume. On the other hand, the achievable SNR is in a complex relationship with the coil diameter (Lawry *et al.*, 1990). Cross *et al.* (1984) showed that surface coils in combination with the surface nulling technique provide a spectroscopic selection of the liver of better than 95%.

The DEPT (distortionless enhancement by polarization transfer) pulse train is a development based on the concept of surface nulling as just described (Bendall and Gordon, 1983). Instead of a single 180° pulse, a $\Theta\text{-}(2\Theta)_n$ pulse train with concomitant phase cycling is applied. This pulse sequence eventually leads to a further restriction of the sensitive volume of the surface coil by attenuating signals from regions close to the coil (David *et al.*, 1990).

Similar to the sensitivity profile, the excitation profile of surface coils rapidly decays in regions further apart from the coil's center. With the implementation of adiabatic pulses, the region of homogeneous excitation can be extended to these outer regions, provided enough rf power is delivered to the coil (Silver *et al.*, 1984; Bendall and Pegg, 1986). The main advantage is the increased contribution of signal from deeper lying organs. However, spectral contributions from tissue near the surface are not suppressed. Nevertheless, since the adiabatic condition is only fulfilled for a limited spectral region, it is possible to obtain a considerable signal reduction off-resonance, while on-resonance adiabaticity is preserved over the sensitive volume of the surface coil. Van Cauteren *et al.* (1992) showed that with this approach the signals from liver glycogen could be enhanced, while at the same time lipid resonances from adipose tissue were suppressed.

Volume selection based on polarization transfer relies on the premise that polarization transfer is only effective for certain well-defined pulse angles in both the proton and carbon frequencies. For surface coil spectroscopy where the excitation field is very inhomogeneous, this condition is clearly not fulfilled and optimum polarization transfer occurs only within a very limited region. Advantage has been taken of this property to spatially select tissue within the animal using concentric proton and ^{13}C surface coils with different diameters (Bendall and Pegg, 1984). By varying the flip angles of the proton and ^{13}C pulses and by including extensive phase cycling, different volumes could be selected.

Polarization transfer from protons to ^{13}C nuclei using DEPT has also been used for both spectral editing and increased sensitivity in the *in vivo* ^{13}C MRS experiment. Theoretically, polarization transfer from protons to ^{13}C could provide a signal gain of more than four. (The net magnetization of the carbon nuclei is increased by a factor of 3.98 and the T_1 relaxation becomes entirely dependent on the relaxation properties of the protons, which are in general somewhat shorter than the corresponding carbon relaxation times.) However, spectra acquired through polarization transfer do not contain the spectral information obtainable from nonprotonated carbons. Furthermore, the signal gain achieved by polarization transfer *in vivo* is much less than the theoretical maximum. On the one hand, the well-defined pulses in both the proton and carbon frequencies necessary for an optimum polarization transfer are not available with surface coils. On the other hand, fast T_1 and T_2 relaxation as found under *in vivo* conditions further reduce the efficacy of the polarization transfer (B. Künnecke and J. Seelig, unpublished data; see also Chapter 6). Xue *et al.* (1990) reported a sensitivity improvement of 1.7 for the DEPT sequence applied to tumor tissue *in situ* as compared to the signal acquired with a plain spin-echo sequence.

Chen and Ackerman (1989, 1990) developed an alternative method for suppressing surface signals by applying spoiling gradients during the surface coil experiment. Basically, a grid of parallel wires driven by a low current is placed between the animal and the surface coil. The current induces magnetic inhomogeneities close to the animal's surface that decrease the coherence. Hence, the intensity of signals originating from superficial tissues is attenuated, while at the same time the acquisition of signals from the deeper lying tissues is fully retained.

All the localization schemes described here have the drawback that the MR signal originates from an ill-defined spatial region. This has led to the development of gradient-based localization techniques such as ^{13}C chemical-shift imaging and localization in the proton frequency range together with polarization transfer (see Chapter 6). However, these techniques have not found widespread application to ^{13}C MRS on animals. A more pragmatic approach involves surgery and exposure of the organ of interest. This method is in stark contrast to the noninvasive techniques, which rely entirely on gradients and rf pulses. Nevertheless, the acquisition of signals from a selected tissue is greatly facilitated by exposing the organ to the receiver coil. Spectra with a very good SNR and excellent tissue selection can be acquired. To further enhance the sensitivity of the experiment, surface coils have been replaced by solenoids in a way such that the tissue is completely surrounded by the coil (Neurohr *et al.*, 1984; Laughlin *et al.*, 1990). To reduce the impact of surgery, coils have also been implanted permanently and the animals were allowed to recover before the actual experiment took place.

In contrast to the *in vivo* situation, ^{13}C MRS on excised tissues, cells, and tissue extracts is far less constrained by the technical setup of coils. Localization is not needed and the entire sample can be placed in a probe tuned for sensitivity and resolution. Spectroscopy *ex vivo* is generally carried out at high field strength where full advantage can be taken of the increased sensitivity and the longer relaxation times. In particular for tissue extracts, the full range of homonuclear and heteronuclear pulse techniques is thus potentially available for chemical analyses.

5.4.3 Decoupling and Nuclear Overhauser Effect

Broadband proton decoupling during acquisition is routinely used in ^{13}C MRS to reduce the complexity of spectra and to increase signal intensities. Such decoupling procedures are straightforward for *in vitro* samples. In the *in vivo* situation, decoupling is achieved either with a second rf field induced by a second surface coil concentric to the ^{13}C surface coil (e.g., Canioni *et al.*, 1983) or by a whole-body resonator (Cross *et al.*, 1984; Behar *et al.*, 1986; Hammer *et al.*, 1990). In this respect, local and total power deposition and concomitant tissue heating, as well as the homogeneity of the decoupling over the volume of interest, become important issues when dealing with living biological material.

Resonators provide a homogeneous decoupling field but require high decoupling power because of their large volume. On the other hand, decoupling by a second surface coil limits the power deposition to a confined area. Nonetheless, areas of very high flux occur close to the coil with a concomitant high local power deposition. To avoid these high-flux regions within the sample, spacers between the decoupling coil and the animal's body have been used. The concentric arrangement of the proton decoupling coil and the ^{13}C receiver coil has the disadvantage that the two coils are strongly coupled, which leads to additional noise induction in the ^{13}C receiver coil. To reduce this electrical crosstalk, proton coils with approximately twice the diameter of the carbon coil have been used together with an array of frequency filters (see also Chapter 6). This latter geometry furthermore provides a relatively homogeneous decoupling field over the sensitive volume of the carbon coil.

In striving for an increased SNR performance of ^{13}C MRS, low-power broadband proton decoupling during the relaxation delay has been used for inducing the NOE. Theoretically, a maximal signal increase of $\eta = 2$ is obtainable for ^{13}C nuclei in the vicinity of protons. However, under *in vivo* conditions, the NOE is most often drastically reduced because the motional narrowing limit does not apply to molecules in the viscous environment of cells and because of the major contribution of relaxation mechanisms other than dipolar relaxation (i.e., the presence of paramagnetic substances).

Broadband decoupling is achieved by means of a modulated rf field. Alternatively, pulsing methods such as WALTZ and MLEV have been used that feature a much lower power dissipation than needed with modulated rf fields

(Levitt and Freeman, 1981). Even though these latter pulse sequences were developed for homogeneous rf fields, they are relatively insensitive to maladjusted pulse angles and are thus also used in conjunction with surface coils. Bilevel decoupling has been used to achieve decoupling and NOE with a minimum of power deposition. The decoupler power is set to a level high enough to sustain NOE buildup and only during the acquisition period is the decoupler power raised to a level that enables full decoupling over the entire spectral range. The average power deposition by the decoupler is thus controlled by the duty cycle of the acquisition period.

5.4.4 Quantification

The quantitative assessment of biochemical pathways with ^{13}C MRS in combination with ^{13}C-labeled precursors involves the determination of fractional enrichments, positional enrichments, ^{13}C isotopomer distributions, and metabolite concentrations (see Section 5.3). The fact that ^{13}C MR signal intensities are simultaneously dependent on a variety of factors such as concentration, NOE, degree of MR visibility, saturation, and positional enrichment might be, at first sight, confusing and could lead to the conclusion that no quantitative assessment is possible at all. However, a step-by-step deconvolution of the different contributions to the detected signal can resolve this problem. We focus first on the quantification of ^{13}C MR spectra obtained *in vitro*.

Long T_1 values, corresponding to a slow spin-lattice relaxation of ^{13}C nuclei (typically of the order of 3 to 30 s for carboxylic groups), impede rapid pulsing or alternatively lead to significant saturation of the resonances when the repetition rate of the spectra acquisition is set for a maximal signal gain per time. Furthermore, T_1 relaxation and NOE of carbons in different chemical and physical environments vary over a wide range. Time constraints may preclude the acquisition of fully relaxed ^{13}C MR spectra and thus, for quantification purposes, every resonance has to be corrected individually for NOE/saturation. Correction factors for saturation/NOE have been determined in model solutions, mimicking the physicochemical properties of the sample to be measured. The model solution is assessed under fully relaxed conditions and again under the acquisition conditions prevailing for the measurement of the actual sample. Under the assumption that saturation/NOE is adequately represented in the model solution, the signal reduction/enhancement as detected between these two spectra can be used to correct signal integrals for the combined effect of saturation and NOE.

^{13}C magnetic resonances, corrected for NOE and saturation, intrinsically render relative measurements of ^{13}C nuclei only. However, the addition of an external or internal standard of known ^{13}C concentration ultimately enables absolute quantification of the amount of MR-visible ^{13}C present at every carbon position. For natural abundance ^{13}C MRS, the values obtained directly relate to the total metabolite concentration (under the assumption of full MR visibility). However, in ^{13}C tracer experiments, information on metabolite concentrations and on ^{13}C enrichments is still intermingled.

Several different approaches have been used to extract data on the ^{13}C enrichments:

1. Due to the topology of the pathway under investigation, a particular carbon position might be known to be void of ^{13}C enrichment. The corresponding ^{13}C resonances can hence be used as an intramolecular enrichment standard whereby the enrichment in other positions can be inferred (Shulman *et al.*, 1985; Cline and Shulman, 1991).

2. The total amount (^{13}C + ^{12}C) of the metabolite is derived from an alternative measurement such as ^{1}H MRS, HPLC, GC-MS, enzymatic assay, etc., and is related to the amount of ^{13}C. In particular, enzymatic assays have also been used to determine the degree of MR visibility of metabolites labeled at the natural abundance level (Gruetter *et al.*, 1991).

3. A third method, referred to as the *satellite technique*, is entirely based on proton MRS (Shulman *et al.*, 1985; Behar *et al.*, 1986). In ^{1}H MR spectra, protons covalently linked to ^{13}C nuclei exhibit, in addition to the usual ^{1}H–^{1}H homonuclear spin couplings, heteronuclear ^{13}C–^{1}H couplings ($^1J_{CH}$ ~ 100 to 200 Hz). These "satellites" flank the central resonance of the corresponding ^{12}C-bound protons. Hence, the positional fractional enrichment is calculated as the ratio of the integrated satellites versus the total intensity (satellites and central resonance) of the corresponding proton resonance. It is evident that this method only provides enrichments for carbons that are spin-coupled to protons with well-resolved resonances. Furthermore, the accuracy of the satellite technique is limited by the dynamic range between the signals of ^{13}C-bound and ^{12}C-bound protons. Thus, fractional enrichments well above 1.1% (and well below 100%) are most accurately detected.

4. Mass spectrometry is a convenient method for measuring the relative amount of isotopomers with different molecular masses due to the different mass number of ^{12}C substituted with ^{13}C (Kalderon *et al.*, 1986). Mass spectrometry is limited insofar as it only provides fractional enrichments of molecules or their fragments.

5. In contrast to the analysis of metabolic pathways by determining fractional and positional enrichments, isotopomer analysis is based on probability calculations of the intramolecular and intermolecular distribution of ^{13}C nuclei. Ultimately, isotopomer analysis involves the deconvolution of spin-coupling patterns into the spectral contributions of MR-distinguish-

able groups of isotopomers. After appropriate correction for intensity distortions (e.g., NOE, saturation, etc.), a quantitative estimate of the occurrence of these isotopomers is obtained that can be related to specific metabolic pathways and their relative contributions to the overall metabolism. Entirely based on ^{13}C MRS, isotopomer analysis is a very powerful technique. It also provides positional enrichments through an appropriate summation of isotopomer contributions to a particular carbon resonance. By taking advantage of spin couplings, isotopomer analysis is capable of detecting enrichments-in-excess below 0.1% (Künnecke and Seelig, 1991). Once the positional enrichments are known, relative concentrations and, with the addition of a concentration standard, absolute metabolite concentrations can be inferred. Isotopomer analysis has also been used to determine the relative MR visibility of cellular metabolites (Künnecke and Seelig, 1991). However, isotopomer analysis has the drawback that it requires assumptions about the topology of the metabolic pathways involved in the formation of the metabolites in question.

The power of MRS is its ability to provide biochemical analyses without destroying the sample. Thus, the same quantitative information as just described is potentially available from MR experiments *in vivo*. However, further complications may arise in the *in vivo* situation. On the one hand, T_1 relaxation times of metabolites in biological tissues are most often drastically reduced (e.g., due to the accumulation of paramagnetic ions), which alleviates the problem of saturation considerably. On the other hand, the inhomogeneous distribution of the rf field and inhomogeneous reception characteristic of surface coils, as well as a changing filling factor due to sample motion, impose severe constraints on absolute quantifications (see Bovée, 1992, and references therein). Furthermore, inhomogeneities of the organ structure and composition with accompanying magnetic and electric susceptibility changes and a variable degree of MR visibility of metabolites in the *in vivo* environment might prevent absolute quantification of metabolite concentrations even in the presence of a reliable concentration standard. Finally, the generally lower SNR of spectra acquired *in vivo*, the peak overlap of broad lines with unknown lineshapes, and the fact that MR spectra represent a time average of the metabolism are limiting factors for the accuracy of direct quantification *in vivo*. As a result, most ^{13}C MR data acquired *in vivo* have been interpreted only qualitatively or quantified relatively to control spectra. For this purpose, a variety of internal and external concentration and chemical-shift standards have been used. Most prominent of the internal standards were the methyl resonance of choline and the methylene resonances of fatty acids (Pahl-Wostl and Seelig, 1986; Jehenson et al., 1992). Alternatively, external standards have been employed that consist of small glass spheres filled with a standard solution (e.g., CS, formic acid doped with Gd^{2+}) and located adjacent to the surface coil (Siegfried et al., 1985), or an intraventricular balloon in the case of heart preparations (Brainard et al., 1986). Absolute quantification of natural abundance metabolites by ^{13}C MRS *in situ* has been attempted by comparing the MR signals from a tissue sample with those acquired from a phantom simulating the physical and chemical conditions of the tissue (Gruetter et al., 1991).

The focus of labeling experiments *in vivo*, however, has been on the elucidation of metabolic pathways and their activities. Thus, the accumulation of ^{13}C label at certain carbon positions was of primary interest. Advantage has been taken of the nondestructive trait of MRS, which offers the possibility of repetitive measurements and the acquisition of time sequences of metabolism. The animal/organ under investigation can therefore serve as its own control and thus reduce artifacts due to intersubject biological variability. Relative resonance increases, as compared to a spectrum acquired before administration of the labeled precursor, have been interpreted as ^{13}C labeling. However, the problem arises as to whether the observed changes of signal intensities are purely due to label incorporation or also from an increase of the total amount of the corresponding metabolite. Complementary measurements including *in vivo* ^1H MRS and more invasive biochemical assays have been used to ascertain that the metabolite concentration remained constant. Alternatively, multilabeled precursors with two or more contiguously arranged ^{13}C nuclei together with spin-coupling analysis have been successfully employed for discriminating signal increases due to ^{13}C enrichment versus changes of the total metabolite concentration (Cerdan et al., 1988).

5.5 Metabolism in the Liver

The metabolism in liver is associated with the capacity of this organ to serve as a metabolic mediator between the dietary source of nutrients and the metabolic requirements of extrahepatic tissue. The liver has perhaps the greatest metabolic flexibility of all the organs because it must constantly adapt to the intermittent ingestion and highly variable composition of nutrients, and to the variable demands of extrahepatic tissues. Storage and synthesis of metabolic fuels, disposal of waste products, and homeostasis of essential nutrients in blood are thus the major tasks fulfilled by the liver. Despite the wealth of data available, liver metabolism is not completely understood and still represents a field of active research. MRS assessments, and in particular ^{13}C MRS, have certainly made significant contributions to the attempt of disentangling the intricately interwoven metabolism of faccharides, amino acids, and lipids. Some of the relevant results from ^{13}C MR studies are outlined in this section.

The metabolic scheme presented in Figure 2 is used as a guide for the following discussion of metabolic events and their assessment by MRS in the liver as well as in other organs. Glycolysis, gluconeogenesis, and the TCA cycle play a central role in cellular metabolism by providing energy and the means for the anabolism and catabolism of a vast variety of metabolites. The pentose phosphate pathway (PPP) directly links hexoses and trioses in the glycolysis with the synthesis of pentoses and the production of reduction equivalents. Ancillary pathways of glycolysis/gluconeogenesis include (1) glucose disposal by its storage as glycogen, a feature that is predominantly observed in liver (this section) and skeletal muscles (Section 5.10); (2) the polyol pathway with the formation of sorbitol, which is discussed for the metabolism in ocular lenses (Section 5.8) and kidney (Section 5.9); and (3) the glycerol pathway that provides and disposes of the glycerol backbone in lipids. Aspartate, glutamate, and its amination product glutamine originate directly from intermediary metabolites of the TCA cycle and connect the carbon and nitrogen metabolism. Glutamine, glutamate, and the decarboxylation product 4-aminobutyric acid (GABA) are of eminent importance for brain metabolism as outlined in Section 5.7. The supply of propionate to the TCA cycle is an example of anaplerotic reactions that replenish TCA cycle intermediates. Propionate originates from the β-oxidation of fatty acids. β-Oxidation, in turn, has a major role in the production of acetyl-CoA, the fuel of the TCA cycle. β-Oxidation of fatty acids is discussed for the metabolism in hepatic (this section) and myocardial (Section 5.6) tissue where this pathway can contribute significantly to the overall metabolism. Instead of pursuing the catabolic sequence of the TCA cycle, acetyl-CoA may be channeled into ketogenesis, a process that is well documented for liver (this section). Alternatively, acetyl-CoA is used in the *de novo* synthesis of fatty acids, which is described next for liver and adipose tissue. Finally, we must point out that the metabolic scheme presented in Figure 2 is oversimplified and highlights features of the metabolism of different organs in the same sketch.

5.5.1 Lipogenesis and Fatty Acid Stores

Triglycerides and carbohydrates are accumulated in the liver for the purpose of storage of nutrients that have been ingested in excess. Storage lipids such as triglycerides are accumulated in the liver in small droplets. Fast and isotropic molecular tumbling together with the high intracellular concentration render these lipids highly detectable as sharp magnetic resonances. Figure 3a shows a characteristic natural abundance ^{13}C MR spectrum obtained

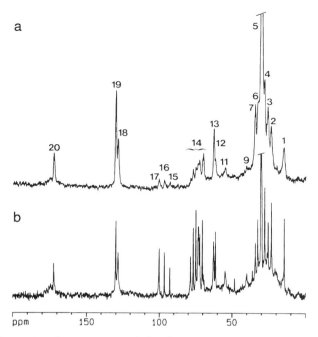

Figure 3. Comparison between a natural abundance ^{13}C MR spectra (at 90 MHz) obtained from a mouse liver *in situ* in (a) the intact animal and (b) a spectrum of the same liver after excision. The *in situ* spectrum was acquired from the liver of a fed mouse using a surface coil. Pieces of the excised liver were submersed in phosphate-buffered saline and assessed in a standard high-resolution probe. Strong resonances from mobile fatty acids (resonances 1–7, 18–20), choline (resonance 11), glycerol (resonance 13), glucose (resonances 12, 14–16) and glycogen (resonances 13, 14, 17) are clearly discernible in both spectra (B. Künnecke, unpublished data, 1992).

with a surface coil from the liver *in situ* in an intact mouse. For the sake of comparison, a spectrum of the excised liver from the same animal is presented in Figure 3b. Clearly, the spectra consist of a number of sharp resonances that have been assigned to various lipid and carbohydrate metabolites (Canioni et al., 1983). The spectra are dominated by the various methylene resonances of lipids (resonances 2–7). Smaller resonances at 15 ppm (resonance 1), 129 ppm (resonances 18, 19) and 174 ppm (resonance 20) are attributable to methyl, olefinic, and carboxylic groups of fatty acid chains. Signals arising from glycerol (resonance 13) and choline (resonance 11) in the backbone and headgroup of triacylglycerols and phospholipids, respectively, are also detected. In addition, resonances from glucose (resonances 12, 14–16) and glycogen (resonances 13, 14, 17) are visible in the spectral region between 61 and 101 ppm.

The high spectral resolution provided by ^{13}C MRS has been used to assess the triglyceride composition in liver for different nutritional conditions (Canioni et al., 1983; Gallis and Canioni, 1992). Unsaturated fatty acids are readily distinguishable from their saturated counterparts by the detection of distinct resonances at approximately 129 ppm. The specific chemical structure of polyunsaturated fatty acids, where two or several double bonds are separated by a single methylene group, allows a further direct distinction between monounsaturated and polyunsaturated fatty acids in the ^{13}C MR spectrum. In polyunsaturated fatty acids the olefinic carbons adjacent to the intermittent methylene group ("inner" olefinic carbons) resonate at ~128.5 ppm (resonance 18). The two "outer" olefinic carbons in polyunsaturated fatty acids as well as the olefinic carbons in monounsaturated fatty acids give rise to resonances at ~130 ppm (resonance 19). A comparison between the intensities of these two groups of resonances can then be used to determine the degree of polyunsaturation. Canioni et al (1983) showed that chronic restriction of essential polyunsaturated fatty acids in the diet results in a decrease of the polyunsaturated fatty acids in adipose tissue of the living rat. Similarly, a chronic supplementation of essential fatty acids leads to an increase of polyunsaturated fatty acids in liver and adipose tissue (Canioni et al., 1983; Gallis and Canioni, 1992).

Fatty acids of different chain length are distinguishable by the cluster of methylene resonances in the spectral region between 20 and 35 ppm. Hence, a quantitative analysis of the fatty acid composition on the basis of nondestructive ^{13}C MRS is possible as was shown for adipocytes in suspension (Sillerud et al., 1986) and for excised liver and liver extracts (Deslauriers et al., 1988; Cunnane, 1992). Direct carbon-by-carbon discrimination of carbon positions within different fatty acid chains offered further insight into lipogenesis *in vivo*. Cunnane (1992) analyzed lipid synthesis from [2-^{13}C]acetate in rat liver. Lipid extracts prepared from livers of rats injected with [2-^{13}C]acetate revealed that ^{13}C enrichment was greatest at C1 (carboxylic group) and decreased toward the methyl terminal. The high enrichment of C1 indicates a very active redistribution of the ^{13}C label from C2 of [2-^{13}C]acetate to C1 of the ultimate fatty acid precursors acetyl-CoA and malonyl-CoA. The gradation of the ^{13}C enrichment in the fatty acyl chains from a very high label incorporation at C1 to a much lower enrichment at the methyl end provides evidence that exogenous acetate not only provided the total carbon skeleton of freshly synthesized fatty acids but also contributed to chain elongations of fatty acids already partially synthesized. Furthermore, it was shown that the synthesis of monounsaturated fatty acids exceeded that of polyenes.

Similar labeling experiments performed with labeled ethanol and pyruvate in perfused rat liver revealed that chain elongation of fatty acids with labeled acetyl-CoA was minor under these conditions. The *de novo* synthesis of fatty acids from pyruvate was found to require the presence of ethanol (Cohen, 1987c). Therefore, Cohen inferred that the lipogenesis from the oxidized substrate pyruvate required the reduction equivalents provided by the oxidation of the cosupplied ethanol in additon to those produced by the PPP. Finally, the labeling data indicated that approximately 6% of the product of the *de novo* lipogenesis were precursors for monoenic fatty acid chains.

Infusion of the medium-chain fatty acid [1,3–$^{13}C_2$]octanoate into rats shed light on novel aspects of the triacylglycerol synthesis in liver. In spectra obtained *in vivo*, the ^{13}C MR chemical shifts for the infused [1,3–$^{13}C_2$]octanoate indicated that all octanoate was bound as glycerolester (Pahl-Wostl and Seelig, 1987). This latter finding is in contrast with previous experimental data, which showed that medium-chain fatty acids do not serve directly as substrates for triacylglycerol formation. However, corroborative evidence was obtained by incubating rat liver slices with lipase, an enzyme that selectively cleaves the carbonyl ester linkage in triacylglycerols. After addition of lipase a decrease of the esterified [^{13}C]octanoate and a concomitant increase of free [^{13}C]octanoate could be observed by ^{13}C MRS *ex vivo*. The esterification of octanoate observed in liver *in vivo* was only transient, and it remains to be seen whether the ester bond is split in the liver or whether the ester is first exported from the liver tissue.

5.5.2 Glycogen, the Major Carbohydrate Store

Glucose and glycogen are the two most abundant saccharides detected in the liver by ^{13}C MRS (see Figure 3b). Carbon C1 of glycogen and of β- and α-glucose gives rise to three well-separated resonances at 100.4 ppm (resonance 17), 96.7 ppm (resonance 16), and 92.6 ppm (resonance 15), respectively. Resonances from C2–C5 are clustered together in the 65- to 80-ppm range, whereas the resonances arising from C6 are segregated in the 60- to 65-ppm region. The $\alpha(1 \rightarrow 4)$ linkage of glycosyl moieties in glycogen is responsible for the characteristic shift of the C1 resonance of glucose in glycogen by 7.8 ppm downfield from the corresponding C1 resonance of unbound α-glucose (Sillerud and Shulman, 1983). For the same reason, carbon C4 of glycogen resonates 7.46 ppm downfield from C4 in α-glucose. Glycogen in its natural form is branched with $\alpha(1 \rightarrow 6)$ bonds between chains of $\alpha(1 \rightarrow 4)$-linked D-glucose homopolymers. Spectra recorded from extracted liver glycogen showed a small shoulder on the C1 resonance at 99.22 ppm that was assigned to carbon C1 in $\alpha(1 \rightarrow 6)$-linked glucose moieties (Sillerud and Shulman, 1983). A similar peak was visible at 70.03 ppm for the C4 resonance of $\alpha(1 \rightarrow 6)$-linked glucose residues. Quantification of the relative contributions of signals from $\alpha(1 \rightarrow 6)$- and $\alpha(1 \rightarrow 4)$-linked C1 carbons indicated that only 7.4% of the glucose units were $\alpha(1 \rightarrow 6)$-linked.

The detection of sharp ^{13}C magnetic resonances for glycogen is somewhat unexpected because glycogen is a very large and highly branched polymer of glucopyranose residues reaching a molecular mass of up to 10^9 Da. For a rigid molecule of the size of glycogen particles, molecular tumbling is severely restricted, which would normally lead to very short T_2 relaxation, dramatically increased linewidths of the magnetic resonances, and ultimately to their disappearance from the high-resolution MR spectrum. In the course of experiments on the glucose/glycogen metabolism, the question has arisen as to whether the macromolecule glycogen is fully visible by ^{13}C MRS. Detailed investigations on the relaxation properties of glycogen *in situ* and *in vitro* were carried out giving credit to this latter issue and to the importance of accurate determination of signal areas for the quantification of metabolite concentrations.

The T_1 values for the carbons of glycogen were found to be strongly field dependent and values ranging from 65 ms at 2.1 T to 310 ms at 8.4 T have been reported (Sillerud and Shulman, 1983; Zang *et al.*, 1990). The T_2 relaxations were found to be less field dependent and values between 5 and 32 ms were measured. The NOE with a value of approximately 1.3 ($\eta + 1$) was fairly constant over the whole field range. The T_2 relaxation of carbons in glycogen is much longer than would be anticipated for a rigid molecule of this large size. This led Zang *et al.* (1990) to propose a model where internal motion of glucose moieties in the glucopyranose chains of glycogen is nearly unrestricted, thus permitting the observation of narrow ^{13}C resonances for glycogen. Because all relaxation parameters were found to be very similar *in vivo* and *in vitro*, this implies that there is little difference in motional properties of the glycogen in the two environments.

Many different approaches have been chosen to assess glycogen MR visibility and to test the possibility that only a fraction of more mobile glycosyl residues contribute to the MR signal, while an immobile portion constitutes an MR-invisible core. A pulse-chase experiment was devised where the relaxation properties of labeled glycosyl units could be followed as the glycogen particles increased in size (Shalwitz *et al.*, 1987). Fasted rats were initially infused with [1–^{13}C]glucose, which was subsequently replaced with natural abundance glucose for a further infusion period. This experimental design enabled the synthesis of glycogen with an inner core of ^{13}C-labeled glucose moieties, which was successively covered by layers of unlabeled residues. It was hypothesized that reduced mobility of the inner core as synthesis progressed would be accompanied by changed ^{13}C MR relaxation parameters (an increase in T_1 and decrease in T_2). Signal acquisition was performed with a repetition rate close to the expected T_1 relaxation of glycogen. Under this condition of partial saturation, it was expected that changes in T_1 would affect the signal intensity, and a decrease of T_2 would lead to an increased linewidth for the glycogen resonances. Neither of these anticipated effects was observed experimentally and hence it was concluded that glycogen visibility does not change during active glycogen synthesis.

The reverse approach was pursued by glycogen hydrolysis *in situ* in excised livers obtained from fed rats. Glycogen hydrolysis was induced by injections of glucagon shortly before the livers were removed from the animals. After excision the time course of glycogen and glucose C1 resonances was followed by ^{13}C MRS (Sillerud and Shulman, 1983). Similar experiments were performed *in vitro* with extracted liver glycogen from fed rats and rabbits where the glycogen was completely hydrolyzed with amyloglucosidase. In both cases it was found that the integrals of the glucose C1 resonances (α and β anomers) increased at the same rate as the glycogen C1 resonance decreased, and it was concluded that hepatic glycogen is fully MR visible *in vivo* and in solution.

Despite the large body of evidence indicating that glycogen is fully visible by ^{13}C MRS, contradictory results have been reported. Cohen (1983) estimated from measurements of metabolite fluxes in perfused livers that only 45% of the synthesized glycogen is MR visible. Further indications that glycogen is not fully MR visible were brought forward by analyses of myocardial glycogen (Brainard *et al.*, 1989a) (Section 5.6.5). More recent experiments on liver (Künnecke and Seelig, 1991) provided compelling evidence that glycogen visibility is dependent on the nutritional condition of the liver. This study deserves a more detailed description because a novel approach based on the analysis of homonuclear spin-coupling patterns was used. Fed and fasted rats were given a bolus of [1,2–^{13}C$_2$]glucose intraperitoneally and after 90 min the livers were excised and extracted for glycogen. The unique homonuclear spin coupling in the doubly labeled [1,2–^{13}C$_2$]glucose is maintained upon glucose incorporation into glycogen since the chemical composition of the glycosyl moieties remains unchanged. As a consequence, freshly synthesized glycogen (doublets in C1 and C2) can be readily distinguished from the preexisting core of natural abundance glycogen (singlets) due to the different spectral patterns. Figure 4 (left panels) shows typical spectral patterns for the C1 resonance of glycogen obtained from fed (Figure 4a) and fasted (Figure 4b) animals. Lorentzian-Gaussian resolution enhancement further clarified the spectral contributions of the central singlet and flanking doublet resonances (middle panels). The use of [1,2–^{13}C$_2$]glucose opens a new avenue to study glycogen visibility by comparing the intensities of doublet and singlet C1 resonances of intact glycogen (middle panels) with those of the corresponding glycogen hydrolysate (right panels). Provided that all glycosyl moieties in glycogen are MR visible, the intensity ratio between doubly labeled and monolabeled (natural abundance) glucose should remain constant. Conversely, a change in the

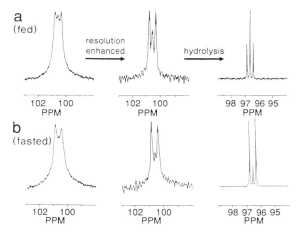

Figure 4. Comparison of the C1 ^{13}C magnetic resonance (at 100.2 MHz) of extracted hepatic glycogen before (left and middle panels) and after hydrolysis (right panel). Glycogen was obtained from (a) fed and (b) fasted rats after the metabolism of [1,2-^{13}C$_2$]glucose in the intact animal. The C1 resonance of extracted hepatic glycogen was analyzed before (left) and after (right) total hydrolysis of the glycogen by amyloglucosidase. The spectra in the middle panel depict the same glycogen resonances as shown on the left but are resolution enhanced by a Lorentzian-Gaussian transformation. In the case of glycogen obtained from fed animals, the intensity ratio between the outer doublet and central singlet decreases on hydrolysis, indicating that not all glycogen is ^{13}C MR visible. (Reproduced with permission from Künnecke and Seelig, 1991. Copyright © 1991 Elsevier Science Publishers BV.)

doublet-to-singlet ratio indicates that a fraction of glycogen is MR invisible and becomes visible upon breakdown of the polymer. Figure 4 reveals that such a change does indeed occur for glycogen isolated from the liver of well-fed rats (Figure 4a) but not for fasted rats (Figure 4b). A quantitative analysis of the spin-coupling patterns indicated that hepatic glycogen from fasted rats is fully MR visible, whereas glycogen of well-fed rats has a reduced visibility of 33%.

5.5.3 Regulation of Glycogen Metabolism

With the advent of ^{13}C MRS, it has not only been possible to monitor changes in glycogen concentration, but it also has become feasible to follow glycogen metabolism *in situ* and noninvasively (Alger *et al.*, 1981; Reo *et al.*, 1984; Siegfried *et al.*, 1985; Chen and Ackerman, 1989; David *et al.*, 1990; Künnecke and Seelig, 1991; Jehenson *et al.*, 1992; for a review, see Shalwitz and Becker, 1990). With a simple labeling experiment, the time course of

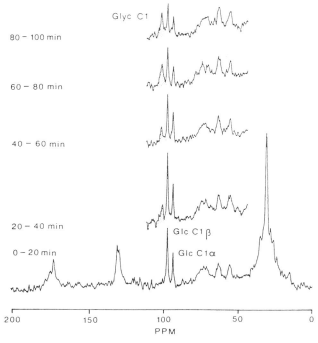

Figure 5. *In vivo* ^{13}C MR spectra (at 20 MHz) of the liver of a fed rat after intraperitoneal injection of [1–^{13}C]glucose. A single bolus of [1–^{13}C]glucose was injected intraperitoneally and its metabolism was followed *in situ* in the liver of an intact animal using a surface coil. Each spectrum represents 20 min of data accumulation. A 5-Hz line-broadening was applied before Fourier transformation. The resonance assignment is as follows: Glc, glucose; Glyc, glycogen. (Reproduced with permission from Künnecke and Seelig, 1991. Copyright © 1991 Elsevier Science Publishers BV.)

glucose incorporation into glycogen can be followed. Figure 5 displays a typical sequence of liver spectra acquired after the intraperitoneal injection of [1–^{13}C]glucose into a live intact rat. At the very beginning, the natural abundance signals of lipids and the resonances of glucose are visible. In later spectra, the glucose resonances decay and a new resonance is observed that can be attributed to C1 of freshly synthesized glycogen.

After ingestion of a larger dose of glucose, the liver changes from a glucose-releasing to a glucose-storing tissue. Among the controlling factors are the release of insulin and a concomitant suppression of glucagon secretion. The effect of these endocrine hormones was followed in the rat liver *in situ* in order to reevaluate the long-standing controversy concerning insulin's role in hepatic glycogen metabolism. The extent to which insulin is able to enhance hepatic glycogenesis *in vivo* has remained unclear. Reo *et al.* (1984) showed that injected glucagon, the major counterregulatory hormone of insulin, leads to a rapid depletion of the hepatic glycogen stores. A more detailed study on rats with somatostatin-suppressed insulin secretion revealed that after a bolus administration of [1-^{13}C]glucose, coinjected glucagon did not significantly change blood glucose concentration compared with control experiments without hormone injection (Siegfried *et al.*, 1985). Glucagon, however, caused a dose-dependent loss of [1-^{13}C]glycosyl units from glycogen and inhibited glucose incorporation into glycogen. In contrast, insulin did not alter glucose incorporation into glycogen. The data presented by Siegfried *et al.* (1985) are consistent with a model where glycogen synthesis increases linearly with hepatic glucose concentration above a threshold concentration. Insulin has no effect on this threshold or on the rate constant of glycogen synthesis.

Intracellularly, the transition from hepatic glycogenolysis to glycogen synthesis is preceded by deactivation of the enzyme glycogen phosphorylase (GP) and simultaneous activation of glycogen synthase (GS), a process that was thought to be well understood. However, recent *in vivo* studies have suggested that with physiological loads of glucose, liver GP may not be completely inactivated and GS not significantly activated, thus implying that futile cycling of glucose in and out of glycogen may occur during the net synthesis of glycogen. With a series of elegant pulse-chase labeling experiments, Shulman *et al.* (1988) and David *et al.* (1990) showed that futile cycling does indeed occur in perfused livers and livers *in situ* of fasted and fed rats. Animals were continually infused with [1-^{13}C]glucose, which was subsequently replaced with either natural abundance or [2-^{13}C]glucose. During the pulse phase with [1-^{13}C]glucose, the C1 resonance of glycogen grew linearly after a short lag period. After switching to the chase phase, the glycogen C1 signal due to the incorporated [1-^{13}C]glucose decayed monotonically. Although the chase experiment with natural abundance labeled glucose suggested that glycogen breakdown occurs within minutes after active synthesis, only the detection of a decaying C1 resonance in the presence of a growing glycogen C2 resonance due to the incorporation of the [2-^{13}C]glucose isotopomer could confirm that glycogen breakdown takes place simultaneously with further net synthesis of glycogen.

5.5.4 Metabolic Pathways for Glycogen Synthesis and the "Glucose Paradox"

The classical concept that ingested glucose is directly incorporated into hepatic glycogen (glucose → glucose-6-phosphate → glucose-1-phosphate → uridinediphosphateglucose → glycogen) has been challenged for more than a decade. It has been suggested that glycogen synthesis requires an active gluconeogenesis, implying that hepatic glycogen is mainly formed by an indirect pathway from C_3 compounds (glucose → → C_3 unit → → glucose-6-phosphate → → glycogen). This so-called "glucose paradox" has attracted a lot of interest. Tracer experiments with specifically labeled metabolic precursors have been an essential prerequisite for elucidating the contribution of the different pathways to glycogen synthesis (for a review, see Shulman and Landau, 1992).

Although the direct incorporation of glucose into glycogen is straightforward and leaves the initial labeling intact, rearrangements of the carbon skeleton of glucose occur in the course of the indirect pathways. Glucose will undergo glycolysis where it is split in several steps to dihydroxyacetone-3-phosphate and glyceraldehyde-3-phosphate. The two molecular species are in equilibrium and isomerize rapidly. Hence, upon labeling, glucose resynthesis from these metabolites results in glucose and glycogen with symmetrical labeling in C1–C3 and C6–C4. Furthermore, C_3 units (either produced intrahepatically or extrahepatically in the Cory cycle) entering gluconeogenesis via carboxylation to oxaloacetate will have their labels scrambled between C2 and C3 due to the symmetrical intermediate fumarate involved in the oxaloacetate ↔ malate ↔ fumarate equilibrium (see Section 5.5.8). The resulting glucose and ultimately the resulting glycogen thus have labels equally distributed between C1 and C2, and C5 and C6. Apart from the latter two indirect pathways, glucose may be recycled through the pentose phosphate pathway. From the point of view of label analysis, the major features of the PPP are the initial decarboxylation of glucose at C1 and the extensive carbon rearrangements in further steps of the pathway (see Section 5.5.7).

On the basis of the above-mentioned considerations, Shalwitz *et al.* (1989) assessed the relative contributions of the indirect and direct pathways in rat liver *in situ*. Fasted, halothane anesthetized rats were given intraduodenally a bolus of an equimolar mixture of glucose and alanine. Two experimental procedures were followed where either [1–^{13}C]glucose in combination with unlabeled alanine or [3–^{13}C]alanine with unlabeled glucose was administered. Glycogen synthesis was followed *in situ* in the surgically exposed liver (Figure 6). In a first series of experiments where [3–^{13}C]alanine was used, resonances from freshly synthesized glycogen labeled at carbons C1, C2 and C5, and C6 could be detected (Figure 6b). Hence it was concluded that despite the relatively large amount of simultaneously injected glucose potentially available for direct synthesis, a significant amount of glycogen was formed from the gluconeogenic precursor alanine. In the reciprocal experiment where [1–^{13}C]glucose was injected together with alanine, a very strong signal for C1 and less intense resonances for C2 and C5, and C6 of

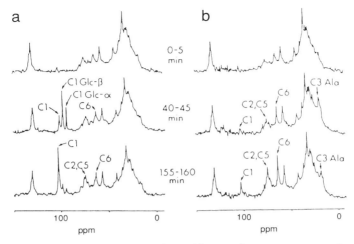

Figure 6. ^{13}C MR spectra (at 90.5 MHz) obtained from rat livers *in situ* representing the time course of hepatic metabolism of (a) [1–^{13}C]glucose and (b) [3–^{13}C]alanine. Representative spectra are shown for each protocol where the first spectrum (0 to 5 min) in each case depicts the background signal contribution before the duodenal injection of the substrates. Spectra were obtained with a surface coil positioned against the surgically exposed liver. Resonance assignments are as follows: Ala, alanine; Cx, glycogen; Glc, glucose. (Reproduced with permission from Shalwitz *et al.*, 1989. Copyright © 1989 The American Society for Biochemistry and Molecular Biology.)

glycogen were detected (Figure 6a). The increase in labeling at the C2, C5, and C6 positions of glycogen indicates recycling of glucose via gluconeogenesis in the indirect pathway. Similar data were obtained from baboon livers *in situ* (Jehenson *et al.*, 1992). Finally, a quantitative analysis of the rather small spectral changes provided estimates of 30% and 35% for the contribution of the indirect pathway to the synthesis of glycogen from labeled glucose (Shalwitz *et al.*, 1989; Jehenson *et al.*, 1992).

Complementary studies were carried out where hepatic glycogen was analyzed *in vitro* after hydrolysis to glucose (Kalderon *et al.*, 1986; Shulman *et al.*, 1985, 1987, 1991; Cline and Shulman, 1991; Künnecke and Seelig, 1991; Moore *et al.*, 1991). Shulman *et al.* (1985) assessed the influence of blood glucose concentration on the pathway selection of hepatic glycogen synthesis. Two different doses of [1–^{13}C]glucose, 1 and 6 g/kg body weight, were administered by gavage into awake fasted rats. Positional enrichments for all the carbons of glucose derived from the hydrolysis of hepatic glycogen as well as of glucose and lactate from blood samples of the portal vein were assessed by ^{13}C MRS. The authors deduced from the high and uniform enrichments at the secondary labeling sites C2, C5, and C6 of the glucosyl moieties in glycogen that only one-third of liver glycogen repletion occurred via the direct pathway and that only a very small amount of glycogen was formed by the conversion of triose-phosphates back to glucose and glycogen. The C_3 units (such as alanine and lactate) accounted for a minimum of 7% and 20% of the glycogen synthesized in the high and low dose groups, respectively. A similar dependency on blood glucose levels was detected in a study where glycogen synthesis was compared in hyperglycemic, euglycemic, and euglycemic/hyperinsulinemic conscious rats after a 24-hour fast (Shulman et al., 1991). The contribution of the direct pathway was clearly increased for the hyperglycemic rats only (41% compared to the two controls with 18% and 17%, respectively), hence excluding a significant effect of insulin on the selection of the synthetic pathway.

The three pathways—direct, indirect, and recycling at the triose-phosphate level—only accounted for a fraction of the total glycogen synthesized (Shulman et al., 1985). This finding led to the conclusion that either (1) different pathways that had not been examined contributed a sizable amount to glycogen synthesis or (2) that label dilution at the level of oxaloacetate was much higher than had been previously appreciated. Hence, experiments were performed where awake fasted rats and dogs were given a continuous intraduodenal infusion of [1-^{13}C]glucose to achieve metabolic and isotopic steady state (Shulman et al., 1987; Moore et al., 1991). The degree of label dilution in the first step of gluconeogenesis was estimated by comparing the enrichments of lactate (or alanine) C2 and C3 with those of the glutamate carbons C2 and C3, which reflected the enrichments in oxaloacetate under the prevailing conditions (Moore et al., 1991). Taking into account this dilution by a factor of 2.4, the calculated contribution of the indirect pathway for glycogen synthesis soars to more than 50% of the total glycogen production. In contrast, Kalderon et al. (1987) showed with similar measurements on liver extracts that in awake, fasted rats infused with [2-^{13}C]acetate only 7% of the freshly synthesized hepatic glycogen originated from labeled acetate. Since in higher animals acetate cannot serve as a precursor for gluconeogenesis the observed labeled glycogen arose from label exchange between oxaloacetate produced in the TCA cycle and oxaloacetate derived from carboxylation of pyruvate. The low contribution of oxaloacetate produced in the TCA cycle to gluconeogenesis supported previous findings by the same research group. In glucose-infused animals, 35% of the freshly synthesized glycogen originated from the direct incorporation of glucose, whereas gluconeogenic precursors accounted for the larger remainder without being significantly diluted by TCA cycle intermediates (Kalderon et al., 1986).

An alternative approach based on the analysis of homonuclear spin-coupling patterns was proposed to further elucidate the glucose paradox (Kalderon et al., 1986; Künnecke and Seelig, 1991). Anaesthetized well-fed and fasted rats were given intraperitoneally a bolus of $[1,2-^{13}C_2]$glucose (Künnecke and Seelig, 1991). The advantage of the contiguous doubly labeled substrate is that the characteristic spin coupling between carbons C1 and C2 distinguishes infused glucose from that present endogenously either as glycogen or unbound glucose. Furthermore, contributions from the direct, the indirect, and the PPP to the hepatic glycogen synthesis can be discriminated in a single ^{13}C MR spectrum. Glucose incorporated into glycogen via the direct pathway retains its spin coupling between the labels at carbons C1 and C2. In the PPP, however, the label at C1 of glucose is lost due to decarboxylation and hence no spin coupling can be detected. The fate of the second label is complex but it reappears at positions C1 and C3 of recycled glucose, notably giving rise to singlet resonances. Finally in the indirect pathways, the two labels appear simultaneously also at positions C5 and C6 with characteristic spin couplings for the corresponding resonances. Figure 7 shows typical spectral patterns of glucose obtained after hydrolysis of hepatic glycogen accrued in fed (Figure 7a) and fasted (Figure 7b) rats upon ingestion of $[1,2-^{13}C_2]$glucose. In both spectra natural abundance contributions of glucose give rise to singlet resonances. The singlet resonances of carbons C1 and C2 are flanked by doublets indicating glucose synthesis via the direct pathway. Closer inspection of the ^{13}C MR spectrum obtained from a fasted rat reveals additional doublets for the resonances of carbons C5 and C6 (see insets of Figure 7b) providing direct evidence for glycogen synthesis via the indirect pathway (either triose-phosphate recycling or gluconeogenesis via oxaloacetate). In contrast, no such line splittings were detected for well-fed rats. Further analysis of the spectral patterns also excluded a significant contribution of the PPP during glycogen synthesis in fasted and well-fed animals. The minimum relative flux through the indirect pathway was calculated from a comparison of the integrated intensities of the C1 and C6 doublets and was found to be 9% of the total glycogen synthesis in fasted rats. However, the actual contribution of the indirect pathway must be larger because of label dilution at the triose level and at oxaloacetate, the crosspoint of gluconeogenesis and the TCA cycle.

Values reported for the relative contributions of different pathways for glycogen repletion vary considerably, which is not surprising since data obtained from a wide variety of experimental conditions are compared with each other. Furthermore, some calculations give a minimum estimate, whereas others give a maximum contribution of the corresponding pathway. However, MR studies revealed that the relative contribution of the indirect pathway is dependent on the venous glucose concentration (Shulman *et al.*, 1985, 1991) and on the dietary state of the liver (Künnecke and Seelig, 1991). Measurements of hepatic uptake and release as well as net glycogen synthesis provided evidence that the relative contribution via the indirect

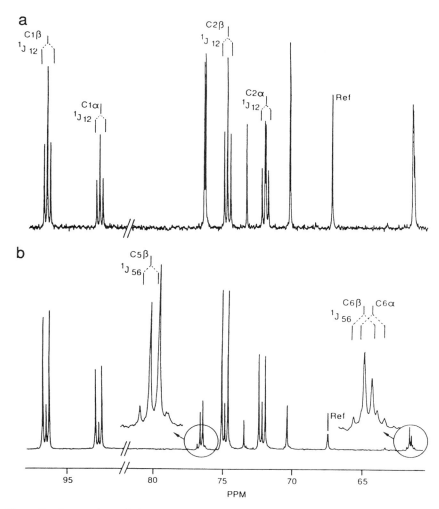

Figure 7. Proton-decoupled ^{13}C MR spectra (at 100.2 MHz) of extracted and enzymatically hydrolyzed liver glycogen obtained from (a) well-fed and (b) fasted rats after the metabolism of [1,2-^{13}C$_2$]glucose. Whereas singlet resonances are due to natural abundance glycosyl units, doublet resonances at C1 and C2 arise from the incorporation of contiguously labeled [1,2-^{13}C$_2$]glucose into glycogen. The insets in the lower panel depict additional homonuclear spin couplings for carbon C5 and C6 in the case of glycogen obtained from fasted animals. These latter spin couplings are evidence for the operation of an "indirect" synthesis of glycogen. (Reproduced with permission from Künnecke and Seelig, 1991. Copyright © 1991 Elsevier Science Publishers BV.)

pathway determines the total mass of glycogen synthesized (Moore et al., 1991). Furthermore, it has been suggested that anaesthesia as well as the way of glucose administration influences the relative contribution of different pathways during glycogen synthesis. On the other hand, metabolic zonation of the liver has almost no effect on the pathway selection for glycogen synthesis in periportal and perivenous hepatocytes (Cline and Shulman, 1991).

5.5.5 Gluconeogenesis

A large number of studies have been undertaken to analyze the gluconeogenic pathway and the involvement of the TCA cycle in synthetic reactions. Of great interest were the hormonal and dietary influences on gluconeogenesis. Cohen et al. (1979a, 1981b) investigated the gluconeogenesis from glycerol and the influence of thyroid hormones on glucose synthesis. Suspensions of primary rat hepatocytes and perfused mouse livers from starved donors were incubated with [2–^{13}C]- and [1,3–^{13}C$_2$]glycerol and label incorporation into glucose was followed. Continuous monitoring of the hepatic metabolism by *in vivo* ^{13}C MRS revealed that hepatocytes from experimentally hyperthyroid animals metabolized the substrate faster by a factor of 2 and had a concomitant faster increase of glucose resonances than in euthyroid controls. Furthermore, the level of glycerol-phosphate dropped to approximately one-third of the level detected in controls.

Glycerol directly feeds into the dihydroxyacetone-3-phosphate (DHAP) pool by activating glycerol to glycerol-phosphate followed by oxidation. DHAP isomerizes to glyceraldehyde-3-phosphate (GAP), leaving the original carbon skeleton and labeling of glycerol intact. Ultimately, a molecule each of GAP and DHAP are linked together in the gluconeogenic pathway to form glucose. If gluconeogenesis from glycerol were the only pathway for glucose synthesis, the labeling in C1 to C3 of glucose would be equal to the labeling of the glucose carbons C6 to C4. In contrast, the synthesis of unlabeled GAP from endogenous sources would dilute the label of the GAP pool, which, in turn, would lead to an asymmetric labeling in glucose unless the GAP ↔ DHAP isomerization step was rapid and close to equilibrium. However, equal labeling was measured experimentally for C3 and C4 of glucose after the incubation of liver tissue with [1,3–^{13}C$_2$]glycerol (Cohen et al., 1979a).

The detection of spin-coupling patterns for the glucose resonances C3 and C4 enabled the determination of label dilution due to glycolysis of unlabeled endogenous substrates (Cohen et al., 1979a; Cohen, 1984). Exclusive synthesis of glucose from [1,3–^{13}C$_2$]-labeled glycerol leads to [1,3,4,6–^{13}C$_4$]glucose with typical homonuclear spin couplings between carbons C3 and C4, which give rise to doublet resonances for C3 and C4. Conversely, label dilution with unlabeled GAP or DHAP leads, apart from other isotopomers, to [1,3-$^{13}C_2$]- and [4,6-$^{13}C_2$]glucose, both of which lack spin couplings for C3 and C4 and hence give rise to singlet resonances. From a quantitative analysis of the doublet-to-singlet ratios at glucose C3 and C4, Cohen *et al.* (1979a) estimated that 22% of the flux into glucose came from unlabeled precursors (compare with Section 5.5.4).

Gluconeogenesis from C_3 precursors such as alanine and pyruvate involves several more steps than the corresponding pathway for glycerol. One of the first reactions in this sequence is the carboxylation of pyruvate via pyruvate carboxylase to mitochondrial oxaloacetate, a product in common for the gluconeogenic pathway and the TCA cycle. Subsequently, oxaloacetate is decarboxylated via the phospho-*enol*-pyruvate carboxykinase reaction (PEPCK) to phospho-*enol*-pyruvate (PEP) from which the same reaction sequence is followed as discussed for glycerol. In the case of labeling experiments, further complications arise. Oxaloacetate is in equilibrium with malate and the symmetric metabolite fumarate. Label entering oxaloacetate at C1 and C2 is randomized to a certain degree to C3 and C4, respectively (see Section 5.5.8). Hence, PEP originating from [2-^{13}C]pyruvate or [3-^{13}C]alanine is labeled at C3 and C2 and, subsequently, glucose emerging from gluconeogenesis should have high ^{13}C enrichments at C1, C2, C5, and C6. This was borne out experimentally in perfused livers and isolated hepatocytes of hyperthyroid, diabetic, and normal animals (Cohen *et al.*, 1979a, 1981a; Cohen, 1983, 1987a,b,c). Glucose synthesis from labeled alanine and pyruvate could be readily observed (e.g., see Figure 8). Rat livers from diabetic donors perfused with [3-^{13}C]alanine or [2-^{13}C]pyruvate plus ammonia together with ethanol had a twofold and fourfold increase in the rate of gluconeogenesis compared with 24-h-fasted and 12-h-fasted normal controls (Cohen, 1987a). An enhanced rate of glucose production from alanine was also observed in hepatocytes of hyperthyroid animals (Cohen *et al.*, 1981a). This increase in the rate of gluconeogenesis is in agreement with the increased activities of hepatic PC, PEPCK, and glucose-6-phosphatase reported previously for hyperthyroid animals.

Cohen (1987b) investigated the effect of insulin on livers perfused with labeled alanine or pyruvate. Treatment of livers from fasted donors with insulin did not significantly affect gluconeogenesis and glycogenesis. However, transiently elevated levels of PEP and glycerol-3-phosphate, both labeled at positions C2 and C3, were seen in the presence of insulin. Insulin had a more pronounced effect on livers from diabetic rats, even partially reversing changes characteristic for the onset of diabetes. Insulin decreased the gluconeogenic rate and led to observable levels of aspartate and N-carbamoylaspartate similar to those detected in controls. However, the most striking effect was a large increase of the flux through the pyruvate kinase reaction.

Gluconeogenesis from alanine and pyruvate was also investigated in animals suffering from diseases other than diabetes and hyperthyroidism. Liu *et al.* (1990, 1991) reported on hepatic gluconeogenesis from [3-^{13}C]alanine in rats bearing tumors. Livers from fasted tumor-bearing hosts showed a more rapid production of [3-^{13}C]oxaloacetate than fasted controls as evidenced by the accumulation of the transamination product [3-^{13}C]aspartate. This augmented oxaloacetate production was attributed to an increased flux through PC in the presence of a distant tumor. Furthermore, the effect of parasite infection by *Plasmodium berghei* (Geoffrion *et al.*, 1985) and *Trypanosoma brucei rhodensiense* (Hall *et al.*, 1988) was assessed in isolated mouse livers perfused with [2-^{13}C]pyruvate and [3-^{13}C]alanine, respectively. *P. berghei* infection caused a marked reduction of hepatic gluconeogenesis with a concomitant lactate accumulation. Similarly, livers isolated from mice infected with *T. b. rhodensiense* formed substantially more labeled lactate from the substrate [3-^{13}C]alanine than livers from uninfected mice. In a variation of the experiments, the metabolism of [3-^{13}C]alanine was followed in isolated reperfused livers in order to monitor liver viability after cold preservation (Nedelec *et al.*, 1989, 1990a). Livers for organ transplantation are routinely stored in preservation solutions at low temperature. Nedelec *et al.* (1989, 1990a) assessed the efficiency of several preservation solutions for hypothermic storage of mouse livers up to 48 h. The amount of labeled glucose, glutamate, and glutamine per consumed [3-^{13}C]alanine was used as an index of metabolic capacity and liver viability.

5.5.6 Substrate Cycling/Futile Cycling

It is generally accepted that futile cycling, or the cycling of metabolites causing adenosine-5′-triphosphate (ATP) hydrolysis without changes in the corresponding reactants, frequently occurs in liver metabolism. An essential prerequisite for futile cycling is the availability of two opposing pathways where the forward and reverse reactions are catalyzed by different enzymes. Futile cycling of glucose in and out of glycogen during net synthesis of hepatic glycogen has already been dealt with in Section 5.5.3. The requirements for futile cycling are also fulfilled for the combined action of the enzymes PC, PEPCK, and pyruvate kinase (PK) in the sequence pyruvate → oxaloacetate → phospho-*enol*-pyruvate → pyruvate in hepatic gluconeogenesis. Cohen and coworkers (1981a; Cohen, 1987b) assessed the extent of futile cycling in hepatocytes in suspension and in perfused livers from hyperthyroid rats, from experimentally diabetic rats and from fasted controls. With [3-^{13}C]alanine as substrate, ^{13}C MRS allowed the simultaneous observations of the recycling of PEP to pyruvate and the flux of PEP to glucose. Briefly, [3-^{13}C]alanine is transaminated to [3-^{13}C]pyruvate, which enters the TCA cycle through the pyruvate carboxylase reaction. The resulting [3-^{13}C]oxaloacetate is scrambled in the malate/fumarate equilibrium (see Section 5.5.8) and hence gives rise to [2-^{13}C]- and [3-^{13}C]PEP and labeled glucose in the subsequent steps of gluconeogenesis. Alternatively, PEP is dephosphorylated to pyruvate in the PK reaction and, in turn, pyruvate is transaminated to alanine by the very active hepatic alanine aminotransferase. Thus, [2-^{13}C]-labeled alanine is detectable in the ^{13}C MR spectra among the original labeling pattern of the substrate [3-^{13}C]alanine. In the presence of a large pool of unscrambled [3-^{13}C]alanine, the steady-state enrichment at position C2 (and C1) of alanine directly reflects the activity of the PK flux. The labeling at C5 and C6 of glucose is a measure of the flux through the gluconeogenesis under this steady-state condition with a prevailing gluconeogenic activity.

Using the method outlined earlier, Cohen (Cohen et al., 1981a; Cohen, 1987b) estimated the PK flux in 24-h-fasted rats to be 74% of the gluconeogenic rate. For 12-h-fasted animals, this value increased to equal contributions of PK and gluconeogenesis. No PK flux could be detected for diabetic livers. However, on treatment of the latter livers with insulin, the PK activity largely increased to a value close to that of the 24-h-fasted controls. Measurements on suspensions of liver cells from 24-h-fasted hyperthyroid and euthyroid rats revealed a 60% and 25% contribution of the PK reaction, respectively, for the slightly different experimental conditions. Cosupply of ethanol to the hepatocytes of hyperthyroid rats reduced the PK flux to that measured in normal controls. Shulman et al. (1987) assessed the extent of pyruvate recycling for 24-h-fasted rats infused with [1-^{13}C]glucose. Based on input–output equations for the label at the C2 position of pyruvate under steady-state conditions, the flux through PK was deduced from the labeling detected for lactate C2 and glutamate C2 and C3 in liver extracts. Using this method, the PK flux relative to the total pyruvate utilization was found to be approximately 40% for fasted rats. Finally, the possibility that malic enzyme (ME) could contribute to the recycling of pyruvate was tested. Malic enzyme catalyzes the decarboxylation of malate directly to pyruvate. On the grounds of experiments with [3-^{13}C]alanine in the presence of the ME inhibitor 2,4-dihydroxybutyrate, ME activity could be excluded because no significant alteration of the labeling patterns was observed on addition of the inhibitor (Cohen et al., 1981a).

5.5.7 The Pentose Phosphate Pathway

Labeling experiments carried out with [2-^{13}C]glycerol on rat hepatocytes and perfused mouse livers revealed strong resonances for C2 and C5 of freshly synthesized glucose. However, in contrast to the prediction that the label of glycerol is equally incorporated into C1–C3 and C4–C6 (see Section

5.5.5), the extent of labeling at position C2 was always smaller than at C5 (Cohen et al., 1979a, 1981b). Similarly, with [1,3-$^{13}C_2$]glycerol as substrate, C1 of glucose was less enriched than C6. Furthermore, significant ^{13}C-enrichments were detected at the glucose positions C1 and C3 that were not directly traceable to the substrate [2-^{13}C]glycerol. These enrichments at secondary labeling sites and the concomitant loss of ^{13}C at the directly labeled C1 and C2 position without corresponding effects at C4 to C6 is an indication of glucose recycling through the complex reaction scheme of the PPP. The PPP fulfills the essential task of providing the cell with reduction equivalents and pentoses. Because no labeled pentoses were detected, it was concluded that the flux through the PPP maintained the level of the reduced form of nicotinamide adenosyl dinucleotide-phosphate (NADPH). Corroborative evidence was obtained from experiments where the NADPH oxidizing agent phenazine methosulfate was added together with labeled glycerol (Cohen et al., 1981b). A twofold increase of PPP activity was detected on the addition of phenazine methosulfate.

In the PPP, glucose is oxidatively decarboxylated at carbon C1 producing ribulose and reducing equivalents in the form of NADPH. In the second, nonoxidative part of the PPP, ribulose is converted back in a series of complex carbon rearrangements to glucose-6-phosphate and glyceraldehyde-3-phosphate. In the absence of pentose synthesis, six molecules of glucose are recycled to five hexoses. Quantitative analyses of the glucose labeling patterns provided estimates of 10% to 17% for the contribution of the glucose recycling through the PPP to the total gluconeogenic flux (Cohen et al., 1979a, 1981b). Similar results (15%) were obtained from experiments on mouse livers perfused with alanine and ethanol (Cohen et al., 1979b) and a negligible contribution of the PPP was reported for the livers of intact rats infused with glucose (Künnecke and Seelig, 1991).

In addition to the estimates on the flux through the PPP, the relative activities of transaldolase and transketolase reactions, which catalyze a complex reaction scheme of carbon rearrangements between C3 to C7 sugars in the nonoxidative part of the PPP, could be evaluated from the labeling patterns in hepatic glucose (Cohen et al., 1979a). With [2-^{13}C]glycerol as the substrate, one of three tentative reaction schemes for the PPP could be verified. An irreversible flux of pentoses formed in the oxidative part of the PPP to recycled hexoses would lead to glucose labeling at C1 and to a smaller extent at C3. A partial reversibility of the transketolase reactions, however, would decrease the contribution of the label at the C3 position. This latter labeling pattern was borne out experimentally and led to the conclusion that the transketolase reactions in the nonoxidative part of the PPP are reversible *in vivo*. Reversibility of both the transketolase and transaldolase could be excluded because such a reaction scheme would give rise to label mainly at the C1 position of glucose and only to a negligible contribution at C3.

5.5.8 Malate–Fumarate Equilibrium

Of particular importance for the accurate interpretation of metabolic fluxes through the TCA cycle is the understanding of the label scrambling occurring in the succinate/fumarate ↔ malate reaction. Succinate and fumarate are symmetrical molecules and if not bound to other cellular components, a simple molecular rotation relocates label from the upper half of the molecules to the lower half. The extent to which this kind of label redistribution occurs between chemically equivalent parts of succinate and fumarate has been assessed. Label following the normal TCA cycle sequence from α-ketoglutarate to succinate, fumarate, and malate in isolated rat liver mitochondria was found to be scrambled to the full extent in freshly synthesized malate (Bernhard and Tompa, 1990).

The reaction sequence oxaloacetate ↔ malate ↔ fumarate is nearly at equilibrium. Therefore, label entering the TCA cycle via oxaloacetate or malate can be randomized at the level of fumarate by a partial reversal of the flux through the TCA cycle. However, substantial label scrambling only occurs under conditions where the reverse fluxes in the equilibrium are large compared with the net forward flux through the TCA cycle. Cohen and coworkers (1979b, 1981a; Cohen, 1983, 1987c) addressed this issue for hepatic metabolism of alanine and pyruvate under various dietary and hormonal conditions. Label introduced by the substrates [^{13}C]alanine or [^{13}C]pyruvate was followed into glutamate, aspartate, and glucose. It was assumed that the ^{13}C distribution of intramitochondrial oxaloacetate is reflected in its transamination product aspartate. Alternatively, glutamate reveals the labeling in α-ketoglutarate, which, in turn, is a probe for oxaloacetate. On the other hand, it was anticipated that the labeling of glucose reflects that of cytosolic oxaloacetate (Cohen et al., 1981a). From the topology of the TCA cycle and gluconeogenesis, it can be deduced that the label distribution in C2 and C3 of glutamate and aspartate, and C5 and C6 of glucose, respectively, mirror the label arrangement in oxaloacetate carbons C2 and C3. The reported ratios of ^{13}C enrichments in glutamate C2 and aspartate C2 versus C3 range between two and close to one, indicating a high degree of randomization. A comparison between the labeling in C5/C6 of glucose and C3/C2 of glutamate revealed an even higher degree of label scrambling in glucose. This finding was explained by the additional label scrambling due to the activity of cytosolic fumarase (Cohen et al., 1979b; Cohen, 1987c). The degree of label randomization measured in glutamate gives a further clue to the flux rates through the mitochondrial malate dehydrogenase and fumarase exchange. The data presented by Cohen (1983, 1987c) suggest a recycling rate of approximately 1.3 to 1.5 times the net flux rate of the TCA cycle.

5.5.9 The Tricarboxylic Acid Cycle

Pyruvate is at the branch point of the catabolic sequence of oxidative glucose metabolism in the TCA cycle and the gluconeogenic pathway from C_3 precursors. In both pathways, pyruvate enters the TCA cycle where it is either oxidatively decarboxylated to acetyl-CoA via pyruvate dehydrogenase or carboxylated to oxaloacetate in the pyruvate carboxylase reaction. Since PC and PDH are competing for pyruvate, they are in key positions for the regulation of the anaplerotic (i.e., replenishing) and oxidative metabolism, respectively. Cohen and coworkers (1979b; Cohen, 1983, 1987a,b,c) used combinations of labeled and unlabeled alanine or pyruvate, and ethanol to investigate hepatic substrate competition for entry into the TCA cycle under various hormonal and dietary states of perfused rat and mouse livers.

When [3-^{13}C]alanine was the sole hepatic substrate, C2, C3, and C4 of glutamate and glutamine were readily labeled as expected from the topology of the TCA cycle (Cohen et al., 1979b; Cohen, 1987a). The [3-^{13}C]alanine is transaminated to pyruvate, which is either decarboxylated to [2-^{13}C]acetyl-CoA or carboxylated to [3-^{13}C]oxaloacetate. The [2-^{13}C]acetyl-CoA is incorporated into citrate and subsequently into [4-^{13}C]α-ketoglutarate, which, in turn, may be transaminated to [4-^{13}C]glutamate. On the other hand, [3-^{13}C]oxaloacetate is in equilibrium with malate and fumarate (see Section 5.5.8) and the label is scrambled into C2 and C3 of oxaloacetate. Condensation of such oxaloacetate isotopomers with acetyl-CoA in the citrate synthase reaction subsequently gives rise to [2-^{13}C]- and [3-^{13}C]α-ketoglutarate and hence to [2-^{13}C]- and [3-^{13}C]glutamate. The relative proportion of pyruvate entering the TCA cycle via PC and PDH can therefore be estimated by the relative enrichments in glutamate C2 and C4. Under the assumption of a first-order model where only one turn of the TCA cycle is considered and continuous recycling of label is neglected (see Section 5.3.7), Cohen (1983, 1987c) calculated ratios of PC versus PDH flux rate that ranged from 1.2 to 2.6 and 7.7 in livers from well-fed, diabetic and 24-h-fasted donor rats, respectively. In diabetic liver, the relative proportion of pyruvate channeled into the TCA cycle through the two alternative pathways was not altered by the *in vitro* incubation with insulin.

In contrast to experiments where [3-^{13}C]alanine was the sole substrate for hepatic metabolism, coaddition of unlabeled ethanol resulted in glutamate C4 and glutamine C4 essentially devoid of label (Cohen et al., 1979b). Ethanol, which is readily oxidized by alcohol dehydrogenase to acetaldehyde and then further to acetate, serves as an additional source for acetyl-CoA. The results indicated that acetyl-CoA derived from ethanol efficiently competes with acetyl-CoA entering the TCA cycle via PDH. Thus, in the presence of an alternative source of acetyl-CoA, alanine entered the TCA cycle almost exclusively via the PC reaction. Furthermore, cosupply of alanine with ethanol increased the rate of alanine consumption via the PC reaction by almost 50% compared with livers perfused with alanine alone. This result is not unexpected in the light of a PC reaction that is regulated by the level of acetyl-CoA. An elevated concentration of acetyl-CoA, as was available from the metabolism of ethanol, increased the flux through PC in order to enlarge the oxaloacetate pool for an accelerated consumption of acetyl-CoA in the TCA cycle. Similar experiments with labeled pyruvate + NH_4Cl and ethanol as substrates were performed and provided equivalent results (Cohen, 1987a,b). Furthermore, Nedelec et al. (1990b) used the model proposed by Cohen (1983) for assessing the effect of virus-induced myeloproliferative leukemia on liver metabolism. The [3–^{13}C]alanine metabolism in infected livers of fasted donors was found to be similar to fed controls. Livers of both groups had a reduced contribution of PC versus PDH to the flux of alanine into the TCA cycle when compared with fasted controls. The altered ratio of PC/PDH flux in leukemic fasted livers was explained by the drastically reduced availability of triglycerides and hence acetyl-CoA in these organs.

^{13}C MRS offers the unique possibility of following the metabolism of ethanol and alanine simultaneously (see Figure 8). In experiments where [3–^{13}C]alanine and [1,2–^{13}C$_2$]ethanol were used as carbon sources, acetyl-CoA with two different labeling patterns was produced (Cohen, 1987b). On entry of alanine into the TCA cycle via the PDH reaction, [2–^{13}C]acetyl-CoA is formed and [1,2–^{13}C$_2$]acetyl-CoA is derived from ethanol. In addition to labeling citrate, α-ketoglutarate, glutamate, and glutamine at positions C4 and C5, the incorporation of [1,2–^{13}C$_2$]acetyl-CoA gives rise to typical spin couplings between the neighboring carbons C4 and C5 of glutamate and glutamine, which are readily distinguishable from the singlet resonances generated by the incorporation of [3–^{13}C]alanine. Hence, a direct estimate for the competition between alanine and ethanol can be obtained from the analysis of the singlet-to-multiplet ratio in C4 of glutamate. The analysis of spin-coupling patterns was also useful for the assignment of resonances from glutathione observed in ^{13}C MR spectra obtained from perfused livers (Cohen, 1987a). The spin-coupling patterns detected for the glutamyl moiety in glutathione upon liver perfusion with ^{13}C-labeled precursors closely corresponded to the labeling pattern observed in glutamate. Thus, it was concluded that the carbon skeleton of glutamate remained intact and was directly incorporated into *de novo* synthesized glutathione. Chemical shifts measured in PCA extracts of livers indicated that glutathione is present in its reduced form.

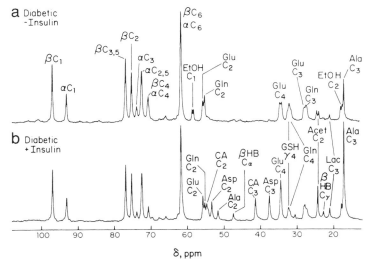

Figure 8. ^{13}C MR spectra (at 90.5 MHz) of isolated livers from diabetic rats. Livers were perfused with labeled alanine and labeled ethanol in the (a) absence or (b) presence of insulin. Spectrum (a) is taken from a series of spectra obtained from the liver of a diabetic rat and represents the accumulation between 170 and 180 min after the initial addition of 10 mM [3-^{13}C]alanine and 7.3 mM [1,2-^{13}C$_2$]ethanol. Spectrum (b) is part of a similar time sequence of spectra and was accumulated 170 to 180 min after the initial addition of 10 mM [3-^{13}C]alanine and 7.3 mM [2-^{13}C]ethanol. Spectra of the ^{13}C natural abundance background of these livers were accumulated under identical conditions before the addition of the labeled substrates. These background spectra were subtracted from the spectra shown. Resonance assignments are as follows: Ala, alanine; Acet, acetate; Asp, aspartate; βHB, 3-hydroxybutyrate; C, glucose; CA, N-carbamoylaspartate; EtOH, ethanol; Gln, glutamine; Glu, glutamate; GSH, glutathione. (Reprinted with permission from Cohen, 1987b. Copyright © 1987 American Chemical Society.)

Apart from glutamate, labeling of hepatic glutamine was observed in many experiments. This is not surprising because glutamine is readily formed in the glutamine synthase reaction. Cohen et al. (1979b) followed the time course of label incorporation into hepatic glutamate and glutamine in mouse livers perfused in the presence of [3-^{13}C]alanine and labeled or unlabeled ethanol. Under all conditions investigated, the enrichment in glutamine lagged behind that of glutamate as evidenced by the ratio C4/C2 of the ^{13}C MR signal intensities in glutamate and glutamine, respectively. The C4/C2 ratio observed in glutamine at any given time was not the same as the corresponding ratio in glutamate but instead reflected the glutamate ratio of about 1 h earlier. This time difference was used as a measure of the flux through the allosterically controlled glutamine synthase reaction in vivo. Similar results were obtained from rat livers perfused with [2-^{13}C]acetate (Desmoulin *et al.*, 1985; Canioni *et al.*, 1985). However, the interpretation of the labeling in glutamate and glutamine was carried one step further. The authors suggested that glutamate and glutamine might take part in a futile cycle, which comprises the activities of cytosolic glutamine synthase, mitochondrial glutaminase, and glutamate/glutamine exchange across the mitochondrial membranes.

In fasted livers perfused with [3–^{13}C]alanine and [2–^{13}C]ethanol or [2–^{13}C]pyruvate, ammonium chloride and [2–^{13}C]ethanol, aspartate and N-carbamoylaspartate labeled at C2 and C3 could be observed (Cohen, 1987a). The detection of labeled aspartate is not unexpected because it is readily formed by the transamination of the TCA cycle intermediate oxaloacetate. N-carbamoylaspartate, however, is produced in the liver in one of the first steps of the *de novo* pyrimidine nucleotide synthesis. N-carbamoylaspartate incorporates the intact aspartate moiety, thus leading to the observed labeling pattern. The further detection of ^{13}C label in the pathway's end products, such as uridine and its phosphorylation products, indicates that a net synthesis of N-carbamoylaspartate, as opposed to turnover, was observed in perfused livers (Cohen, 1987a).

5.5.10 β-Oxidation of Fatty Acids and Ketogenesis

Ketogenesis was observed in perfused livers and isolated hepatocytes under a variety of conditions including diabetes and a high supply of acetate or ethanol as hepatic fuels (Cohen *et al.*, 1979b, 1981a; Canioni *et al.*, 1985; Cohen, 1987a,b). Liver metabolism gave rise to the synthesis of 3-hydroxybutyrate and acetoacetate in these model systems. In contrast, Seelig and coworkers (Cross *et al.*, 1984; Pahl-Wostl and Seelig, 1986, 1987) investigated the β-oxidation and ketogenesis of short- and medium-chain fatty acids in rat liver *in situ*. In a first series of experiments where [1–^{13}C]-, [3–^{13}C]-, and [1,3–^{13}C$_2$]butyrate were chosen as metabolic probes, the influence of hormonal and dietary states on hepatic ketogenesis was assessed (Cross *et al.*, 1984; Pahl-Wostl and Seelig, 1986). Butyrate, like other short-chain fatty acids, does not require the carnitine acyl transferase system to enter the mitochondrial matrix, thus bypassing a major control element of the metabolism of long-chain fatty acids. Once in the mitochondria, butyrate effectively competes with long-chain fatty acids for β-oxidation. This was borne out experimentally. Figure 9 displays representative ^{13}C MR spectra of the liver of an intact, fasted rat before (Figure 9a) and during (Figure 9b) the intravenous infusion of [1,3–^{13}C$_2$]butyrate. In the difference spectrum (b − a, shown in Figure 9c), resonances from the infusate butyrate (resonances 1 and 13) and the labeled metabolites glutamate (resonances 2 and 4), glutamine (resonance 4), carbonate (resonance 8), 3-hydroxybutyrate (resonances

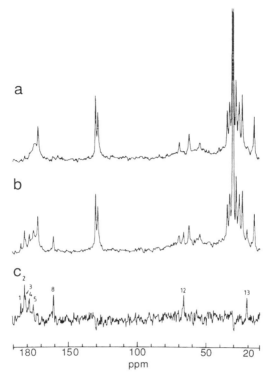

Figure 9. ^{13}C MR spectra (at 20 MHz) acquired *in situ* from the liver of a fasted rat (a) before and (b) during the infusion of [1,3-^{13}C$_2$]butyrate. Trace (c) displays the spectral differences between (a) and (b), which arise from the metabolism of [1,3-^{13}C$_2$]butyrate. Spectra were acquired over 40 min in the intact animal using a surface coil. An artificial line broadening of 4 Hz was applied before Fourier transformation. Resonances are as follows: 1, butyrate C1; 2, glutamate C5; 3, 3-hydroxybutyrate C1; 4, glutamine C5; 5, acetoacetate C1; 8, carbonate; 12, 3-hydroxybutyrate C3; 13, butyrate C3. (Reprinted with permission from Pahl-Wostl and Seelig 1986. Copyright © 1986 American Chemical Society.)

3 and 12), and acetoacetate (resonance 5) are clearly discernible. The [1-^{13}C]acetyl-CoA generated during the metabolism of butyrate may be channeled either into the TCA cycle or into the ketogenesis. The detection of [5-^{13}C]glutamate, [5-^{13}C]glutamine, and a strong resonance from labeled bicarbonate confirmed the operation of the TCA cycle. The presence of [1-^{13}C]acetoacetate and [1-^{13}C]3-hydroxybutyrate indicated active ketogenesis.

The same labeling experiments were repeated with diabetic rats. The recorded spectra showed a simpler spectral pattern compared with that of fasted rats. Well-resolved resonances of the substrate, bicarbonate, and 3-hydroxybutyrate were present, while the resonances of the amino acids and acetoacetate were missing. Instead, the resonances of bicarbonate, butyrate, and 3-hydroxybutyrate were distinctly enhanced in intensity. The spectral differences detected for animals under different metabolic conditions suggested that acetyl-CoA was predominantly channeled into the TCA cycle and into ketogenesis in the liver of fasted and diabetic rats, respectively. Furthermore, in diabetic rats where an increased amount of free fatty acids was available, butyrate accumulated to a higher extent due to an increased substrate competition with other fatty acids for β-oxidation. Similar to the assessment of butyrate metabolism in liver, experiments were carried out using the medium-chain fatty acid [1,3–$^{13}C_2$] octanoate as substrate (Pahl-Wostl and Seelig, 1987). The data obtained from these experiments essentially confirmed the findings derived from the metabolism of butyrate.

The simultaneous observation of 3-hydroxybutyrate and acetoacetate in fasted animals allowed an assessment of the mitochondrial redox state of the liver, which may be defined as the ratio [3–hydroxybutyrate]/[acetoacetate] (Pahl-Wostl and Seelig, 1986). A ratio of 1.4 was measured for fasted rats and a corresponding value of 10 for diabetic rats (the latter value was measured in spectra of liver extracts) which provided evidence for a more reduced state of the diabetic liver as compared to fasted liver. The differences in short-chain fatty acid metabolism between fasted and diabetic animals together with the changed mitochondrial redox state of the liver supports the previous view that the mitochondrial redox state of the liver is of primary regulatory importance. Thus, the overproduction of NADH and acetyl-CoA in diabetic rats due to disease-related accelerated β-oxidation is counteracted by the less oxidative ketogenesis. In addition, the oxidative TCA cycle is suppressed and further reduction equivalents are used up in an enhanced gluconeogenesis in the diabetic state (see also Section 5.5.5).

Labeling experiments with [1–^{13}C]- and [3–^{13}C]butyrate further clarified the pathway selection for 3-hydroxybutyrate synthesis in liver (Pahl-Wostl and Seelig, 1986). Hepatic 3-hydroxybutyrate is exclusively produced by reduction of acetoacetate with NADH. Acetoacetate, however, can be synthesized from butyrate by the activity of two different pathways. In the first, butyrate undergoes β-oxidation in which two molecules of acetyl-CoA are produced that are fed into a common pool of acetyl-CoA. With both [1–^{13}C]- and [3–^{13}C]butyrate, [1–^{13}C]acetyl-CoA is formed apart from natural abundance acetyl-CoA. Acetoacetate, which is synthesized by condensation of acetyl-CoA via the rather complex hydroxymethylglutaryl pathway, would hence be labeled equally at carbons C1 and C3 and the same labeling pattern would be expected for 3-hydroxybutyrate. In the second pathway, butyrate is transformed to acetoacetyl-CoA, which still retains the original labeling pattern. However, acetyl-CoA from the common pool is added in a following step and the resulting hydroxymethylglutaryl-CoA is split into acetoacetate and acetyl-CoA. In this mechanism carbons C1 and C2 of butyrate are exchanged against acetyl-CoA whereas carbons C3 and C4 are retained. Thus for [1-^{13}C]- and [3-^{13}C]butyrate as substrates, this pathway leads to an asymmetric label distribution in C1 and C3 of 3-hydroxybutyrate. Quantitative analysis of the C1/C3 ratios detected by ^{13}C MRS in 3-hydroxybutyrate for [1-^{13}C]- and [3-^{13}C]-labeled substrates led to the conclusion that 40% to 50% of 3-hydroxybutyrate is generated by direct conversion of labeled butyrate to acetoacetyl-CoA, thus omitting β-oxidation. Complementary experiments where hepatocytes in suspension were perfused with unlabeled 3-hydroxybutyrate and [3-^{13}C]alanine provided corroborative evidence for the operation of the direct pathway for the synthesis of 3-hydroxybutyrate in liver (Cohen et al., 1981a). Again, an asymmetrical labeling was detected for this metabolite.

5.5.11 ω-Oxidation and Dicarboxylic Fatty Acids

Apart from β-oxidation, fatty acids in the liver can undergo ω-oxidation, a process in which the methyl ends of monocarboxylic fatty acids are oxidized and 1,ω-dicarboxylic acids emerge. Elevated tissue levels and increased urinary excretion of medium- and short-chain dicarboxylic acids have been found in situations of impaired or stimulated fatty acid metabolism such as inherited β-oxidation defects, high fat diets, and diabetes. These findings led to the hypothesis that dicarboxylic acids formed by ω-oxidation could be subjected to bidirectional β-oxidation, a process that was suggested to be far more efficient than conventional monodirectional β-oxidation. Cerdan et al. (1988) addressed this aspect of liver metabolism with a study on the in situ metabolism of 1,ω-medium-chain dicarboxylic acids in the liver of intact rats. [1,12-^{13}C$_2$]- and [1,2,11,12-^{13}C$_4$]dodecanedioic acid was infused into well-fed, fasted, and diabetic rats, thus bypassing ω-oxidation of the corresponding monocarboxylic acids. ^{13}C MRS of the rat liver in situ revealed the accumulation of [1-^{13}C]- and [1,2-^{13}C$_2$]adipic acid, respectively. Furthermore, high-resolution spectra of liver extracts revealed the presence of labeled sebacic, suberic, and adipic acid, the products of subsequent steps of β-oxidation on dodecanedioic acid.

Homonuclear spin coupling introduced by adjacent ^{13}C labels in [1,2,11,12-^{13}C$_4$]dodecanedioic acid was retained in the products of subsequent β-oxidation steps, hence providing evidence that at least one carboxylic end of the initial substrate stayed intact during β-oxidation. Heteronuclear spin coupling between protons and ^{13}C as detected in the proton spectra of liver extracts offered a further means of distinguishing between two different pathways for β-oxidation of medium-chain dicarboxylic acids. In principle, ^{13}C-labeled adipic acid can be produced from [1,12-^{13}C$_2$] and [1,2,11,12-^{13}C$_4$]dodecanedioic acid by a monodirectional or bidirectional three-step β-oxidation mechanism. The monodirectional degradation pathway would involve three consecutive β-oxidation steps starting at position one and proceeding monodirectionally toward position ω. Conversely, the bidirectional β-oxidation mechanism would involve a three-step β-oxidation starting and progressing randomly from either of the carboxyl groups. Whereas a monodirectional β-oxidation would yield [^{13}C]adipic acid, the bidirectional pathway generates a 1:3 mixture of [^{13}C]- and unlabeled adipic acid. A quantitative analysis of the ^{13}C fractional enrichment in adipic acid using the satellite technique in proton spectra revealed that all adipic acid was labeled at C1 (Cerdan et al., 1988). Consequently, it was concluded that the β-oxidation of medium-chain dicarboxylic acids is monodirectional, which contrasts with the view held previously.

5.6 Metabolism in the Heart

Heart muscle is very similar in structure to skeletal muscles. However, the continuous although cyclic work performed by the heart requires a very efficient supply of energy. Myocardial adaptation to the increased energy requirement has led to a very high abundance of mitochondria, which can occupy up to 40% of the intracellular space of myocytes. In stark contrast to skeletal muscles and other tissues where glucose is among the preferred fuels, myocytes almost exclusively rely on the energy derived from the oxidation of fatty acids. Under normal conditions, glycolysis has a secondary function in the heart muscle and it is only further activated during metabolic emergencies such as ischemia and hypoxia.

^{13}C MRS of heart metabolism has evolved considerably over the last decade since the first publication about ^{13}C MRS tracer studies on the perfused heart (Bailey et al., 1981). Starting with the qualitative assignment of a few resonances, currently the instruments for assessing myocardial metabolism include complex isotopomer analyses for the quantitative evaluation of metabolic fluxes. Assessment of myocardial metabolism with ^{13}C MRS has primarily been carried out on isolated perfused heart. The ease with which the physiology of the isolated beating heart can be followed and controlled makes the perfused heart an attractive experimental system for metabolic investigations. Predominantly, the Langendorff method is adopted where the heart is retrograde perfused through the pulmonary artery and is working against a balloon placed in the left ventricle. In contrast, in situ ^{13}C MRS of the heart has been carried out on open-chest preparations where the heart is surgically exposed before being placed into a receiver coil.

5.6.1 Substrate Selection in the Myocardial Tricarboxylic Acid Cycle

Chance *et al.* (1983) assessed the metabolic fluxes through the TCA cycle in intact beating rat hearts. Hearts were perfused for different time intervals with either [2-^{13}C]acetate or [3-^{13}C]pyruvate, each in combination with unlabeled glucose. Perchloric acid extracts subsequently obtained from these hearts were analyzed with ^{13}C MRS for the positional enrichments of the individual carbons in glutamate, aspartate, and alanine. A mathematical model based on differential equations (see Section 5.3.4) was developed for the interpretation of the labeling kinetics in the myocardial TCA cycle under metabolic steady state (Chance *et al.*, 1983). The ^{13}C MR data together with analytical determinations of metabolite concentrations allowed the calculation of absolute flux rates of metabolites into and through the TCA cycle.

More recently, a mathematical model has been proposed that takes full advantage of the additional metabolic information provided by ^{13}C homonuclear spin couplings (Malloy *et al.*, 1988, 1990a). Based on isotopomer analysis (see Section 5.3.6) applied to key metabolites such as glutamate, aspartate, and alanine, the relative flux rates of metabolites into the TCA cycle were evaluated for metabolic and isotopic steady-state conditions. In particular, the competition of glycolytically and lipolytically derived acetyl-CoA for entry into the TCA cycle was explored in detail (Sherry *et al.*, 1985; Malloy *et al.*, 1987, 1988). Figure 10 shows representative ^{13}C MR spectral patterns of glutamate carbons C2 (left column), C4 (middle column), and C3 (right column) in perchloric acid extracts obtained from rat hearts perfused with (a) [2-^{13}C]acetate, (b) [2-^{13}C]acetate and [3-^{13}C]propionate, (c) [2-^{13}C]acetate and unlabeled pyruvate, (d) unlabeled acetate and [3-^{13}C]pyruvate, and (e) [1,2-^{13}C$_2$]acetate and [3-^{13}C]pyruvate. All resonances are of multiplet character and are, in fact, superpositions of the spectral patterns from different isotopomers. From a first qualitative comparison of the labeling patterns detected for the different substrates, it becomes evident that the observed ^{13}C MR spectra of glutamate are sensitive to the substrates available to the heart and to the position of the ^{13}C enrichment within these substrates.

Detailed isotopomer analyses provides additional quantitative information on the metabolism in perfused hearts. The combined fluxes from labeled acetate and labeled pyruvate into acetyl-CoA accounted for more than 95% of the total acetyl-CoA consumption in the TCA cycle of perfused hearts (Malloy *et al.*, 1988). Hence, approximately 5% of the acetyl-CoA was derived from unlabeled endogenous sources, presumably triglycerides. During

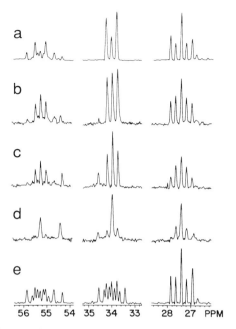

Figure 10. Proton-decoupled ^{13}C MR spectra (at 75.45 MHz) of glutamate in perchloric acid extracts obtained from perfused rat hearts. Spectra were obtained from hearts perfused with the following substrates: (a) [2-^{13}C]acetate; (b) [2-^{13}C]acetate and [3-^{13}C]propionate (5:2); (c) [2-^{13}C]acetate and pyruvate (1:1); (d) acetate and [3-^{13}C]pyruvate (1:1); (e) [1,2-^{13}C$_2$]acetate and [3-^{13}C]pyruvate (1:1). Only the expansions of the spectral regions containing the resonances from glutamate C2 (left column), C4 (middle column), and C3 (right column) are shown. Other metabolites were also detected: the trimethylamino resonance of carnitine and acetylcarnitine at 54.3 ppm, the C2 and C3 resonances of succinate at 34.6 ppm, and the C3 resonance of pyruvate at 26.9 ppm. (Reproduced with permission from Malloy *et al.*, 1988. Copyright © 1988 The American Society for Biochemistry and Molecular Biology.)

cosupply of acetate and pyruvate, acetate was clearly favored as the source for acetyl-CoA entering the TCA cycle with a contribution of 65% to 70% to the total acetyl-CoA supply in perfused rat hearts (Malloy *et al.*, 1988). Similar results were obtained for the competition of lactate and acetate (Sherry *et al.*, 1985). A complete inhibition of the influx of acetyl-CoA derived from glucose was reported for coperfusion with the short-chain fatty acid hexanoate (Weiss *et al.*, 1989a). Furthermore, glucose, even in the presence of insulin, did not compete significantly with lactate and to an even lesser extent with pyruvate and acetate. These findings were attributed to the inhibitory actions of elevated levels of citrate and acetyl-CoA. In pyruvate perfused hearts, a relatively high concentration of citrate was detected, which is known to lower the activity of the phosphofructokinase reaction and hence decrease the flux through glycolysis (Sherry et al., 1985). Similarly, β-oxidation of fatty acids provides increased amounts of acetyl-CoA, which, in turn, can act as an inhibitor of the PDH complex (Weiss et al., 1989a).

More recently, models for isotopomer analysis under non-steady-state conditions (metabolic and isotopic) were developed (Malloy et al., 1990b; Sherry et al., 1992) (see Section 5.3.6). These methods have allowed a rapid and easy quantification of substrate selection for entry as acetyl-CoA into the TCA cycle and have opened a new avenue for assessing dynamic metabolic changes. Malloy et al. (1990b) followed changes of substrate selection in perfused hearts that were offered a mixture of [1,2-$^{13}C_2$]acetate, [3-^{13}C]lactate, and natural abundance glucose. Only minor variations in the relative flux rates of the different substrates were detected for perfusion periods of 5 and 30 min. Similar labeling experiments were performed to assess transient metabolic changes in hearts after a short ischemic insult (Sherry et al., 1992). During the early period of recovery, acetate was clearly favored over other substrates (see Section 5.6.6).

5.6.2 Anaplerosis

Isotopomer analysis not only offers the possibility of simultaneously discriminating fluxes from differently ^{13}C-labeled precursors, but it has the potential to detect fluxes of unlabeled metabolites by their indirect influence on the label distribution. The ^{13}C labeling patterns detected in perchloric acid extracts obtained from hearts perfused with [^{13}C]acetate, [^{13}C]pyruvate, and [^{13}C]propionate provided evidence for the operation of anaplerotic reactions (Malloy et al., 1987, 1988). Anaplerosis, i.e., all the fluxes into the TCA cycle other than through the condensation of acetyl-CoA with oxaloacetate to citrate, has the function of replenishing TCA cycle intermediates that were used up in synthetic pathways. Depending on the substrate combination, the contribution of these anaplerotic pathways was found to range from as low as 5% in purely acetate-perfused hearts to 24% in the case of combined pyruvate and acetate perfusion. The origin of the carbon skeletons for anaplerosis was dependent on the substrate available in the perfusion medium. In hearts perfused with pyruvate and acetate, a significant amount of pyruvate was metabolized through PC to oxaloacetate. In contrast, hearts perfused with pyruvate as the sole carbon source, or with pyruvate together with glucose, derived most of the substrate for anaplerosis from metabolites other than pyruvate or glucose. These observations are in line with the known stimulation of PC by acetyl-CoA, which is available at increased amounts from acetate in hearts perfused with the latter substrate.

Sherry et al. (1988) and Malloy et al. (1988) used propionate in conjunction with ^{13}C-MR-based isotopomer analysis to further elucidate anaplerosis in perfused hearts. Propionate, which is not a usual substrate supplied to the heart by blood, is derived in myocardial tissue from the β-oxidation of odd-chain-length fatty acids and the degradation of certain amino acids. However, exogenously supplied propionate is readily metabolized in the propionate pathway, which involves carboxylation and carbon rearrangements to form succinyl-CoA, which ultimately enters the TCA cycle. Hearts perfused with acetate plus propionate showed an increased (up to 40%) contribution of anaplerosis to the TCA cycle for a 1:1 mixture of acetate and propionate as substrate. Furthermore, isotopomer analyses on extracts of hearts perfused with pyruvate and propionate provided evidence that approximately 27% of the metabolized pyruvate was derived from propionate via the synthesis of oxaloacetate and the subsequent decarboxylation by malic enzyme. This carbon flow was further substantiated by experiments where [3-^{13}C]propionate was the sole exogenous substrate. Under these conditions, all the propionate ultimately reentered the oxidative pathway as pyruvate.

An indirect flux parameter representing the extent of anaplerosis was calculated from the ratio of glutamate C2/C4 enrichment at isotopic steady state (Malloy *et al.*, 1987; Weiss *et al.*, 1989b; Lewandowski, 1991; Lewandowski and Hulbert, 1991). The topology of the TCA cycle predicts that the ratio C2/C4 of the enrichments in α-ketoglutarate approaches 1.0 for the metabolism of [2-^{13}C]acetyl-CoA under steady-state conditions. Any influx of unlabeled material via anaplerotic pathways reduces this value. Anaplerotic fluxes as low as 6% and 9% were measured for working hearts perfused with glucose and acetate, respectively, whereas a 39% contribution was detected for hearts perfused with acetate plus propionate (5:2 mixture) (Weiss *et al.*, 1989b; Lewandowski, 1991). The relatively low contribution of anaplerotic pathways in the acetate-perfused, working heart, and a high degree of anaplerosis in propionate-perfused hearts are in accordance with the 5% and 40% contributions derived from isotopomer analyses reported by Malloy *et al.* (1988) for rat hearts perfused under similar conditions.

5.6.3 Metabolic Consequences of Altered Workload

The heartbeat easily adapts to a wide range of output volumes and pressures according to the body's requirements for blood supply. Such dramatic alterations of the cardiac workload obviously demand effective adjustments in the cellular energy metabolism. Weiss *et al.* (1992) and Lewandowski (1992) addressed this issue by *in vivo* ^{13}C MRS on the intact perfused heart of rats and rabbits. Figure 11 shows a typical ^{13}C MR spectrum of an intact rat heart functioning at a high developed pressure. The spectrum was ac-

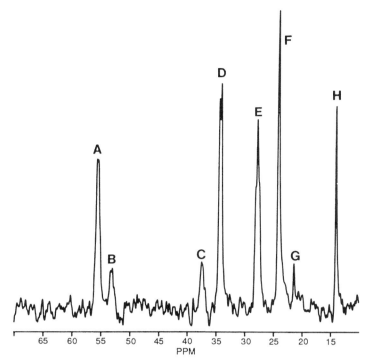

Figure 11. ^{13}C MR spectrum of an isolated rat heart functioning at a high developed pressure during perfusion with [2-^{13}C]acetate. The spectrum represents the acquisition over 5 min at 30 min after the substitution of natural abundance acetate with [2-^{13}C]acetate. A spectrum acquired over an equivalent time period before the introduction of [2-^{13}C]acetate has been subtracted to remove signals due to natural abundance contributions of ^{13}C. Detected resonances are: A, glutamate C2; B, aspartate C2; C, aspartate C3; D, glutamate C4; E, glutamate C3; F, acetate C2; G, acetylcarnitine C2; H, hexanoic acid C6, contained in an intraventricular balloon. (Reproduced with permission from Weiss *et al.*, 1992. Copyright © 1992 American Heart Association.)

quired during 5 min after the heart had been perfused with [2-^{13}C]acetate for 30 min. Resonances from the myocardial metabolites glutamate (resonances A, D, and E), aspartate (resonances B and C), and acetylcarnitine (resonance G) are clearly visible. To quantify TCA cycle activity from such *in vivo* spectra, an empirical flux parameter K_T was calculated from the metabolite pool sizes divided by the time difference Δt_{50} between the half maximal enrichment in glutamate C4 and glutamate C2 (Weiss *et al.*, 1992). The parameter K_T is based on the premise that the flux through the TCA cycle is inversely proportional to the time it takes for label to be scrambled from the primary labeling site at carbon C4 of α-ketoglutarate (and due to transamination in glutamate C4) to C2 and C3 in the latter compounds after recycling in the TCA cycle. Furthermore, the flux index K_T relies on the presumptions that the observed glutamate accurately reflects the labeling of intramitochondrial TCA metabolites, that the pool sizes of the TCA cycle intermediates remain constant, and that the contribution of anaplerotic fluxes is relatively small compared to the total flux through the TCA cycle. All three prerequisites were shown to be adequately fulfilled for KCl arrested hearts, hearts working at normal level, and hearts in an increased contractile state (Weiss et al., 1992; Lewandowski, 1992). The flux index K_T strongly correlated with myocardial oxygen consumption over a wide range of developed pressures. This finding suggests that the regulation of the respiratory activity in heart is mediated by changes in the rate of production of reducing equivalents in the TCA cycle (Weiss et al., 1992). Furthermore, the C2/C4 ratio in glutamate of KCl-arrested hearts was significantly lower than in working hearts. This led to the conclusion that in KCl-arrested hearts the relative contribution of anaplerotic fluxes increased to 32% of the total flux through the TCA cycle compared to a contribution as low as 9% in the working heart (Lewandowski, 1991). However, the simultaneously reduced flux through the TCA cycle in arrested hearts supports the interpretation that anaplerosis remains fairly constant for arrested and working hearts.

5.6.4 Ischemia, Hypoxia, and the Tricarboxylic Acid Cycle

Weiss et al. (1989b) analyzed the effects of reduced coronary flow on myocardial performance and metabolism. Figure 12 displays representative in vivo ^{13}C MR (middle trace) and ^{31}P MR (lower trace) spectral assessments of rat hearts perfused with [1-^{13}C]glucose at normal coronary flow (Figure 12a), at reduced flow (Figure 12b), and at severely reduced flow (Figure 12c). On reduction of the coronary flow, ^{13}C MR spectra indicated a lowered level of TCA cycle activity, as detected by a decreased labeling of glutamate and an increased labeling of lactate. These changes closely coincided with the depression of the contractile function (top trace) caused by the decreased coronary flow rates. In contrast, ^{31}P MR spectra obtained from the same hearts remained very constant over a wide range of coronary flows and only for severely restricted flows did the levels of high-energy phosphates drop significantly. Measurements of the elapsed time Δt_{50} for half-maximum enrichment in carbon C2 of glutamate significantly increased for hearts perfused with [1-^{13}C]glucose under reduced coronary flow (Weiss et al., 1989b). Furthermore, [3-^{13}C]lactate increased and the ratio of steady-state labeling in C2/C4 of glutamate fell. With Δt_{50} as an indicator for the meta-

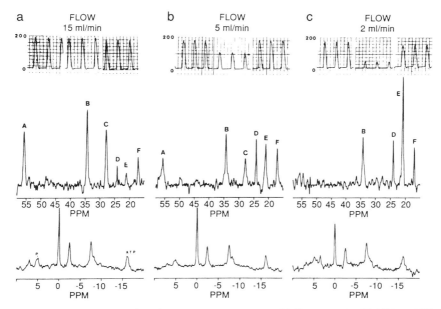

Figure 12. The mechanical performance (upper trace), ^{13}C MR spectra (at 90 MHz, middle trace) and ^{31}P MR spectra (lower trace) obtained from rat hearts perfused with [1-^{13}C]glucose under (a) normal, (b) reduced, and (c) severely restricted coronary flow. At the top, representative tracings of the left ventricular pressure (mmHg) before, during and after the experimental period are shown for each heart. The following resonances are discernible: A, glutamate C2; B, glutamate C4; C, glutamate C3; D, acetate C2 (contained in an intraventricular balloon); E, lactate C3; F, alanine C3. Resonances in the ^{31}P MR spectra are P$_i$, phosphate; PCr, phosphocreatine. (Reproduced with permission from Weiss et al., 1989b.)

bolic flux through the TCA cycle, the data attest to a decreased TCA cycle activity in hearts with decreased coronary flow. The relative contribution of anaplerosis, as indicated by the ratio of labeling in C2/C4 of glutamate, increased from 6% in hearts with normal coronary flow to 35% in corresponding hearts with severely restricted flow. Together, these results indicate a strong correlation between TCA flux and changes in cardiac work due to reduced coronal flow. The remarkably invariable levels of phosphocreatine (PCr) and ATP for a wide range of coronary flows and the correlation of TCA flow rate with the contractile function suggest that the heart was in a "hibernating" state where the myocardial workload was in balance with the reduced oxygen supply.

In vivo ^{13}C MRS of guinea pig hearts perfused with [3-^{13}C] pyruvate revealed the appearance of [2-^{13}C]succinate (and [3-^{13}C]succinate) on induction of anoxia by replacing oxygen with nitrogen (Brainard et al., 1986). Simultaneously, anoxia caused a decrease of labeled glutamate and aspartate. Reoxygenation reversed this effect and succinate disappeared. Similar results were obtained from experiments on rat hearts perfused with [2–^{13}C]acetate (Bernard et al., 1990; Bernard and Cozzone, 1992). During low-flow ischemia, labeled succinate was synthesized concurrently with a loss of labeled glutamate. Brainard et al. (1986) suggested that anaerobic metabolism of glutamate and aspartate occurred in the anoxic heart as a complementary pathway to glycolysis for yielding ATP without the participation of oxygen. Briefly, aspartate and glutamate are transaminated to their respective 2-oxoacids. α-Ketoglutarate enters the mitochondria and is converted to succinyl-CoA and succinate with the production of guanosine-5'-triphosphate (GTP) by succinyl-CoA synthase. Oxaloacetate, on the other hand, is reduced to malate which enters the mitochondria. Malate is reduced to fumarate and subsequently to succinate and ATP is formed. The occurrence of such an anaerobic pathway was substantiated with a further series of experiments in which the transaminase inhibitor cycloserine was added to the perfusion medium. Because the proposed pathways for the anaerobic metabolism of glutamate and aspartate involve transamination reactions, the addition of cycloserine should inhibit succinate synthesis during anoxia. This was borne out experimentally in guinea pig hearts perfused with [3–^{13}C]pyruvate (Brainard et al., 1986).

5.6.5 Ischemia, Hypoxia, and Glycogen Synthesis

Heart muscle can store excessive glucose as glycogen, which, in turn, provides an essential source of energy for maintaining myocardial performance during periods of reduced substrate and oxygen supplies. Neurohr et al. (1983, 1984) and Laughlin et al. (1988) followed myocardial glycogen synthesis in the intact heart in situ (open chest preparation) by infusing guinea pigs intravenously with [1–^{13}C]glucose and insulin. Due to the incorporation of [1–^{13}C]glucosyl moieties into glycogen, the glycogen resonance C1 grew linearly over a period of approximately 50 min and then levelled off. Similar experiments on hearts of diabetic and fasted rats showed that glycogen synthesis proceeded at a slower rate than in fed controls (Laughlin et al, 1990). Additional measurements of enzyme activities in tissue extracts indicated that glycogen synthase is rate limiting for glycogen synthesis in insulin-treated hearts (Laughlin et al., 1988).

Analogous to experiments on liver, glycogen turnover was assessed with a series of glucose pulse-chase experiments in situ on live guinea pigs and rats. Two different glucose isotopomers were infused for the pulse and the chase phase, respectively (Neurohr et al., 1984; Laughlin et al., 1988). Infusion of [1–^{13}C]glucose followed by the infusion of unlabeled glucose revealed that the glycogen resonance increased during the administration of labeled glucose and stayed constant during the chase phase. These *in vivo* ^{13}C MRS measurements suggest that in contrast to liver, myocardial glycogen phosphorylase is completely suppressed in the intact heart and hence no futile cycling of glucose in and out of glycogen occurs during active glycogen synthesis (Laughlin *et al.*, 1988). However, this finding is in apparent contrast to biochemical measurements of enzyme activities *in vitro*, which indicated a very high activity of the glycogen catabolizing enzyme glycogen phosphorylase (Laughlin *et al.*, 1988).

Similar pulse-chase experiments were carried out on the isolated hearts of guinea pigs (Brainard *et al.*, 1989a). Hearts perfused with [1–^{13}C]glucose and subsequently with [2–^{13}C]glucose (or vice versa) accumulated labeled glycogen. In contrast to the findings *in situ*, however, the integrated glycogen resonance due to the first substrate was found to decrease during the chase experiment. At the same time an increase of the linewidth of the glycogen C1 resonance was detected. The decrease of the glycogen resonance together with a concomitant increase in linewidth during net glycogen synthesis was explained by myocardial glycogen molecules, which lose their MR visibility while growing in size. These findings are at odds with the experiments carried out by Neurohr *et al.* (1984), which showed a constant intensity of the glycogen resonance during the chase phase. Furthermore, extracted myocardial glycogen and glycogen in excised myocardial tissue was shown by enzymatic degradation to glucose to be completely ^{13}C MR visible (Neurohr *et al.*, 1984; Garlick and Pritchard, 1993). In this context, a comparison of relaxation parameters of myocardial glycogen with those of hepatic glycogen might be of interest particularly since hepatic and myocardial glycogen have different higher structural organizations known as α- and β-particles, respectively. Despite this structural difference, T_1 and NOE values of myocardial glycogen *in vivo* and *in vitro* (at 1.9 and 8.5 T, respectively) were found to be very similar to the corresponding values obtained for liver glycogen (Neurohr *et al.*, 1984).

During episodes of hypoxia and ischemia, accumulated glycogen is readily mobilized (Neurohr *et al.*, 1983, 1984; Lavanchy *et al.*, 1984; Hoekenga *et al.*, 1988; Laughlin *et al.*, 1991). On previous labeling of glycogen, hypoxia and ischemia lead to the detection of labeled lactate as well as labeled glutamate originating from glycolysis and TCA cycle metabolism. Isolated guinea pig hearts initially perfused with [1–^{13}C]glucose and later perfused with [2–^{13}C]glucose released, on experimentally induced ischemia, first [2–^{13}C]lactate and later [3–^{13}C]lactate (Brainard *et al.*, 1989a). This finding demonstrated an ordered synthesis and subsequent ordered release of glycosyl moieties. In postischemic perfused hearts, continuing glycogenolysis was observed during the early stage of reperfusion (Kalil-Filho et al., 1991). Glycogenolysis at this stage was attributed to a persistent activation of glycogen phosphorylase by an elevated, albeit falling, phosphate level in the early postischemic heart. In the later phase of postischemia, glycogen synthesis is readily resumed and enhanced by activation of the glycogen synthase (Laughlin et al., 1991). In hearts from diabetic animals, this response was markedly blunted and only detectable in the presence of insulin. It was suggested that this impairment of rapid repletion of the glycogen stores might reduce the ability of the diabetic myocardium to withstand repeated hypoxic or ischemic stress.

5.6.6 Postischemic Recovery, Cardioprotectants, and Cardiotoxins

During severe myocardial ischemia such as that found in cardiac infarction, the oxidative metabolism rapidly ceases, levels of high-energy phosphates decline, and contractile failure of the heart ensues. Recovery of heart function during reperfusion after an ischemic insult not only depends on restoring the ATP level, but it has been suggested that a successful resumption of the TCA cycle activity is of vital importance. Labeling experiments with [3-^{13}C]pyruvate and [2-^{13}C]acetate on postischemic reperfused hearts with a remaining impaired contractility revealed a reduced flux through the TCA cycle of these hearts compared to nonischemic controls (Lewandowski and Johnston, 1990). This finding was put into perspective by recent work (Weiss et al., 1993). Postischemic perfused hearts with persistent contractile dysfunction were found to have a higher TCA flux rate than nonischemic control hearts that developed the same left ventricular pressure. This relative increase of the TCA flux indicated a higher energy demand for the same workload of hearts with contractile impairment.

In recent years, attention has been directed to possible therapeutic treatments that may aid postischemic recovery after an acute myocardial infarction. Substrate availability and selection are among the factors that have been suggested to have a major influence on the recuperation of postischemic contractility. Lewandowski and coworkers (Johnston and Lewandowski, 1991; Lewandowski et al., 1991a) compared postischemic recovery of the heart during perfusion with fatty acids and substrates not requiring β-oxidation. The metabolite flux through the TCA cycle and the contribution of anaplerotic fluxes were assessed by ^{13}C MRS. The dynamics of the label appearance at C4 and C2 of glutamate and the labeling ratio C2/C4 in glutamate, respectively, were used as flux indicators. Butyrate as well as palmitate were found to maintain a continued contractile dysfunction in the reperfused heart even though both substrates sustained full contractility in nonischemic hearts. In contrast, glucose and acetate provide an accelerated recovery of the contractile performance and support an improved respiratory efficiency of the heart metabolism. Malloy and coworkers (1990b; Sherry et al., 1992) addressed the issue of intrinsic substrate selection of postischemic hearts in the early phase of reperfusion. By means of isotopomer analysis under non-steady-state conditions (see Section 5.3.6), the relative contributions of different sources of acetyl-CoA to the metabolism in the myocardial TCA cycle were assessed. Postischemic rat hearts that had been given mixtures of [2-^{13}C]acetate/glucose or [1,2-^{13}C$_2$]acetate/glucose/[3-^{13}C]lactate clearly favored acetate as the primary fuel for the TCA cycle. This finding is consistent with an inactivation of the PDH complex, which is known to occur in ischemic tissue.

In addition to the influence of different metabolic fuels on myocardial metabolism during and after ischemia, the possible alleviatory effect of various drugs has been investigated. One of them, inosine, is a naturally occurring adenosine catabolite in heart metabolism. Inosine has been used to improve left ventricular function without a profound knowledge of its role in benefiting the stressed myocardium. To assess the effect of inosine, rabbit hearts were perfused with [1-^{13}C]glucose or [3-^{13}C]pyruvate each in combination with inosine. ^{13}C MRS revealed that the cardioprotective effect of inosine in hypoxia was due to an increased glycolytic activity (Lewandowski et al., 1991b). At the same time, inosine favored the transamination of pyruvate to alanine and hence, avoided the further accumulation of lactate during ischemia. The contrasting approach was followed by the treatment of myocardial infarction with a mixture of glucose, insulin and potassium (GIK) (Hoekenga et al., 1988). In experiments where [1-^{13}C]GIK or [1-^{13}C]glucose and insulin were administered to perfused guinea pig hearts, glycogen synthesis and labeling of glutamate could be readily observed during the preischemic period. On ischemia, glycogen decreased and labeled lactate and alanine were observed. GIK treatment resulted in a slower decrease of the glycogen store with no reduction in the glycolytic flux as inferred from [3-^{13}C]lactate accumulation. These results indicate that GIK had a glycogen-saving effect and at the same time enabled the ischemic heart to utilize exogenous glucose.

Chatham et al. (1990, 1992) compared the glucose metabolism in perfused hearts with adriamycin-induced heart failure and untreated controls. Adriamycin, a potent antineoplastic agent, causes a dose-dependent cardiomyopathy. While ^{31}P MRS detected only minor metabolic changes on treatment of hearts with adriamycin, ^{13}C MR spectra of the same hearts perfused with [1-^{13}C]glucose revealed decreased incorporation of label into lactate, alanine, and glutamate compared with hearts from untreated animals. At reduced coronary flow, the controls and the drug-treated hearts showed similar labeling of glutamate. However, labeling of lactate and alanine was significantly higher in the adriamycin-treated hearts. A decline of cardiac function was detected concomitantly with the reduced flow in treated and untreated hearts. Whereas in control hearts the cardiac performance was restored on reperfusion, cardiac function was irreversibly impaired in treated hearts. This finding indicated that adriamycin is more toxic under reduced flow conditions. The depressed level of ATP in ischemic adriamycin-treated hearts together with the fairly constant level of labeling in glutamate then suggested that adriamycin toxicity may be due to uncoupling of the oxidative energy metabolism from cardiac work.

5.6.7 Metabolic Compartmentation

The heart metabolism discussed so far is based on the assumption that the metabolism is divided into the mitochondrial and cytosolic contributions without further compartmentation of the metabolite pools. However, some evidence for metabolic heterogeneity of the pyruvate pool arose from labeling experiments where perfused rabbit hearts were incubated with [3-^{13}C]pyruvate (Lewandowski, 1992). The fractional enrichment found for glutamate C4 was much higher and that of [3-^{13}C]lactate was lower than the enrichment of the common precursor [3-^{13}C]pyruvate. This apparent discrepancy between individual fractional enrichments led to the conclusion that one pool of pyruvate within the cytosol of mainly exogenously supplied [3-^{13}C]pyruvate preferentially contributes to mitochondrial TCA cycle activity, while a second pool of mostly unlabeled endogenous pyruvate is used for lactate formation.

5.7 Metabolism in the Brain

The mammalian brain is a highly specialized organ for data processing. This specialization makes it potentially vulnerable to all sorts of exogenous influences. However, the skull protects the brain well against mechanical damage, and the blood-brain barrier with its impermeability for many metabolites is an effective asset for maintaining the brain in a constant chemical surrounding. Even though the brain is well shielded from its environment, its location in the protruding head leads to easy accessibility for *in situ* MRS. Cerebral metabolism has been studied extensively with ^{31}P and proton MRS. In contrast, ^{13}C MRS has received much less attention and only in the last few years have numerous studies been published that deal with carbon metabolism and its compartmentation in cerebral tissue.

Lajhta *et al.* (1959) provided first evidence of metabolic compartmentation in mammalian brain. Two classes of [^{14}C]-labeled substrates were identified, one of which led to an unaccountably higher enrichment of glutamine compared to that measured for its direct precursor glutamate. In subsequent

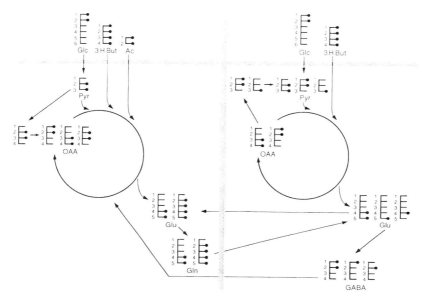

Figure 13. Scheme of cerebral metabolism of [1,2-$^{13}C_2$]glucose, [U-$^{13}C_4$]3-hydroxybutyrate, and [1,2-$^{13}C_2$]acetate. Two kinetically different TCA cycles exist in mammalian brain, located presumably in the glial cells (left) and the neurons (right). The glial compartment is characterized by the access of exogenous glucose, 3-hydroxybutyrate, and acetate to the TCA cycle, as well as by the activity of glutamine synthase and pyruvate carboxylase. The neuronal compartment lacks acetate metabolism and glutamine synthase activity, but synthesizes GABA and contains a pyruvate recycling system. Both compartments are interconnected by the fluxes of glutamate, glutamine, and GABA. Filled circles indicate ^{13}C labeling after 1.5 turns of the TCA cycles. Abbreviations: Glc, glucose; 3HBut, 3-hydroxybutyrate; Ac, acetate; Pyr, pyruvate; OAA, oxaloacetate; Glu, glutamate; Gln, glutamine. (Reproduced from Künnecke et al., 1993. Reprinted by permission of John Wiley & Sons, Ltd.)

radiolabeling experiments, this apparent inconsistency was resolved by the detection of two kinetically different glutamate pools, with only one of these pools being accessible to glutamine synthesis. The two pools of glutamate have been associated with two separate TCA cycles located in the two major cerebral constituents, glial cells and neurons, respectively. Neurons are responsible for the signal transmission, whereas glial cells have a variety of maintenance functions including insulation of axons (oligodendrocytes) and metabolic support (astrocytes).

A wealth of radio tracer studies—and more recently ^{13}C MRS assessments—has led to a fairly complex picture of the metabolism in cerebral tissue that is not adequately represented by the metabolic scheme given in Figure 2. Figure 13 outlines the basic features of the current understanding of the TCA cycle metabolism and compartmentation in brain, providing a framework for the following discussion of ^{13}C MRS experiments. The model considers two compartments that are presumably located in the glial cells (in particular astrocytes) (left) and in the neurons (right). Both compartments contain glycolytic and TCA cycle activity and are interconnected by exchange reactions at the level of the C4 and C5 units. The TCA cycles are fed by the influx of acetyl-CoA derived from exogenous substrates such as acetate, which enters the glial cells preferentially, and glucose or 3-hydroxybutyrate, which are thought to be metabolized by glial cells and neurons. The glial compartment is further characterized by the presence of a small pool of glutamate and the enzymatic activities of pyruvate carboxylase and glutamine synthase. The activity of glutamine synthase then leads to a relatively large pool of glutamine. The neurons, on the other hand, locate a large pool of glutamate from which 4-aminobutyrate (GABA) is derived by decarboxylation of glutamate at C1. Furthermore, the neurons contain a pyruvate recycling system and are virtually devoid of glutamine synthase activity.

Research on brain metabolism has concentrated on the issues of metabolic compartmentation *in vivo*, substrate selection, and metabolic interaction between glial cells and neurons. However, the inseparability of glial and neuronal tissue *in vivo* defies the direct assessment of glial and neuronal contributions to the metabolism. By taking advantage of the particular distribution of enzyme activities between glial cells and neurons, this problem can be resolved partially and has led to the adoption of the following qualitative notion for labeling experiments: Since cerebral glutamine is produced predominantly in the glial cells, its labeling pattern mainly reflects metabolic activity in this compartment. On the other hand, the labeling pattern of glutamate is due to a superposition of neuronal and glial metabolism. However, because the glial pool is much smaller than the neuronal pool, the observed labeling pattern of glutamate can be mainly attributed to the biochemical reactions in the neuronal TCA cycle.

5.7.1 Uptake and Metabolism of Cerebral Fuels *in Situ*

The brain almost exclusively uses glucose as its sole metabolic fuel. Because cerebral tissue is neither capable of synthesizing glucose nor has large glucose stores, the brain relies on a constant supply of glucose via the blood stream. Even small changes of the blood glucose level severely impede brain function. Glucose uptake and metabolism in brain was followed with *in vivo* ^{13}C MRS (Behar *et al.*, 1986; Fujiwara *et al.*, 1989; Hammer *et al.*, 1990; Künnecke *et al.*, 1993). Figure 14 shows spectra of a rabbit brain *in situ* prior (Figure 14a) and during (Figure 14b) the intravenous infusion of [1–^{13}C]glucose. The intense signals near 30, 129, and 180 ppm detected in both

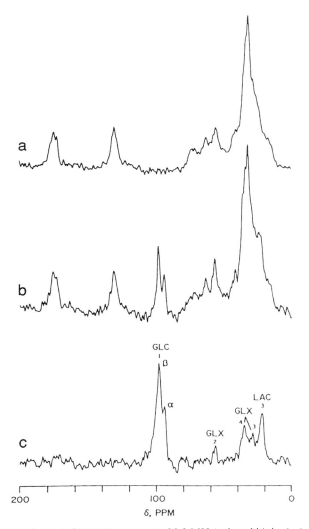

Figure 14. Proton-decoupled ^{13}C MR spectra (at 20.2 MHz) of a rabbit brain *in situ* (a) prior and (b) during the infusion of [1–^{13}C]glucose. Spectra of the brain were acquired in the intact animal using a surface coil placed on top of the skull. Spectrum (a) represents the spectral accumulation over 50 min. Spectrum (b) was accumulated for 25 min during a 39-min period of hypoxia following the infusion of [1–^{13}C]glucose. Spectrum (c) is the spectral difference between (b) and (a). Exponential line broadening of 14 Hz was applied to spectrum (c) before Fourier transformation. Resonances from labeled cerebral metabolites are as follows: GLC, glucose C1α and C1β; GLX (2, 3, 4), glutamate/glutamine C2, C3, and C4; LAC (3), lactate C3. (Reproduced with permission from Behar *et al.*, 1986. Copyright © 1986 Academic Press.)

spectra are from natural abundance ^{13}C in fatty acid chains. A comparison with spectra of excised brains and lipid extracts of brain tissue has shown that an appreciable portion of these signals arises from sufficiently mobile lipids (Bárány et al., 1985; Bárány and Venkatasubramanian, 1987) in brain tissue, whereas the rest is due to lipid deposits in the skin and bone marrow (Behar et al., 1986). Shortly after the start of the [1–^{13}C]glucose infusion, the resonances of carbon C1 of the two anomers of glucose became apparent and increased up to a steady-state level. Delayed with respect to the appearance of the glucose signals, further less intense resonances became detectable in the spectral region between 20 and 60 ppm. This is borne out in the difference spectrum (b − a) displayed in Figure 14c where three distinct resonances due to the labeling of glutamate/glutamine C2, C3, and C4 are discernible. The time course of the intensity of the glutamate/glutamine resonances indicated initial labeling at position C4 followed by C2 and C3, albeit at a lower enrichment (Behar et al., 1986; Fujiwara et al., 1989). During experimental hypoxia, an additional resonance from lactate C3 was observed. Together, these labeling patterns are characteristic for the generation of [3–^{13}C]pyruvate in glycolysis and its subsequent entry into the TCA cycle via PDH.

The anticipated linear sequence of cerebral glucose metabolism comprising uptake → phosphorylation → subsequent metabolism came under scrutiny when arteriovenous difference measurements of cerebral glucose uptake by radio tracer methods showed conflicting results. During periods in which the arterial radioactivity of ^{14}C fell rapidly, unlabeled glucose was apparently used in preference over the labeled glucose. The contrary behavior was observed during rapidly increasing specific activity of arterial glucose (Sacks et al., 1985). From these apparently inconsistent findings, it was concluded that cerebral tissue is capable of releasing glucose back into the blood stream. Mason et al. (1992) focused on this issue and measured the dependence of the cerebral glucose level on alterations of the blood glucose concentration. The glucose concentrations in brain *in situ* were measured by ^{13}C MRS during a steady-state infusion of variable amounts of [1–^{13}C]glucose. A symmetric Michaelis-Menten model, which allowed for mediated glucose influx and outflux, was proposed to correlate the glucose concentration in cerebral tissue with that measured in blood. Applied to the experimental data, this model predicted that the glucose influx into brain is 2.4-fold the basal glycolytic rate in cerebral tissue, indicating that ample glucose-6-phosphate is dephosphorylated and excreted back into the blood stream. Corroborative evidence was obtained from experiments with the glucose analog [6–^{13}C]2-deoxyglucose (Kotyk et al., 1989). Spectroscopic distinction of [6–^{13}C]2-deoxyglucose from its phosphorylated counterpart [6–^{13}C]2-deoxyglucose-

6-phosphate allowed simultaneous measurements of the two pools. A relatively rapid decay of the resonance from [6-^{13}C]2-deoxyglucose-6-phosphate was observed and ascribed to dephosphorylation of the otherwise barely metabolized 2-deoxyglucose.

During periods of reduced glucose availability, such as prolonged fasting, diabetes, or in the early postnatal phase, a significant portion of the brain's energy requirement can be met by the uptake and subsequent oxidation of ketone bodies in the cerebral tissue. The question as to how the brain tissue of well-fed animals can cope with such alternative fuels substituting glucose has been investigated with a series of experiments *in situ* in intact rats (Cerdan *et al.*, 1990; Künnecke *et al.*, 1993). During the infusion of [1,2-^{13}C$_2$]acetate or the physiologically more relevant [U-^{13}C$_4$]3-hydroxybutyrate, strong resonances from these substrates were detected. Smaller resonances from labeled bicarbonate and of labeled glutamate/glutamine were also discernible. The detection of these labeled metabolites demonstrates that ketone bodies are readily metabolized even in well-fed rats. This finding is in accordance with previous *in vitro* experiments, which showed that the enzymes necessary for the metabolism of ketone bodies are present and active in the brain tissue of well-fed, adult animals.

5.7.2 The Tricarboxylic Acid Cycle

The concept of compartmentation between glial cells and neurons as derived from ^{14}C tracer experiments has been further substantiated with labeling experiments that use synaptosomically enriched fractions (synaptosomes) of rat cerebrum (Petroff *et al.*, 1991), neuron-enriched fractions (Petroff *et al.*, 1993), and cultures of primary astrocytes and neurons (Portais *et al*, 1991; Sonnewald *et al.*, 1991, 1993). ^{13}C MR spectra of perchloric acid extracts of such primary cultures of neurons incubated with [1-^{13}C]glucose exhibited distinct peaks for glutamate and GABA, whereas PCA extracts of astrocytes showed resonances from glutamate and glutamine in the corresponding ^{13}C MR spectra. Analogous experiments with [2-^{13}C]acetate revealed strongly labeled glutamate and glutamine in astrocytes but virtually no label incorporation into neuronal metabolites. These results are consistent with the neuronal location of the GABA synthesizing enzyme glutamate decarboxylase and the location of glutamine synthase in astrocytes. Furthermore, the selective metabolism of acetate in astrocytes was confirmed.

Astrocytes were found to excrete labeled citrate and glutamine (Sonnewald *et al.*, 1991, 1993). Studies on cocultures of astrocytes and neurons showed that under these conditions [2-^{13}C]acetate labels GABA, suggesting that labeled citrate or/and glutamine excreted by the astrocytes was taken up by the neurons and metabolized to GABA (Sonnewald et al., 1993). The selective inhibition of glutamine synthase with methionine sulfoximine drastically reduced the labeling of GABA. Together with the experiments on cultures of neurons alone, this finding indicates that astrocytal glutamine is a major but not unique precursor for neuronal GABA synthesis.

Badar-Goffer et al. (1990) followed the time course of label incorporation from different ^{13}C isotopomers of glucose and acetate into superfused cerebral-cortical slices obtained from guinea pigs. ^{13}C MR spectra of PCA extracts of such brain slices incubated with [1–^{13}C]glucose showed strong labeling of glutamate C4 and GABA C2, whereas no label was observed in glutamine. It was also found that GABA C2 always had a higher enrichment with ^{13}C than its direct metabolic precursor glutamate C4. In contrast, [2–^{13}C]acetate labeled glutamine C4 to a higher extent than GABA C2 and glutamate C4, respectively. These experiments and analogous experiments carried out with [2–^{13}C]glucose and [1–^{13}C]acetate essentially confirmed labeling patterns reported from ^{14}C tracer studies and further support the compartmentation concept outlined earlier and depicted in Figure 13. [^{13}C]acetate, which preferentially enters glial cells, is metabolized in the glial TCA cycle to [^{13}C]α-ketoglutarate and subsequently to [^{13}C]glutamate and [^{13}C]glutamine. Due to exchange reactions between glial cells and neurons, some label is transferred into a neuronal metabolite pool whereby [^{13}C]GABA is formed after label dilution by unlabeled metabolites. On extraction, the labeled glial glutamate pool and the much larger unlabeled neuronal glutamate pool mix, resulting in a lower fractional enrichment for glutamate than is observed in glutamine. The reverse situation is encountered for the metabolism of glucose. [^{13}C]glucose, which is thought to be metabolized in glial cells and neurons, exclusively enters the neuronal TCA cycle under the prevailing conditions, leading to labeled glutamate and GABA. On extraction, the two glutamate pools mix with an ensuing reduction of the fractional enrichment of glutamate as compared to GABA. These metabolic interactions were found to be dramatically changed on depolarization of the cerebral-cortical slices with KCl (Badar-Goffer et al., 1992). First, steady-state labeling was attained faster than in the resting condition. Secondly, [^{13}C]glucose labeled glutamate, GABA, and also glutamine. Because no significant changes of metabolite pool sizes were detected, we can conclude that glucose metabolism by glial cells was selectively stimulated and that in general the TCA cycle activity in cerebral tissue was increased as a consequence.

In further labeling experiments, metabolic compartmentation and substrate selection were assessed quantitatively in the intact rat brain by taking advantage of the additional information provided by ^{13}C homonuclear spin couplings. Rats were infused intravenously with either [1,2–^{13}C$_2$]glucose,

[U-^{13}C$_4$]3-hydroxybutyrate, or [1,2-^{13}C$_2$]acetate until steady-state labeling was achieved (Cerdan et al., 1990; Künnecke et al., 1993). Figure 15 shows representative spectra of PCA extracts obtained from the brains of rats infused with saline (Figure 15a), [U-^{13}C$_4$]3-hydroxybutyrate (Figure 15b), and [1,2-^{13}C$_2$]glucose (Figure 15c), respectively. The majority of the resonances (see figure legend for resonance assignment) exhibit a multiplet character due to homonuclear spin couplings between neighboring ^{13}C nuclei. Closer inspection reveals that each multiplet is a superposition of the spectral contributions from various isotopomers. Careful analysis of the isotopomer distribution and the application of a computer model based on input–output equations (see Section 5.3.6) simulating the metabolic scheme shown in Figure 13 provided novel qualitative and quantitative information on cerebral metabolism (Künnecke et al., 1993). While the proposed mathematical model clearly distinguishes between glial and neuronal metabolism and distinguishes the two glutamate pools, the discussion presented next compromises on this aspect to allow for better comprehension. Hence, it is assumed that the labeling detected in glutamate is entirely due to metabolic activity in the neurons.

5.7.3 Substrate Selection in Neuronal and Glial Tricarboxylic Acid Cycles

Radio tracer experiments and more recently ^{13}C-MR experiments on cerebral tissue have revealed that potential substrates of brain tissue are not equally metabolized in glial cells and neurons. Therefore, cerebral substrates were qualitatively separated into two classes according to their preference to enter the neuronal or glial metabolism, respectively. Künnecke et al. (1993) quantitatively assessed this substrate selection in the intact brain for exogenously supplied glucose, 3-hydroxybutyrate, and acetate. Figure 16 depicts the C4 resonances of glutamate (left column) and glutamine (right column), and the C2 resonance of GABA (left column) in perchloric acid extracts of brains from intact rats infused with [1,2-^{13}C$_2$]glucose (Figure 16a), [U-^{13}C$_4$]3-hydroxybutyrate (Figure 16b), and [1,2-^{13}C$_2$]acetate (Figure 16c), respectively. Firstly, the structure of the multiplets detected for glutamine C4 and its precursor glutamate C4 (as well as the multiplet structures of carbons C1 to C3 and C5) differ considerably, thus supporting the multicompartmental metabolism mentioned previously. Secondly, the positional enrichments at C4 (and C5) of glutamate and glutamine, as determined by isotopomer analysis, delineate the preference of the substrates to enter either the neurons or glial cells, respectively. Glucose was preferentially used by the neurons (96% of the acetyl-CoA was derived from the infusate) and to a smaller extent by the glial cells (52%). For 3-hydroxybutyrate this preference was reversed (neurons, 48%; glial, 72%) and acetate entered the glial compartment almost exclusively (neurons, 5%; glial, 48%). Thus, the black-and-white distinction previously attained between substrates entering the glial and neuronal compartment, respectively, deserves further revision.

5.7.4 Pyruvate Recycling

Referring to the spectra obtained from PCA extracts of rats infused with multiply labeled glucose, 3-hydroxybutyrate, and acetate (see Figures 15 and 16), more information can be obtained on the metabolic differences between these cerebral substrates (Künnecke *et al.*, 1993). From the topology of the TCA cycle it follows that the outer doublet resonances in glutamate and glutamine C4 (and C5) and GABA C2 (and C1) are due to the incorporation of doubly labeled acetyl-CoA into C4 and C5 of α-ketoglutarate and, in turn, into glutamate, glutamine, and GABA. In contrast, the singlet resonances at glutamate C4 (and C5) and at GABA C2 (and C1) arise from the incorporation of singly labeled acetyl-CoA. Since these singlet resonances are clearly above the natural abundance level, they suggest an alternative pathway for the synthesis of monolabeled acetyl-CoA. The production of singly labeled acetyl-CoA can be explained by the combined activity of phospho-*enol*-pyruvate carboxykinase and pyruvate kinase or by malic enzyme. $[1,2-^{13}C_2]$- and $[3,4-^{13}C_2]$oxaloacetate or malate, produced in the first turn of the TCA cycle due to the incorporation of $[1,2-^{13}C_2]$acetyl-CoA, is decarboxylated to

---

Figure 15. Proton-decoupled ^{13}C MR spectra (at 100 MHz) of perchloric acid extracts obtained from brains of rats infused with (a) saline solution, (b) $[U-^{13}C_4]$3-hydroxybutyrate, and (c) $[1,2-^{13}C_2]$glucose. Multiplet patterns for most of the resonances in spectra (b) and (c) indicate significant enrichment of the metabolites and allow a deconvolution of different metabolic pathways contributing to the synthesis of the corresponding cerebral metabolites. Only the aliphatic and carboxylic regions of the spectra are shown. Resonance assignments are as follows: 1, alanine C3; 2, unassigned; 3, lactate C3; 4, 3-hydroxybutyrate C4; 5, NAA C6; 6, GABA C3; 7, glutamine C3; 8, glutamate C3; 9, glutamine C4; 10, glutamate C4; 11, GABA C2; 12, taurine C2; 13, aspartate C3; 14, unassigned; 15, creatine C2; 16, GABA C4; 17, NAA C3; 18, glycine C2; 19, 3-hydroxybutyrate C2; 20, taurine C1; 21, unassigned; 22, alanine C2; 23, aspartate C2; 24, NAA C2; 25, creatine C4, glycerophosphocholine (GPC) C5; 26, glutamine C2; 27, glutamate C2; 28, unassigned; 29, GPC C4; 30, unassigned; 31, 3-hydroxybutyrate C3; 32, unassigned; 33, lactate C2; 34, glucose C4; 35, inositol C4, C6; 36, glucose C2α, C5α; 37, inositol C2; 38, inositol C1, C3; 39, glucose C3α; 40, glucose C2β; 41, inositol C5; 42, glucose C3β; 43, glucose C5β; 44, NAA C5; 45, glutamine C1; 46, glutamate C1, aspartate C1; 47, aspartate C4, 48, glutamine C5; 49, NAA C4; 50, NAA C1; 51, unassigned; 52, 3-hydroxybutyrate C1; 53, glutamate C5; 54, GABA C1; 55, lactate C1. (Reproduced from Künnecke *et al.*, 1993. Reprinted by permission of John Wiley & Sons, Ltd.)

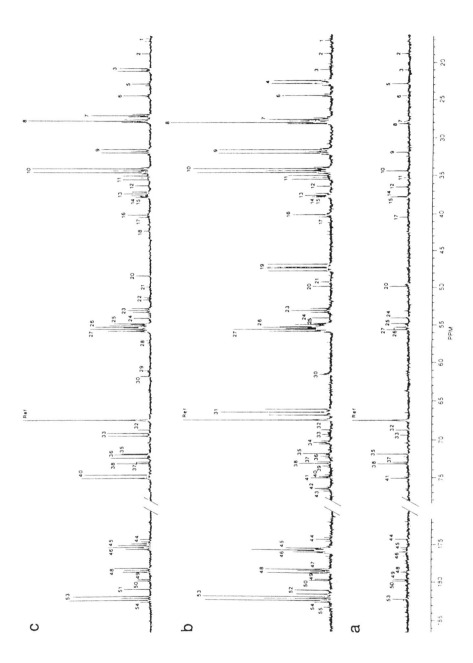

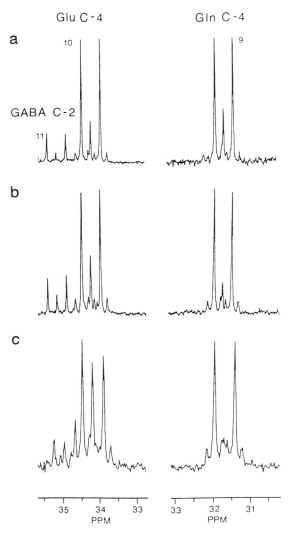

Figure 16. Comparison of the multiplet structures detected for ^{13}C MR resonances at glutamate C4 (left column) and glutamine C4 (right column) measured in extracts of brains obtained from rats infused with (a) [1,2-^{13}C$_2$]glucose, (b) [U-^{13}C$_4$]3-hydroxybutyrate, and (c) [1,2-^{13}C$_2$]acetate. The glutamate and glutamine resonances are scaled to the same height in order to visualize the gradual change of the doublet-to-singlet ratios. In addition, the GABA C2 resonance is displayed for a comparison with the multiplet structures of glutamate C4 and glutamine C4. (Reproduced from Künnecke *et al.*, 1993. Reprinted by permission of John Wiley & Sons, Ltd.)

pyruvate and on the action of PDH to [1–^{13}C]- and [2–^{13}C]acetyl-CoA. The operation of such a decarboxylation sequence is not unexpected because all the necessary enzymes are present in brain tissue, and it has been reported that malic enzyme primarily acts in the direction of decarboxylation.

A clear gradation of the ratio between the outer doublet and central singlet is observed for glutamate C4 in the sequence glucose < 3-hydroxybutyrate < acetate infusion. For the glutamine C4 resonances, the reverse order of the intensity ratios is observed. The predominant appearance of singlets in glutamate resonances leads to the conclusion that the pyruvate recycling system is located in the neuronal compartment with contributions to the total acetyl-CoA flux of 0%, 17%, and 25% in glucose, 3-hydroxybutyrate, and acetate-infused animals, respectively. The physiological relevance of such apparent futile cycling might be found in the supply of the neuronal TCA cycle with acetyl-CoA under conditions of reduced glucose availability.

5.7.5 4-Aminobutyrate Synthesis

The origin of the carbon skeleton for the synthesis of GABA has been subject to a long-standing controversy and it remains inconclusive whether GABA is synthesized from neuronal or glial precursors. Isotopomer analysis on brain extracts obtained after infusion of intact rats with [1,2–^{13}C$_2$]glucose, [U–^{13}C$_4$]3-hydroxybutyrate, or [1,2–^{13}C$_2$]acetate provides further quantitative insight into this matter. The isotopomer distribution in GABA for all three infusion conditions was closely related to glutamate (taking into account that GABA synthesis involves decarboxylation of glutamate at C1 and that carbons C2 to C5 of glutamate become C4 to C1 of GABA) but not to glutamine (see Figure 16) (Künnecke *et al.*, 1993). The strong resemblance of labeling patterns in glutamate and GABA indicates that the direct precursor of GABA is neuronal glutamate. Similar evidence was obtained from spin-coupling analyses of brain extracts from young rabbits infused with [U–^{13}C$_6$]glucose (Gopher *et al.*, 1990).

Isotopomer analysis has shown that the glial and neuronal TCA cycles are interconnected via fluxes of C5 units (presumably glutamine) into the neuronal compartment and of C4 units (presumably GABA) into the glial compartment. Together with the fact that GABA is labeled from acetate, which enters the glial cells exclusively, these latter findings suggest that a significant part of neuronal glutamate and, in turn, GABA is derived indirectly from glial precursors. These findings agree well with those obtained from experiments with brain slices and cocultures of primary neurons and glial cells, which have shown that glial glutamine is a nonexclusive precursor for GABA production (see Section 5.7.2). However, the findings outlined are in stark contrast to the conclusions Brainard *et al.* (1989b) inferred from glucose infusion experiments on intact rat brains. They concluded that GABA precursors are derived from a glutamate pool that is metabolically independent from the small or large glutamate pool in glial cells and neurons, respectively. Brainard *et al.* (1989b) then suggested that this independent glutamate pool would be replenished by a TCA cycle fed through PC activity.

5.7.6 Anaplerosis

Isotopomer analysis also provides quantitative information on the contribution of pyruvate carboxylase to the anaplerotic reactions in brain tissue. Figure 17 exemplifies the spectral deconvolution and isotopomer analysis of glutamate/glutamine C2 resonances for an extract obtained from a rat brain infused with [1,2-$^{13}C_2$]glucose. Incorporation of the two labels into the TCA cycle via PC produces [2,3-$^{13}C_2$]α-ketoglutarate isotopomers (Figure 17c), whereas the activity of PDH leads predominantly to [1,2-$^{13}C_2$] isotopomers (Figure 17d), and only after heavy cycling in the TCA cycle are [2,3-$^{13}C_2$] isotopomers produced. With glutamate and glutamine as probes for neuronal and glial TCA cycle activity, respectively, it was borne out that PC is operative in the glial compartment. A quantitative evaluation of the relative contributions of [2,3-$^{13}C_2$]- and [1,2-$^{13}C_2$]glutamine isotopomers revealed that PC contributed 38% to the total TCA flux in the glial compartment in anaesthetized rats (Künnecke *et al.*, 1993).

Similar experiments on rabbits provided comparable results for the PC activity in brain. A 33% contribution of PC was inferred from the labeling pattern in glutamine on cerebral metabolism of [U-$^{13}C_6$]glucose (Gopher and Lapidot, 1991). Experimental hyperammonia considerably increased the PC activity as well as the cerebral glutamine concentration. When glutamine synthesis was blocked by methionine sulfoximine, PC activity remained at the basal level and alanine accumulated instead of glutamine. Thus, the increased activity of PC was directly related to the increased synthesis of glu-

---

Figure 17. Spectral analysis of the ^{13}C MR multiplet patterns detected for glutamate C2 and glutamine C2 in a brain extract obtained from a rat infused with [1,2-$^{13}C_2$]glucose. Experimental spectrum (a) is deconvoluted into the contributions of different glutamate (left column) and glutamine (right column) isotopomers. (b) [2-^{13}C]-, (c) [2,3-$^{13}C_2$]-, and (d) [1,2-$^{13}C_2$]-isotopomers can be distinguished. The spectral contribution of creatine C4 and glycerophosphocholine C5 are shown in (b), and a simulated spectrum with the superposition of all detectable isotopomers in the appropriate ratios is depicted in (e). The increased relative contribution of [2,3-$^{13}C_2$]glutamine compared with [2,3-$^{13}C_2$]glutamate is indicative of glucose metabolism via pyruvate carboxylase in glial cells. (Reproduced from Künnecke *et al.*, 1993. Reprinted by permission of John Wiley & Sons, Ltd.)

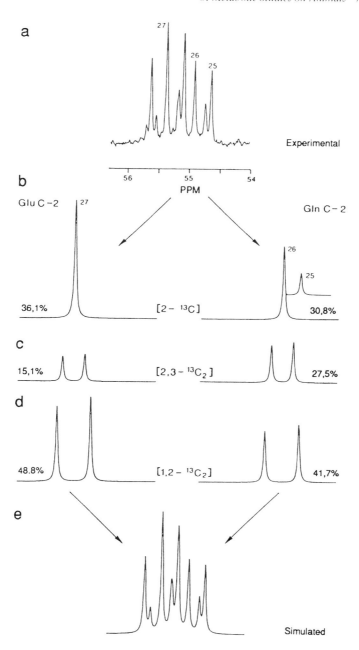

tamine in astrocytes, suggesting that the excess of ammonia was detoxified by glutamine synthase (Gopher and Lapidot, 1991).

5.7.7 N-Acetylaspartate Metabolism

N-acetylaspartate (NAA), or "bound aspartate" as first described by Tallan *et al.* (1956) several decades ago, is found almost exclusively in the nervous system in concentrations as large as 10 mM. NAA has gained renewed interest because of its excellent visibility by means of proton MRS and because of its potential use as an indicator for changes in brain physiology. Even though NAA is involved in neurological disorders such as Canavan's disease, its function and biosynthesis in brain are still elusive. To shed light on the synthesis of NAA in the adult brain, Künnecke *et al.* (1993) followed the label incorporation into NAA after infusing rats with [1,2-$^{13}C_2$]glucose, [U-$^{13}C_4$]3-hydroxybutyrate, or [1,2-$^{13}C_2$]acetate, respectively. It was found that under all infusion conditions the aspartyl moiety of NAA remained unlabeled, whereas the acetyl moiety was strongly labeled and showed a labeling pattern similar to that of the C1–C2 residue in neuronal glutamate. Hence, the acetyl part of NAA is in fast turnover with TCA cycle intermediates, and the aspartyl moiety is not in direct contact with glycolysis or TCA cycle intermediates. NAA has been suggested to be involved in the regulation of protein synthesis, in the production of myelin, and in the synthesis of neurotransmitters. It has also been suggested that NAA may serve as an acetyl carrier in the neuronal compartment. This latter proposal is supported by the peculiar labeling patterns observed for NAA.

5.7.8 Choline Metabolism

Choline is an essential precursor for the synthesis of the neurotransmitter acetylcholine and for the synthesis of phosphatidylcholine and sphingomyelin, the two most abundant phospholipids in brain. Tunggal *et al.* (1990) followed the biosynthesis of choline and its incorporation into phospholipids in rabbit brain. [N-$^{13}CH_3$]choline or [S-$^{13}CH_3$]methionine was injected subcutaneously and the label accumulation in cerebral tissue was monitored *in situ* by ^{13}C MRS. Myelogenesis in the central nervous system is the most intensive period of choline anabolism in the life span of the animals. This was borne out experimentally in newborn rabbits where labeled choline accumulated in the brain after administration of [N-$^{13}CH_3$]choline or of the methyl donor [S-$^{13}CH_3$]methionine, which is involved in choline synthesis. Adult animals treated in the same way showed only natural abundance choline. These findings suggested that during maturation of cerebral tissue the brain utilizes choline entering the brain for the synthesis of phospholipids.

Furthermore, the labeling of cerebral choline from methionine together with the low concentration of choline detected in blood led to the conclusion that *de novo* synthesis of choline occurred to a large extent in the cerebral tissue. Additionally, phospholipid turnover was assessed in the mature brain of animals that had received labeled choline during the myelination phase. After maturation, the supply of labeled choline was stopped and the distribution of the ^{13}C label was monitored during the next 24 days. The relatively fast decrease of the choline resonance over time indicated a remaining very active choline metabolism in adult brains (Tunggal *et al.*, 1990).

5.8 Metabolism in the Ocular Lens

^{13}C MRS in ophthalmic research has focused on assessing the metabolism of the ocular lens. Particular attention has been paid to the carbohydrate metabolism under elevated glucose levels created by metabolic diseases such as *diabetes mellitus*. Exposure of the crystalline lens to increased levels of glucose leads to a rapid opacification of the normally highly translucent lens and a so-called sugar cataract ensues. ^{13}C MRS provides a convenient means of following the metabolic interactions and changes during cataractogenesis in intact lenses. The avascular nature of the tissue allows the isolated lens to be kept viable in organ culture over a long period of time. Due to the noninvasive nature of MRS, repetitive measurements of the same tissue sample are thus within the realm of experiments that aim at determining the long-term effects of physiological changes.

5.8.1 The Polyol Pathway

The sorbitol pathway, an accessory pathway of the glucose metabolism, plays a fundamental role in sugar cataract formation in the ocular lens. Under conditions of high plasma glucose concentration, lenticular hexokinase becomes saturated and the excess glucose is converted into its sugar alcohol, sorbitol. However, the physiological function of this pathway as yet remains unclear. The sorbitol pathway comprises the action of two enzymes: Aldose reductase reduces glucose to sorbitol using NADPH as cofactor and, in the second reaction, polyol dehydrogenase oxidizes sorbitol to fructose by a simultaneous reduction of NAD$^+$ to NADH. Once formed, sorbitol is trapped intracellularly because its penetration through cell membranes and its metabolic degradation are severely restricted. Elevated plasma glucose levels foster the accumulation of sorbitol in the lens and, as a consequence, the cytoplasmic osmolarity rises, water is drawn into the lens fibers, and the lens swells. Eventually the lens fibers rupture, resulting in lenticular opacity.

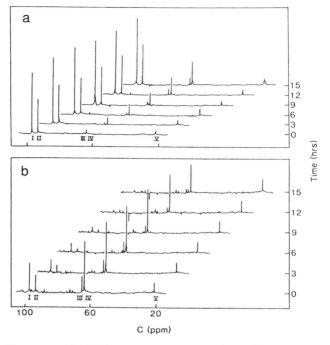

Figure 18. Time course of ^{13}C MR spectra (at 68 MHz) obtained from rat lenses (a) during incubation with a high concentration of [1–^{13}C]glucose and (b) after the removal of labeled glucose and in the absence of exogenous energy sources. Resonances I and II represent the beta and alpha anomers of glucose C1, respectively. Resonances III and IV are derived from fructose C1 and sorbitol C1, respectively, and V is due to lactate C3. (Reproduced with permission from Cheng et al., 1985. Copyright © 1985 Academic Press, Ltd., London.)

Glucose uptake and metabolism have been studied for a wide variety of conditions including normal and increased glucose levels and diabetes. Figure 18a displays a typical time course of hydrocarbon metabolism in rat lenses incubated in a culture medium containing a high concentration (35.5 mM) of [1–^{13}C]glucose. A linear increase in the labeled sorbitol (resonance IV) and fructose (resonance III) is observed over time (Cheng et al., 1985, Willis and Schleich, 1986). Resonances from the substrate (resonances I and II) and labeled lactate (resonance V) are also detectable. Similar experiments were performed with lenses from acute diabetic rats. Lenses excised 1 to 11 days after induction of diabetes showed an increased sorbitol accumulation, which coincided with the progress of the disease.

To assess sorbitol and fructose consumption, lenses incubated with [1–

^{13}C]glucose for 48 h were transferred to a medium containing no exogenous energy source (Figure 18b). During the first few hours, mainly intracellular glucose was utilized for maintaining the lens' energy requirement (Cheng *et al.*, 1985). After a lag period, first fructose and then sorbitol linearly decreased over time according to zero-order kinetics. The sorbitol and fructose turnover did not vary with the onset and progression of diabetes (Cheng *et al.*, 1989). Taken together, it can be concluded that the pronounced accumulation of sorbitol in lenses from diabetic animals is indicative of a more active aldose reductase in acute diabetes.

The deleterious effect of sorbitol accumulation on ocular lenses as a side effect of diabetes mellitus has evoked a quest for efficient aldose reductase inhibitors that could reduce sorbitol synthesis and thus prevent sugar cataracts. ^{13}C MRS has been used extensively to screen a wide variety of experimentally and commercially available drugs for their performance (Williams and Odom, 1986, 1987; Lerman and Moran, 1988). ^{13}C MRS was found to be advantageous over other methods since the influence of the aldolase reductase inhibitors on competing pathways such as glycolysis could be monitored simultaneously.

5.8.2 The Hexose Monophosphate Shunt

Up to one-third of the glucose turnover in the intact ocular lens was found to proceed via the sorbitol pathway under conditions of elevated plasma glucose levels (González *et al.*, 1984). Since the formation of sorbitol is tightly coupled to the oxidation of NADPH, this high metabolic activity of the polyol pathway influences the redox state of the ocular lens. NADPH is replenished via the activity of the hexose monophosphate shunt (HMPS or pentose phosphate pathway, PPP). Cheng *et al.* (1991) followed the activity of the HMPS with ^{13}C MRS. Ocular lenses were incubated with [2–^{13}C]glucose and the label distribution in lactate was monitored. [2–^{13}C]glucose metabolized by either the glycolysis or sorbitol pathway gives rise to [2–^{13}C]pyruvate and hence to [2–^{13}C]lactate. In contrast, glucose recycled through the HMPS has the ^{13}C label rearranged and mainly gives rise to [3–^{13}C]lactate. By analyzing the ratio of [3–^{13}C]-labeled lactate over the total amount of labeled lactate, a quantitative measure of the HMPS can be derived. Cheng and coworkers (1991) estimated that, at a glucose concentration of 20 mM, 7.1% of the consumed glucose in ocular lenses of rats was channelled through the HMPS.

Glutathione participates in a number of antioxidant reactions critical to normal lens function by protecting the lens from potentially deleterious peroxides. Willis and Schleich (1992) followed glutathione synthesis and homeostasis in ocular lenses. Intact rabbit lenses were incubated with

[3-^{13}C]cystein and the incorporation of label into the cysteinyl moiety of reduced glutathione was readily observed by a resonance appearing at 25.5 ppm. Treatment of the lenses with oxidizing agents decreased the content of reduced glutathione and concurrently the corresponding resonance of oxidized glutathione at 38.7 ppm grew. Similarly, a limited NADPH availability due to a competition of glutathione reductase and the polyol pathway for NADPH leads to a decreased oxidative resistance of the ocular lens because of a lowered glutathione reduction rate (Cheng and González, 1986; Cheng et al., 1989).

5.9 Metabolism in the Kidneys

Kidneys filter blood with the aim of removing obsolete catabolites from the blood stream for ultimate excretion in the urine. In a first step, a fluid resembling blood plasma is excreted into the proximate convoluted tubule (PCT). Cells lining these renal tubules provide the essential functions of reabsorbing vital metabolites and water from the tubular fluid for the later use in the body and for further secreting waste products into the tubular fluid. Tubular fluid is concentrated during its passage through the medulla, and the resulting highly hypertonic fluid, urine, is finally gathered in the papillary collecting ducts and transported to the bladder.

^{13}C MRS of kidney has been focused on the metabolism in PCT cells. Metabolic fluxes into and through the TCA cycle were assessed in a manner similar to that for labeling experiments on liver and heart. Predominantly, isolated PCT cells in suspension were used as the experimental system and after incubation with labeled substrates, perchloric acid extracts were prepared and analyzed by ^{13}C MRS.

5.9.1 Gluconeogenesis

In many experiments with PCT cells, glucose was found to be labeled to a significant extent, indicating active gluconeogenesis in these cells (Nissim et al., 1987, 1990; Jans and Willem, 1988, 1989, 1990, 1991; Jans and Leibfritz, 1989; Jans et al., 1989b). Jans and Willem (1988) followed glucose synthesis from [1,3-^{13}C$_2$]- and [2-^{13}C]glycerol. Measurements of positional enrichments revealed that carbons C1–C3 of freshly synthesized glucose were labeled to a lesser extent than the corresponding carbons C6–C4. Furthermore, enrichment was observed at carbon positions of glucose that were not directly traceable to the labeled substrate. These additional enrichments in glucose C1, C3, C4 and C6 from [2-^{13}C]glycerol and in C2 and C5 from [1,3-^{13}C$_2$]glycerol, respectively, were explained by the recycling of glucose through the PPP according to the metabolic scheme developed for the PPP in liver (see Section 5.5.7). A quantitative analysis of the label scrambling from [2-^{13}C]glycerol to glucose C1 and lactate C3 indicated a PPP activity of 10% of the gluconeogenic flux.

Elevated levels of ^{13}C were also detected in alanine and lactate at positions not directly traceable to the initial substrates [2-^{13}C]- and [3-^{13}C]alanine (Jans and Willem, 1989). These findings were interpreted as a partial recycling of pyruvate in the gluconeogenic pathway by the action of pyruvate kinase. Jans and Willem (1989) used a metabolic model based on infinite convergent series to calculate the relative fluxes through PK and the gluconeogenic pathway (see Section 5.3.5). A value of 0.78 was determined for the ratio of the PK flux to the gluconeogenic flux in rat renal cells. A very similar ratio was obtained from experiments where PCT cells were incubated with [3-^{13}C]aspartate (Jans and Willem, 1989). Alternatively, [3-^{13}C]citrate was used as the sole substrate for PCT cells and the label incorporation into alanine and glucose was followed (Jans et al., 1989b). Jans et al. (1989b) made an attempt to derive the relative flux rates through PK and the gluconeogenesis from the label distribution between alanine C2/C3 and glucose C2/C1 in rabbit PCT cells. However, this approach attracted heavy criticism from Katz (1989) who pointed out that the calculated labeling ratios are not a measure of the relative flux rates under the prevailing conditions. Further controversial data were obtained from the metabolism of [3-^{13}C]citrate. In contrast to the metabolism of labeled alanine and pyruvate, glucose synthesis from [3-^{13}C]citrate resulted in a ratio of 2.3 for the labeling at glucose C2/C1 (C5/C6) rather than the expected ratio of 1. Incorporation of [3-^{13}C]citrate into oxaloacetate, and ultimately into glucose, via succinate and fumarate, causes complete label scrambling between C2 and C3 of these acids. Hence, glucose should be labeled to equal extents at C1 and C2 (C6 and C5). Jans et al. (1989b) explained this apparent discrepancy by the operation of a citrate lyase reaction in PCT cells. Citrate lyase reverses the first step of the TCA cycle and produces [2-^{13}C]oxaloacetate from [3-^{13}C]citrate. On removal of [2-^{13}C]oxaloacetate from the TCA cycle by the PEPCK reaction, glucose with additional labeling at C2 and C5 is formed.

Gluconeogenic activity was also followed in PCT cells from animals with metabolic impairments. Streptozotocin-induced diabetes clearly increased the levels of labeled glucose on the metabolism of labeled alanine or aspartate (Jans and Willem, 1989). Metabolic modeling of the label distribution further revealed that streptozotocin reduced the ratio of PK-to-gluconeogenesis from 0.78 in controls to 0.33 in PCT cells obtained from diabetic animals. Together, these data suggest that the gluconeogenic activity in PCT cells is enhanced with diabetes. A similar increase of gluconeogenesis was observed for renal cells obtained from animals with chronic acidosis (Nissim et al.,

1990; Jans and Willem, 1991). By using a metabolic model based on infinite convergent series, Jans and Willem (1991) showed that the flux through the PEPCK reaction in rat PCT cells incubated with [2-^{13}C]- or [2,2'-^{13}C$_2$]succinate was increased in renal tissue from acidotic animals. Concomitantly, the flux of pyruvate through PDH was decreased from 65% to 44% of the amount of PEP formed in the gluconeogenic pathway. Together with the finding that the degree of labeling of lactate and alanine was not affected by acidosis, the data suggest that the gluconeogenic activity was more than doubled in acidotic PCT cells when compared to nonacidotic controls.

5.9.2 The Tricarboxylic Acid Cycle

The competition of pyruvate dehydrogenase and pyruvate carboxylase for the entry of pyruvate into the TCA cycle was evaluated by incubating rabbit PCT cells with [2-^{13}C]pyruvate, [3-^{13}C]pyruvate, [3-^{13}C]alanine or [1,3-^{13}C$_2$]glycerol (Jans and Leibfritz, 1989; Jans and Willem, 1988, 1989). Figure 19 shows a representative spectrum of a perchloric acid extract obtained from rabbit PCT cells incubated with [3-^{13}C]pyruvate. Apart from the dominant resonances of alanine, lactate, and glutamate, a multitude of less intense peaks from renal metabolites including glutamine, aspartate, citrate, malate, succinate, glucose, and glutathione can be observed. The relative fluxes of [3-^{13}C]pyruvate through PC and PDH were estimated from the ratio of labeling at glutamate C3 and C2 versus C4 according to a first-order model proposed for liver metabolism (Cohen, 1987b; see also Sections 5.3.7 and 5.5.9). However, ^{13}C-^{13}C spin couplings detected for glutamate C2 and C3 indicated that appreciable amounts of TCA intermediates were recycled over several turns of the TCA cycle. Therefore, a correction factor was applied in order to compensate for the fact that after two turns of the TCA cycle, label flux through PDH also gives rise to [2-^{13}C]- and [3-^{13}C]glutamate isotopomers. Using this correction procedure, Jans and Leibfritz (1989) calculated a ratio of 1:1.48 for the fluxes through PDH and PC, respectively. A similar result of PC/PDH = 1.4 was obtained by a complementary method where [2-^{13}C]pyruvate was used as substrate for PCT cells (Jans and Leibfritz, 1989). Due to the topology of the TCA cycle, label detected in glutamate C2 and C3 is solely derived from the influx of label via PC, whereas the labeling at glutamate C5 reflects the flux of pyruvate through PDH. The advantage of this labeling scheme is apparent insofar as the flux calculations do not require any correction for isotopomers common to both pathways. The ratio of enrichment at glutamate C5 over C2 and C3 gives a direct estimate of the PC versus PDH flux. The quantitative analysis of carboxylic and methylene groups, however, demands a prudent correction for saturation and NOE. Furthermore, neither of the two presented methods

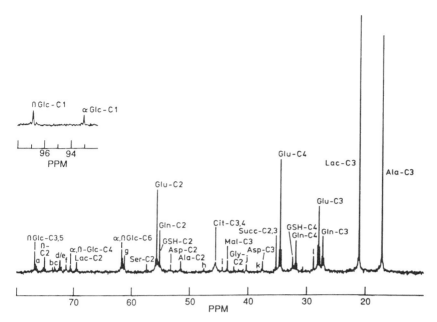

Figure 19. ^{13}C MR spectrum of a perchloric acid extract obtained from rabbit renal cells (proximal convoluted tubule) incubated with [3-^{13}C]pyruvate. The resonance assignment is as follows: a, citrate C2; Ala, alanine; Asp, aspartate; b, α-glucose C3; c, unassigned; Cit, citrate; d, α-glucose C2 and C5; e, glyceraldehyde-3-phosphate C3; f, malate C2; g, triglycerides C1, C3 and serine C3; Glc, glucose; Glu, glutamate; Gly, glycine; GSH, glutathione; h, 3-hydroxybutyrate C4; i, unassigned; j, cis-aconitate C4; k, glutarate C3; l, unassigned; Lac, lactate; Mal, malate; Ser, serine; Succ, succinate. (Reproduced from Jans and Leibfritz, 1989. Reprinted by permission of John Wiley & Sons, Ltd.)

accounts for label recycling in the TCA cycle of labeled substrate entering via the PC reaction, thus leading to an overestimation of the PC reaction.

A more sophisticated mathematical model based on the concept of infinite convergent series was used to interpret positional enrichments at isotopic steady state obtained from experiments on rabbit and rat PCT cells incubated with [2-^{13}C]alanine, [3-^{13}C]alanine, or [3-^{13}C]aspartate (Jans and Willem, 1989). A ratio of 1.92 was calculated for the relative fluxes through PC and PDH in rat PCT cells. Experimental diabetes increased this ratio to PC/PDH=2.27. These data together with an augmented gluconeogenic activity in renal tissue from diabetic animals indicate an absolute increase of the flux through PC (Jans and Willem, 1989). Similarly, chronic acidosis caused an increased flux through PC and a concomitantly decreased flux through PDH and citrate synthase as was evidenced by the label distribution detected in acidotic PCT cells incubated with [2,3-$^{13}C_2$]pyruvate (Nissim et al., 1990). Furthermore, acetyl-CoA and glucose accumulated while the levels of citrate, aspartate, glutamate, and glutamine decreased in the renal tissue of acidotic animals (Nissim et al., 1987, 1988, 1990). These findings suggest that the adaption of renal metabolism to chronic acidosis involves a reduced synthesis of α-ketoglutarate in the TCA cycle, thereby stimulating the degradation of glutamate and glutamine in order to replenish the partially depleted pool of α-ketoglutarate. The overall result is an augmented release of ammonia by the deamination of glutamate and glutamine, and an increased flux of glutamate and glutamine carbons into the gluconeogenesis. The released ammonia, in turn, is used to counteract the renal acidification by the excessive filtration of H^+ into the urine. Corroborative evidence for this metabolic adaptation was obtained from measurements on the utilization of glutamate carbons (Nissim et al., 1987). In control tissue, 44% of the initally applied [3-^{13}C]glutamate was recovered in glutamate and in glutamate-related metabolites such as glutamine and GABA. In PCT cells of acidotic animals, a significantly lower recovery of only 11% was recorded, however, indicating metabolic consumption of glutamate rather than recycling of the carbon skeleton. Furthermore, on the metabolism of [2,3-$^{13}C_2$]pyruvate in PCT cells, chronic acidosis led to aspartate with a labeling at C2/C3 reaching a ratio of 1:2 as opposed to 1:1 in controls (Nissim et al., 1990). The cause for this observation was not immediately apparent and hence metabolic compartmentation of intramitochondrial oxaloacetate was considered. In the same experiments the synthesis of the ketone bodies [^{13}C]3-hydroxybutyrate and [^{13}C]acetoacetate could be observed. Acidosis provoked a shift from a predominant production of 3-hydroxybutyrate to an accumulation of acetoacetate. This finding was attributed to a higher $NAD^+/NADH$ ratio in renal cortical mitochondria of chronically acidotic animals.

Jans and coworkers (1989b; Jans and Willem, 1990, 1991) assessed the extent of metabolite fluxes leaving (as opposed to entering) the TCA cycle at various branch points of the metabolism. With a merely qualitative approach, metabolic cycling through the various metabolic routes of the TCA cycle and ancillary pathways in rabbit PTC cells was elucidated by using [U-$^{13}C_4$]succinate as substrate (Jans and Willem, 1990). Particular advantage was taken of the homonuclear ^{13}C spin couplings caused by the multiply labeled substrate. Spectra of perchloric acid extracts of the cells showed multiplets for the resonances of all observed metabolites, attesting to the direct relationship with the substrate. Furthermore, doublet resonances detected for glutamate and glutamine C5 indicated that a part of malate or oxaloacetate that left the TCA by the operation of malic enzyme or PEPCK gave rise to pyruvate, which, in turn, reentered the TCA cycle via PDH. Jans and Willem (1991) quantified the metabolite efflux at the level of oxaloacetate by using [2-^{13}C]- and [2,2'-$^{13}C_2$]succinate as substrates. A fraction of 41% of the total oxaloacetate synthesized in the TCA cycle was found to leave the TCA cycle. In PCT cells of chronic acidotic animals this efflux increased to 57%. Alternatively, the consumption of α-ketoglutarate due to the synthesis of glutamate via glutamate dehydrogenase was estimated. Jans and Willem (1991) calculated that 27% and 39% of the flux through the TCA cycle accounted for the synthesis of glutamate in controls and acidotic PCT cells, respectively. Similar experiments were also performed with [3–^{13}C]citrate (Jans et al., 1989b). However, this work attracted criticism and Katz (1989) emphasized that the flux ascribed to glutamate synthesis is merely a label exchange in the transamination reaction between glutamate and oxaloacetate. Obviously, net synthesis of glutamate would require ammonia, but none was added to the perfusion medium.

In kidney, the cells of the proximal convoluted tubule are the primary target of the nephrotoxic agent cadmium. Winkel and Jans (1990) evaluated the effect of cadmium on the metabolism of PCT cells. Perchloric acid extracts of PCT cells incubated with [2–^{13}C]succinate and different concentrations of cadmium were assessed by ^{13}C MRS. The flux through PEPCK from oxaloacetate to phospho-*enol*-pyruvate as well as the labeling of alanine and lactate increased on cadmium treatment. Furthermore, cadmium caused an increased synthesis of glutamate and glutamine. In contrast, Fukuoka et al. (1988) focused on the metabolic changes induced by the nephrotoxic agent 2,3-dibromopropyl phosphate. Administration of 2,3-dibromopropyl phosphate to rats is known to cause proximal tubular damage and acute renal failure. Natural abundance ^{13}C MR spectra acquired from excised kidneys of rats at day 1 to 10 after injection of 2,3-dibromopropyl phosphate showed a progressive increase of sialic acid and the osmolyte inositol.

5.9.3 Renal Osmolytes

Cells of the papillary collecting ducts are in contact with the highly hypertonic urine, which contains urea and sodium chloride as the main solutes. It has been suggested that papillary tissue is capable of synthesizing organic osmolytes such as sorbitol, inositol, glycerophosphocholine and betaine in order to sustain and counteract the enormous osmotic stress caused by urine. Jans and coworkers (1988, 1989a) followed the carbohydrate metabolism in renal cells of papillary collecting ducts. Rabbit papillary tissue in superfusion incubated with [1–^{13}C]glucose produced significant amounts of [1–^{13}C]sorbitol. Simultaneous incubation with the aldose reductase inhibitor sorbinil inhibited the synthesis of sorbitol dramatically (see Section 5.8.1). Taken together, these results demonstrated that aldose reductase is active in renal papillary. The concept that gluconeogenesis is occurring in papillary tissue was substantiated with experiments where sorbitol became labeled from the gluconeogenic precursors [3–^{13}C]alanine, [2–^{13}C]pyruvate, and [2–

[^{13}C]glycerol (Jans *et al.*, 1988, 1989a). In these experiments, label was also found in sorbitol at positions not directly traceable through gluconeogenesis. The authors inferred from the detected labeling pattern that label scrambling through the PPP had occurred in the cells of the papillary collecting ducts.

5.10 Metabolism in Skeletal Muscles

In contrast to heart, ^{13}C MRS of skeletal muscles and of smooth muscular tissue has received only minor attention. Despite the good spectroscopic accessibility of skeletal muscles *in vivo*, ^{13}C MRS has not as yet attained the widespread application of ^{31}P MRS.

5.10.1 Metabolite Profile

Doyle (Doyle *et al.*, 1981; Doyle and Bárány, 1982), Bárány *et al.* (1982), and Arús *et al.* (1985) reported on the natural abundance spectra of excised chicken pectoralis muscles and gastrocnemius muscles of frogs. Spectra of excised muscles were dominated by relatively broad resonances from lipids (30, 130, and 175 ppm) and bound choline coresonating with creatine (55 ppm). In addition, distinct sharp resonances from the γ- and δ-carbons of histidine, phenylalanine, tyrosine, and tryptophan residues were observed in the spectral region between 115 to 140 ppm. At 170 to 185 ppm and 15 to 45 ppm, resonances of a variety of small peptides and amino acids, including lactate, were detected. Furthermore, the resonances of guanidino carbons in phosphocreatine (PCr) and creatine (Cr) residues were identified at 157.13 and 158.06 ppm, respectively. The well-separated resonances for creatine and its phosphorylated counterpart enabled the measurement of the ratio PCr/(Cr + PCr) as a means for assessing the state of high-energy phosphate metabolism (Arús *et al.*, 1985). The lactate signals and magnetic resonances from mobile phospholipids were found to increase in anaerobic muscles on contraction induced by caffeine (Doyle and Bárány, 1982; Bárány *et al.*, 1982; Arús *et al.*, 1985). Furthermore, differences in the fatty acid composition of muscles were detected between muscles from active frogs assessed in summer and hibernating frogs assessed during winter. Frog muscles obtained during the cold season had accrued a larger amount of polyunsaturated fatty acids as evidenced by the ratio of the resonances from olefinic carbons at 128.68 and 130.33 ppm (see Section 5.5.1) (Arús *et al.*, 1985).

5.10.2 Glycogen

Muscles of well-fed animals contain an appreciable amount of glycogen, supplementing the liver's capacity to store glucose. Gruetter *et al.* (1991)

quantified glycogen *in situ* in the *biceps femoris* of rabbits. The integrated signals obtained from glycogen C1 *in vivo* were compared with those acquired from solutions of oyster glycogen of known concentrations. The quantification of glycogen in muscles *in situ* revived the question as to what extent glycogen is ^{13}C MR visible. Thus, the glycogen was extracted from the muscles previously assessed by ^{13}C MRS *in situ* and the content of glycogen was measured chemically. Relying on the premises that glycogen in solution is completely MR visible and that glycogen is extracted quantitatively with perchloric acid, Gruetter *et al.* (1991) concluded from their experiments that muscle glycogen *in situ* is fully ^{13}C MR visible (see also Section 5.5.2).

5.11 Metabolism in Parasites

The significance of parasites with mammalian hosts can be inferred from the large number of human diseases evoked by parasites. These include the well-known malaria, trypanosomia, ascardiasis, and schistosomiasis, which together affect more than a billion people. Parasites are found equally spread among animals. Parasites adapt and specialize their metabolism to the physicochemical conditions provided by the host. It enables the parasites to utilize host nutrients in a manner compatible with parasite and host survival. Parasites transferred to surroundings other than those found in the host, however, barely survive, which substantiates the close parasite–host relationship. This integration of host and parasite metabolism and the parasites' dependence on host metabolism certainly justifies the consideration of parasites as an additional organ of the host's body.

The prevalently anaerobic or facultative aerobic metabolism of parasites allows the intact parasites to be kept in a superfusion system without the need for sophisticated oxygenation procedures. This metabolic feature provides an ideal prerequisite for ^{13}C MRS assessments of single or multiple parasites *in vivo* over a long period of time. As a consequence, ^{13}C MRS has provided a large body of experimental data on the metabolism of parasitic protozoas and helminths during the last two decades. This section focuses on the metabolism of multicellular adult helminths only. The review by Thompson (1991) is recommended for further reading on the application of ^{13}C MRS to protozoa.

5.11.1 Glucose Metabolism in Cestodes

Wasylishen and Novak (1983) examined the ^{13}C MR natural abundance spectra of several intact adult tapeworms species including *Hymenolepis diminuta*, *Hymenolepis microstoma*, and *Hymenolepis nana*. For all species, strong signals from lipids dominated the spectra. The relatively large intensity of the olefinic carbons indicated the presence of a high fraction of unsaturated fatty acids. The ratio derived from the intensities of the olefinic resonances at 128.0 and 129.7 ppm approached unity whereby it could be inferred that most of the unsaturated fatty acids were monounsaturated. Smaller resonances from glycerol and choline moieties of esterified lipids and phospholipids were also present in all cestode species investigated. Furthermore, distinct peaks from glycogen and glucose were observed. During prolonged incubation of the cestodes in saline solution, the glucose resonances increased in intensity, which was attributed to the hydrolysis of glycogen.

Behm et al. (1987) have further investigated carbohydrate metabolism in cestodes. Intact adult Hymenolepis diminuta from well-fed hosts were incubated with [U-$^{13}C_6$]glucose in the spectrometer under anaerobic conditions and glucose metabolism followed in vivo. Over time, the glucose resonances decreased and new resonances from glucose-related catabolites, namely, alanine, lactate, acetate, and succinate, appeared. Furthermore, as a consequence of the incorporation of labeled glucose, the resonances from glycogen increased. Starvation of the host depleted the glycogen stores of the parasite and increased the rate of glucose utilization in H. diminuta.

Separate measurements of incubation media and extracts of the tapeworm H. diminuta discriminated between internal metabolites and excretory products. Figure 20 shows spectra of incubation media (Figures 20a and b) and PCA extracts (Figure 20c and d) of two H. diminuta strains, denoted UT (Figure 20, upper traces) and ANU (Figure 20, lower traces), after incubation with [U-$^{13}C_6$]glucose under aerobic conditions. Lactate, acetate, and succinate were identified as excretory products, whereas alanine and malate occurred only intracellularly (Behm et al., 1987). The relative proportion of the excreted products was found to be different for the two strains, which was attributed to differences in the rate of excretion. In contrast to the findings of Behm et al. (1987), Blackburn et al. (1986), and Novak and Blackburn (1988) found substantial amounts of alanine among the excretory products of H. diminuta and Mesocestoides corti incubated with [U-$^{13}C_6$]glucose under anaerobic conditions.

Alanine, lactate, succinate, and acetate are the four expected end products of the currently accepted pathways of glucose metabolism in helminths,

---

Figure 20. ^{13}C MR spectra (at 67.9 MHz) of incubation media (left column) and perchlorate extracts (right column) of two strains of the cestode Hymenolepis diminuta. (a) and (c) UT strain. (b) and (d) ANU strain. Parasites from starved hosts were incubated in vitro for 30 min with [U-$^{13}C_6$]glucose. Resonances detected in the spectra are assigned as follows: A, acetate; Ala, alanine; B, buffer; C, glucose; Glyc, glycogen; L, lactate; M, malate; S, succinate; U, unidentified. (Reproduced with permission from Behm et al., 1987. Copyright © 1987 Australian Society for Parasitology.)

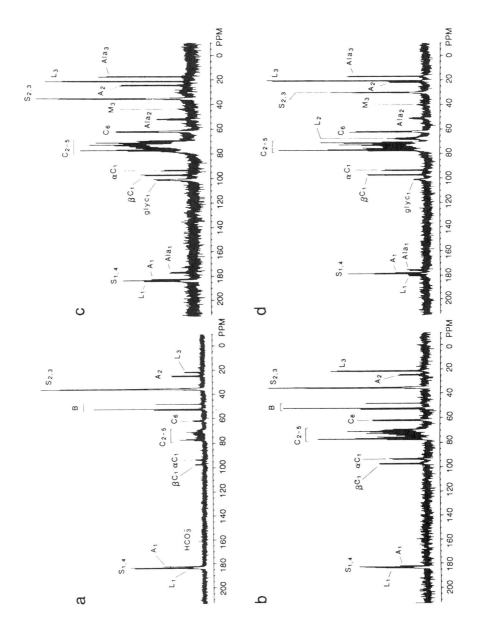

which differs slightly from the metabolism common in mammalian tissue (for a review, see Thompson, 1991). Briefly, glucose is metabolized in glycolysis to phospho-*enol*-pyruvate (PEP), which is dephosphorylated and either reduced to lactate or transaminated to alanine. Alternatively, PEP is carboxylated via cytoplasmic PEPCK to oxaloacetate. Cytosolic malate dehydrogenase reduces oxaloacetate to malate, which enters the mitochondria. By a partial reversal of the TCA cycle sequence, malate is transformed to fumarate and finally to succinate. Malate may also be decarboxylated by mitochondrial malic enzyme to reform pyruvate which, in turn, is the precursor of acetate.

The specific labeling patterns detected in the excretory products after incubation of the cestodes with [U-$^{13}C_6$]glucose gave further essential clues to the glucose metabolism in cestodes (Blackburn *et al.*, 1986). The major isotopomer species were [U-$^{13}C_3$]- and [2,3-$^{13}C_2$]lactate in a ratio of 2:1, [1,2,2'-$^{13}C_3$]- and [2,2'-$^{13}C_2$]succinate, also present at a ratio of 2:1, and [1,2-$^{13}C_2$]acetate. Furthermore, small amounts of [U-$^{13}C_4$]succinate were found. The detection of [2,3-$^{13}C_2$]lactate and [2,2'-$^{13}C_2$]succinate demonstrates that the alternative pathway for pyruvate synthesis via malic enzyme is used extensively and that the precursor malate is rapidly randomized in the fumarase reaction. Blackburn *et al.* (1986) tentatively explained the occurrence of [U-$^{13}C_4$]succcinate by the operation of the glyoxal shunt. However, [U-$^{13}C_4$]succinate may also be formed by carboxylation of [U-$^{13}C_3$]pyruvate with $^{13}CO_2$ derived from decarboxylation reactions.

The occurrence of the sequence PEP → oxaloacetate → pyruvate and randomization in the fumarase step was further substantiated by incubating *H. diminuta* with NaH$^{13}CO_3$ and unlabeled glucose. As expected from the topology of the proposed metabolic network, label was predominantly recovered in lactate C1 and succinate C1/C1'. These labeling patterns are indeed complementary to those revealed in experiments where [U-$^{13}C_6$]glucose and unlabeled CO_2 were the metabolic precursors (Behm *et al.*, 1987). Together, these experimental data suggest that in cestodes the series of reactions between pyruvate and fumarate is much more reversible than previously anticipated and that the equilibrium is fast in comparison to the conversion of fumarate to succinate.

5.11.2 Glucose Metabolism in Trematodes

Kawanaka *et al.* (1989) investigated glucose metabolism in *Schistosoma japonicum*. Incubation of the worms with [U-$^{13}C_6$]glucose under anaerobic conditions revealed excreted labeled lactate, alanine, acetate, and succinate in the incubation medium. Under aerobic conditions, however, succinate was completely lacking and the glucose utilization and lactate production were slightly increased. These observations suggested that the transition from aerobic to anaerobic metabolism reduced the PK activity, whereas the carboxylation of PEP to oxaloacetate and ultimately to succinate was favored. The shift toward the formation of succinate is advantageous to the parasites since it potentially increases the yield of ATP from 2 to 4 moles per mole of glucose depending on the balance between the two pathways. Isotopomer analysis on lactate and succinate recovered from *S. japonicum* incubated under anaerobic conditions disclosed the presence of [2,2'-$^{13}C_2$]succinate and [2,3-$^{13}C_2$]lactate among other isotopomers. This finding clearly indicates a partial reversibility of the metabolic reactions between pyruvate and fumarate, similar to the metabolism in cestodes (see Section 5.11.1).

Matthews *et al.* (1985) examined the excretory products of glucose metabolism in *Fasciola hepatica*. Several liver flukes were incubated with [1-^{13}C]glucose and samples of the incubation medium were taken at various intervals. The distribution and enrichment of ^{13}C in excreted metabolites was assessed by monitoring ^{13}C satellites in proton spectra. Propionate and acetate in a ratio of approximately 2:1 were the major catabolites found in the incubation medium. Propionate with equal ^{13}C labeling at C2 and C3 indicated that its formation occurred via decarboxylation of succinate, which, in turn, was synthesized in a pathway similar to that proposed for cestodes and the trematode *S. japonicum*. Equal label distribution was also found in C1 and C2 of acetate. This observation is not compatible with a direct formation of acetate from [3-^{13}C]PEP involving dephosphorylation and decarboxylation because such an acetate would be labeled only at carbon C2. Hence, it was suggested that acetate excreted by *F. hepatica* is derived via carboxylation of PEP to malate, which is exchanged with fumarate. After label randomization, two subsequent decarboxylation steps would then lead to [1-^{13}C]- and [2-^{13}C]acetate. Enrichments for carbons C2 and C3 of propionate and carbons C1 and C2 of acetate close to the maximal enrichment of 25% finally indicated that both catabolites originated exclusively from labeled exogenous glucose.

Glycogen synthesis from [1-^{13}C]glucose was followed in *F. hepatica* by *in vivo* ^{13}C MRS in the intact fluke (Matthews *et al.*, 1985). Apart from the expected enrichment in C1 of glycogen, C6 contained a significant fraction of the introduced ^{13}C label. Hence, it was suggested that substrate cycling between glucose and the triosephosphates GAP and DHAP occurs at the level of the enzyme couple phosphofructokinase and fructosediphosphatase in the glycolytic and gluconeogenic direction of glucose metabolism, respectively (see also Section 5.5.4). At metabolic steady state, the label distribution between C6 and C1 in freshly synthesized glycogen is a measure of the ratio between net glycolytic flux and total forward flux through phosphofructokinase. Matthews *et al.* (1985) determined the extent of substrate cycling under the assumption that labeling in glycogen accurately reflects the labeling in the fructose-6-phosphate pool, and that the aldolase and triosephosphate isomerase scramble the label in triosephosphates to the full extent. Calculations showed that more than half of the forward glycolytic flux is counteracted by a reverse flux between fructose-6-phosphate and fructose-1,6-diphosphate indicating substrate recycling to a large extent.

5.11.3 Glucose Metabolism in Nematodes

Nematodes provide further astonishing varieties of carbohydrate metabolism in animals. Powell *et al.* (1986a,b) and Christie *et al.* (1987) investigated glucose metabolism in *Dipetalonema viteae* and *Brugia pahangi*. *In vivo* ^{13}C MRS on the intact macrofilariae incubated with [1–^{13}C]glucose showed that glucose is removed from the medium faster during aerobic incubation than under similar anaerobic conditions. This finding is in contrast to the Pasteur effect (accelerated glucose consumption), which occurs in many tissues on oxygen deprivation. The principal product of glucose metabolism in *Dipetalonema viteae* and *Brugia pahangi* was lactate and to a smaller extent succinate, both of which were excreted into the incubation medium. A resonance at 94.3 ppm in spectra of tissue extracts indicated the accumulation of the disaccharide α,α-trehalose in intact worms on incubation with glucose (Powell *et al.*, 1986a). MacKenzie *et al.* (1989) reported acetate and glycogen (in contrast to trehalose) as two further metabolites of aerobic metabolism of [U–^{13}C$_6$]glucose in *B. pahangi*. Powell *et al.* (1986b) found that in the absence of exogenous glucose, trehalose is used up meeting the parasites' energy requirements. Concomitant with reduced glucose availability, a significant shift in *D. viteae* metabolism was observed from predominantly glucose fermentation with lactate as product to a manifold increased synthesis of succinate. *B. pahangi* showed a similar yet not as pronounced alteration in metabolism on deprivation of glucose. Isotopomer analysis of succinate originating from *D. viteae* incubated with [1–^{13}C]glucose in media buffered with [^{13}C]bicarbonate revealed the presence of significant amounts of [1–^{13}C]succinate and [1,2–^{13}C$_2$]succinate, the latter of which gave rise to indicative doublet resonances in the ^{13}C MR spectrum (Christie *et al.*, 1987). The detection of label at the carboxylic group and simultaneous labeling at C2 of succinate indicated succinate synthesis via carboxylation of [3–^{13}C]PEP with ^{13}CO$_2$ followed by a partial reversal of the TCA cycle similar to the metabolism in cestodes and trematodes.

Brugia pahangi was thought to be completely dependent on glucose metabolism. However, *B. pahangi* is able to survive by glutaminolysis as was shown by MacKenzie *et al.* (1989). Filarial worms incubated with [5–^{13}C]glutamine produced [5–^{13}C]glutamate, [1–^{13}C]succinate, [1–^{13}C]acetate, and [1-^{13}C]glyoxylate as detected by ^{13}C MRS. Additional measurements by mass spectrometry revealed that α-ketoglutarate was not labeled. Therefore, it was suggested that glutamine was metabolized to glutamate followed by decarboxylation and deamination via GABA to succinate. Furthermore, the detection of labeled acetate cannot be explained by the sequence [1-^{13}C]succinate → [1-^{13}C]pyruvate → acetate since all label would be lost on formation of acetate. The detection of labeled glyoxylate suggested the synthesis of acetate via the glyoxylate cycle whereby citrate is metabolized to isocitrate, which is split into succinate and glyoxylate. Glyoxylate, in turn, can condense with acetyl-CoA to form malate. Similar experiments with [5-^{13}C]glutamine were performed on *Onchocerca volvulus*, the causative agent of river blindness (MacKenzie *et al.*, 1989). ^{13}C MR spectra of incubation media revealed excreted [5-^{13}C]glutamate and [1-^{13}C]succinate but no labeled acetate or glyoxylate could be detected. On the other hand, *O. volvulus* incubated with [U-^{13}C$_6$]glucose excreted a wide variety of labeled products including lactate, succinate, acetate, and ethanol with a general trend toward a greater variety of energetically favorable anaerobic pathways than in *B. pahangi*. *B. pahangi* has been used as a more convenient laboratory model than *O. volvulus*. However, in light of the distinct metabolic differences between these two species the validity of this model system has to be questioned.

Finally, glucose metabolism in young *Angiostrongylus cantonensis*, a neurotropic parasite in rat and occasionally in man, was investigated by a combination of ^{13}C and proton MRS (Nishina *et al.*, 1988). *A. cantonensis* incubated with [U-^{13}C$_6$]glucose excreted mainly uniformly labeled lactate and minor amounts of uniformly labeled acetate and alanine. Male *A. cantonensis* excreted almost twice as much acetate as females. In contrast to many other helminths, *A. cantonensis* only excreted uniformly labeled products on incubation with [U-^{13}C$_6$]glucose. Hence, Nishina and coworkers (1988) concluded that the alternative pathway of pyruvate synthesis by carboxylation of PEP to oxalate and subsequent decarboxylation of malate to pyruvate is not active in *A. cantonensis*.

5.12 Metabolism in Tumors

The application of ^{13}C-MRS to the analysis of tumor tissue of animal origin has a history dating back over a decade. Despite the early interest and in contrast to studies on animal organs, ^{13}C MRS of tumors is still in its infancy and only a very limited number of publications are available. The major objective of all work has been the elucidation of metabolic differences that characterize and distinguish tumors from their parental tissues. Thus, the quest for alterations to the metabolism of cancer cells is driven by the premise that such cancer-specific modifications of pathways might be a starting point for a more specific and efficient drug treatment. A convenient and widespread approach for investigating the metabolism in tumor cells is the study of cells maintained in culture. The focus of this review, however, is directed at tumors grown *in vivo*.

5.12.1 Metabolite Profile

Natural abundance ^{13}C MRS has been used to assess the metabolite profile in tumors. Evanochko *et al.* (1984) evaluated PCA extracts obtained from a murine radiation-induced fibrosarcoma (RIF-1), a malignant cancer of the connective tissue. Distinct resonances from taurine and a variety of amino acids (alanine, glycine, and glutamate) were observed in RIF-1 tumors grown in mice. Further resonances originating from lactate, phosphorylethanolamine, phosphorylcholine, creatine, and phosphocreatine were also detected. A great deal of variability was found for samples obtained from different tumor specimens. Factors such as tumor size, degree of necrosis, and hypoxic cell fraction were considered responsible for this biological variability.

Xue *et al.* (1990) employed a modified DEPT sequence to obtain CH, CH_2, and CH_3-edited spectra of RIF-1 tumors *in vivo*. The spectra were dominated by the resonances of mobile lipids. Apart from these lipid resonances, *in vivo* spectroscopy revealed a metabolite profile very similar to that in PCA extracts of RIF-1. Comparable *in vivo* spectra were also obtained from multidrug-resistant human breast cancers grown in nude mice (Navon *et al.*, 1989).

5.12.2 Glucose Metabolism

Glucose is one of the main substrates for cellular energy metabolism. In animal tissues, glucose is commonly catabolized in glycolysis and subsequently in the TCA cycle on the availability of oxygen. In contrast, most tumor cells have a low respiration and derive their energy almost exclusively from the less energy yielding glycolysis, even in the presence of high levels of oxygen. As a consequence, the rate of glycolysis (in this case often referred to as aerobic glycolysis) is dramatically increased to meet the tumor cells' energy requirements. A very active glycolysis and a concomitantly reduced flux through the TCA cycle, in turn, are expected to lead to the accumulation of glycolytic end products such as lactate.

Constantinidis *et al.* (1991) and Bhujwalla *et al.* (1991) followed the metabolism of [1–^{13}C]glucose in RIF-1 tumors in mice *in vivo*. Shortly after the injection of [1-^{13}C]glucose directly into the tumors, [3-^{13}C]lactate and to a lesser extent [3-^{13}C]alanine could be detected, indicating a high glycolytic activity. Similar results were obtained with *in vivo* spectroscopy of human colon carcinomas implanted in athymic mice (Lyon *et al.*, 1988). The [2-^{13}C]lactate could be detected after the administration of [2-^{13}C]glucose. None of the *in vivo* spectra revealed any evidence for the metabolism of glucose via the TCA cycle. However, spectra of PCA extracts prepared from the same tumors indicated ^{13}C labeling primarily at the C4 position of glutamate (Constantinidis *et al.*, 1991). This finding led to the conclusion that in RIF-1 tumors, pyruvate enters the TCA cycle primarily via pyruvate dehydrogenase. Mégnin *et al.* (1989) assessed the glycolytic, gluconeogenic, and TCA cycle activity in hematopoietic cells from mice with leukemia induced by the acutely transforming myeloproliferative leukemic retrovirus. Hematopoietic cells were isolated from infected spleens and were embedded in agarose gel threads. *In vivo* ^{13}C MR spectra of cells perfused with [1-^{13}C]glucose revealed that hematopoietic cells were able to store glucose as [1-^{13}C]glycogen and to metabolize it in the glycolysis to [3-^{13}C]lactate. To test the hypothesis that hematopoietic cells are capable of performing gluconeogenesis, cells were perfused with medium containing the gluconeogenic precursors [2-^{13}C]pyruvate or [3-^{13}C]alanine. However, neither of the two substrates led to the labeling of glycogen as would be expected from active gluconeogenesis. Furthermore, neither glucose, alanine, nor pyruvate was capable of labeling TCA cycle intermediates, thus indicating a low TCA cycle activity. To further substantiate this finding, Mégnin *et al.* (1989) conducted another series of experiments in which hematopoietic cells were perfused with [1-^{13}C]glucose in combination with sodium acide, which blocks mitochondrial respiration and hence mitochondrial energy production. Under these "anaerobic" conditions, the glucose consumption by glycolysis increased only by 35% even though the cells remained viable as attested to by ^{31}P MRS.

In contrast to RIF-1 tumors and hematopoietic cells, *in vivo* spectra of rat C6 gliomas disclosed a wider variety of labeled metabolites after infusion of [1-^{13}C]glucose. Strong resonances from [3-^{13}C]lactate, [1-^{13}C]glycogen, [4-^{13}C]-, [3-^{13}C]-, and [2-^{13}C]glutamate/glutamine could be detected (Ross *et al.*, 1988). It is well known that the brain is not gluconeogenic, and that the levels of stored glycogen in cerebral tissue are very low. Thus, the ability of C6 gliomas to store glycogen further exemplifies the diverse metabolic alterations and adaptions that may occur in tumor tissues.

The glucose analogue 2-deoxyglucose was used to monitor the transport kinetics and subsequent phosphorylation of glucose in multidrug-resistant human breast tumors grown in nude mice (Navon *et al.*, 1989). Administration of [6-^{13}C]2-deoxyglucose allowed separate resonances for 2-deoxyglucose and its intracellular phosphorylation product, 2-deoxyglucose-6-phosphate to be followed *in vivo* (see also Section 5.7.1). Administration of glucose to tumor tissues has not only been used to assess glucose metabolism but also to generate lactic acidosis. Excessive availability of glucose, i.e., hyperglycemia, decreases both tumor pH and blood flow in a variety of tumor models. These changes, in turn, can sensitize tumors to hyperthermia. Hwang *et al.* (1992) used the nonmetabolizable glucose analogue 3-O-methyl-glucose to elucidate the involvement of glycolysis and the accumulation of lactate in hyperglycemic tumor acidification. Mice bearing RIF-1 tumors were injected intraperitoneally with a high dose of [1–^{13}C]3-O-methyl-glucose, and ^{13}C MRS was used to verify that the glucose analogue was not metabolized to lactate. Although it was anticipated that the production of lactate was necessary for tumor acidification, a marked drop in pH was observed by ^{31}P MRS in the absence of an increased glucose metabolism and without a change in lactate concentration (measured by ^{1}H MRS in PCA extracts of the tumors). This result suggests an alternative nonglycolytic mechanism for the hyperglycemic pH reduction in RIF-1 tumors.

5.13 Concluding Remarks

^{13}C MRS on animal tissues has often confirmed and sometimes seriously challenged previous findings by providing novel information about the cellular metabolism. Its nonspecificity in detecting and quantifying cellular metabolites, i.e., its capacity to observe metabolites in complex mixtures without specific assays, has kept ^{13}C MRS receptive for the unknown and unexpected. Hence, ^{13}C MRS experiments have allowed the cellular metabolism to be assessed in a comprehensive manner with each spectrum providing an unprecedented wealth of information. Furthermore, ^{13}C MRS in conjunction with specifically ^{13}C-labeled precursors has proven to be a powerful tool for biochemists. The additional exploitation of the unique information obtained from ^{13}C–^{13}C homonuclear spin couplings may rightfully be dubbed the quantum leap in the analysis of labeling data. Finally, the possibility of applying the noninvasive ^{13}C MR techniques *in vivo* has opened up a hitherto unknown avenue for assaying tissue biochemistry of carbon metabolites *in situ* with a minimum of interference from the measurement itself.

The only conspicuous disadvantages of ^{13}C MRS are its relatively low intrinsic sensitivity and the costs of the equipment and ^{13}C labels. Current developments, however, are heading toward increasingly strong magnets with bores large enough to accommodate experimental animals, thus considerably alleviating the sensitivity issue. At the same time, the costs of the equipment are comparatively reduced and the prices for ^{13}C-labeled products have fallen almost exponentially due to more widespread use of ^{13}C MRS.

Indeed, ^{13}C MRS has made its entry into a far wider range of applications for assessing the biochemistry of tissues than the present review might suggest. Exciting developments in areas such as ^{13}C MRS on cultured cells, plants, insects, and a variety of mammalian organs have been addressed only marginally or even completely set aside. The most challenging field, however, is probably the application of ^{13}C MRS to humans. The noninvasive nature of MR methods has allowed many of the ^{13}C MRS examinations reported herein to be readily adapted to the human situation (for a review see Chapter 6). In light of the possible implementation of ^{13}C MRS in human medicine, further developments of ^{13}C MRS on animals may thus proceed along the investigation of animal or cell models of human diseases and their metabolic responses to medications. Cancer biochemistry is one of the most promising fields where ^{13}C MRS may gain a firm footing. However, it is evident that most of the more invasive experimental approaches adopted for animal experiments are not applicable to humans. An interesting, though not yet deeply investigated alternative is the evaluation of body fluids by *in vitro* ^{13}C MRS. Developments along this line would include *in vivo* and *in vitro* ^{13}C MRS on animal tissues in order to establish the relationship between the metabolite and label profiles in body fluids and the metabolic activity in different organs of the body.

Acknowledgments

I thank various authors and publishers for their generous permission to reproduce some of their data. I am particularly indebted to Drs. Sebastian Cerdan, Jim Delikatny, and Wanda Mackinnon for valuable discussions and comments on earlier drafts. Special thanks also go to my wife, Christina Aenishänslin, who contributed in innumerable ways to the final form of this chapter.

References

Alger, J. R., Sillerud, L. O., Behar, K. L., Gillies, R. J., Shulman, R. G., Gordon, R. E., Shaw, D., and Hanley, P. E. (1981). *Science* **214**, 660.
Arús, C., Chang, Y.-C., and Bárány, M. (1985). *Biochim. Biophys. Acta* **844**, 91.
Bachelard, H. S., Cox, D. W. G., and Morris, P. G. (1987). *Gerontology* **33**, 235.
Badar-Goffer, R., and Bachelard, H. (1991). *Essays Biochem.* **26**, 105.
Badar-Goffer, R. S., Bachelard, H. S., and Morris, P. G. (1990). *Biochem. J.* **266**, 133.
Badar-Goffer, R. S., Ben-Yoseph, O., Bachelard, H. S., and Morris, P. G. (1992). *Biochem. J.* **282**, 225.
Bailey, I. A., Gadian, D. G., Matthews, P. M., Radda, G. K., and Seeley, P. J. (1981). *FEBS Lett.* **123**, 315.
Bárány, M., and Venkatasubramanian, P. N. (1987). *Biochim. Biophys. Acta* **923**, 339.

Bárány, M., Doyle, D. D., Graff, G., Westler, W. M., and Markley, J. L. (1982). *J. Biol. Chem.* **257**, 2741.
Bárány, M., Arús, C., and Chang, Y.-C. (1985). *Magn. Reson. Med.* **2**, 289.
Batiz-Hernandez, H. and Bernheim, R. A. (1967). *Prog. Nucl. Magn. Reson. Spectrosc.* **3**, 63–85.
Becker, N. N., and Ackerman, J. H. (1990). *Basic Life Sci.* **56**, 317.
Behar, K. L., Petroff, O. A. C., Prichard, J. W., Alger, J. R., and Shulman, R. G. (1986). *Magn. Reson. Med.* **3**, 911.
Behm, C. A., Bryant, C., and Jones, A. J. (1987). *Int. J. Parasitol.* **17**, 1333.
Bell, J. D. (1989). In "Magnetic Resonance Spectroscopy of Biofluids" (J. D. de Certaines, ed.), pp. 7–32. World Scientific, Singapore.
Bell, J. D. (1992). In "Magnetic Resonance Spectroscopy in Biology and Medicine" (J. D. de Certaines, W. M. M. J. Bovée, and F. Podo, eds.), pp. 529–558. Pergamon, Oxford.
Bendall, M. R., and Gordon, R. E. (1983). *J. Magn. Reson.* **53**, 365.
Bendall, M. R., and Pegg, D. T. (1984). *J. Magn. Reson.* **57**, 337.
Bendall, M. R., and Pegg, D. T. (1986). *J. Magn. Reson.* **67**, 376.
Bernard, M. and Cozzone, P. J. (1992). In "Magnetic Resonance Spectroscopy in Biology and Medicine" (J. D. de Certaines, W. M. M. J. Bovée, and F. Podo, eds.), pp. 387–409. Pergamon, Oxford.
Bernard, M., Confort-Gouny, S., and Cozzone, P. J. (1990). *Proc. Eur. Soc. Magn. Reson. Med. Biol.* **7**, 264.
Bernhard, S. A., and Tompa, P. (1990). *Arch. Biochem. Biophys.* **276**, 191.
Bhujwalla, Z. M., Constantinidis, I., Chatham, J. C., Wehrle, J. P., and Glickson, J. D. (1991). *Int. J. Radiat. Oncol. Biol. Phys.* **22**, 95.
Blackburn, B. J., Hutton, H. M., Novak, M., and Evans, W. S. (1986). *Exp. Parasitol.* **62**, 381.
Bovée, W. M. M. J. (1992). In "Magnetic Resonance Spectroscopy in Biology and Medicine" (J. D. de Certaines, W. M. M. J. Bovée, and F. Podo, eds.), pp. 181–208. Pergamon, Oxford.
Brainard, J. R., Hoekenga, D. E., and Hutson, J. Y. (1986). *Magn. Reson. Med.* **3**, 673.
Brainard, J. R., Hutson, J. Y., Hoekenga, D. E., and Lenhoff, R. (1989a). *Biochemistry* **28**, 9766.
Brainard, J. R., Kyner, E., and Rosenberg, G. A. (1989b). *J. Neurochem.* **53**, 1285.
Canioni, P., Alger, J. R., and Shulman R. G. (1983). *Biochemistry* **22**, 4974.
Canioni, P., Desmoulin, F., Galons, J. P., Bernard, M., Fontanarava, E., and Cozzone, P. J. (1985). *Arch. Int. Physiol. Biochim.* **93**, 119.
Cerdan, S., and Seelig, J. (1990). *Annu. Rev. Biophys. Biophys. Chem.* **19**, 43.
Cerdan, S., Künnecke, B., Dölle, A., and Seelig, J. (1988). *J. Biol. Chem.* **263**, 11664.
Cerdan, S., Künnecke, B., and Seelig, J. (1990). *J. Biol. Chem.* **265**, 12916.
Chacko, V. P., and Weiss, R. G. (1993). *Am. J. Physiol.* **264**, C755.
Chance, E. M., Seeholzer, S. H., Kobayashi, K, and Williamson, R. R. (1983). *J. Biol. Chem.* **258**, 13785.
Chatham, J. C., Cousins, J. P., and Glickson, J. D. (1990). *J. Mol. Cell. Cardiol.* **22**, 1187.
Chatham, J. C., Hutchins, G. M., and Glickson, J. D. (1992). *Biochim. Biophys. Acta* **1138**, 1.
Chen, W., and Ackerman, J. J. H. (1989). *NMR Biomed.* **2**, 267.
Chen, W., and Ackerman, J. H. (1990). *NMR Biomed.* **3**, 147.
Cheng, H.-M., and González, R. G. (1986). *Metab. Clin. Exp.* **35**, 10.
Cheng, H.-M., González, R. G., Barnett, P. A., Aguayo, J. B., Wolfe, J., and Chylack, L. T., Jr. (1985). *Exp. Eye Res.* **40**, 223.
Cheng, H.-M., Hirose, K., Xiong, H., and González, R. G. (1989). *Exp. Eye Res.* **49**, 87.
Cheng, H.-M., Xiong, J., Tanaka, G., Chang, C., Asterlin, A. A., and Aguayo, J. B. (1991). *Exp. Eye Res.* **53**, 363.

Christie, D. A., Powell, J. W., Stables, J. N., and Watt, R. A. (1987). *Mol. Biochem. Parasitol.* **24**, 125.
Cline, G. W., and Shulman, G. I. (1991). *J. Biol. Chem.* **266**, 4094.
Cohen, S. M. (1983). *J. Biol. Chem.* **258**, 14294.
Cohen, S. M. (1984). *Fed. Proc., Fed. Am. Soc. Exp. Biol.* **43**, 2657.
Cohen, S. M. (1987a). *Biochemistry* **26**, 563.
Cohen, S. M. (1987b). *Biochemistry* **26**, 573.
Cohen, S. M. (1987c). *Biochemistry* **26**, 581.
Cohen, S. M., and Shulman, R. G. (1982). *In* "Noninvasive Probes of Tissue Metabolism" (S. J. Cohen, ed.), pp. 119–147. Wiley, New York.
Cohen, S. M., Ogawa, S., and Shulman, R. G. (1979a). *Proc. Natl. Acad. Sci. U.S.A.* **76**, 1603.
Cohen, S. M., Shulman, R. G., and McLaughlin, A. C. (1979b). *Proc. Natl. Acad. Sci. U.S.A.* **76**, 4808.
Cohen, S. M., Glynn, P., and Shulman, R. G. (1981a). *Proc. Natl. Acad. Sci. U.S.A.* **78**, 60.
Cohen, S. M., Rognstad, R., Shulman, R. G., and Katz, J. (1981b). *J. Biol. Chem.* **256**, 3428.
Constantinidis, I., Chatham, J. C., Wehrle, J. P., and Glickson, J. D. (1991). *Magn. Reson. Med.* **20**, 17.
Cozzone, P. J., Canioni, P., Bernard, M., Desmoulin, F., and Galons, J. P. (1985). *Ann. Endocrinol.* **46**, 239.
Cross, T. A., Pahl, C., Oberhänsli, R., Aue, W. P., Keller, U., and Seelig, J. (1984). *Biochemistry* **23**, 6398.
Cunnane, S. C. (1992). *FEBS Lett.* **306**, 273.
David, M., Petit, W. A., Laughlin, M. R., Shulman, R. G., King, J. E., and Barrett, E. J. (1990). *J. Clin. Invest.* **86**, 612.
Deslauriers, R., Somorjai, R. L., Geoffrion, Y., Kroft, T., Smith, I. C. P., and Saunders, J. K. (1988). *NMR Biomed.* **1**, 32.
Desmoulin, F., Canioni, P., and Cozzone, P. J. (1985). *FEBS Lett.* **185**, 29.
Doyle, D. D., and Bárány, M. (1982). *FEBS Lett.* **140**, 237.
Doyle, D. D., Chalovich, J. M., and Bárány (1981). *FEBS Lett.* **131**, 147.
Evanochko, W. T., Sakai, T. T., Ng, T. C., Krishna, N. R., Kim, H. D., Zeidler, R. B., Ghanta, V. K., Brockman, R. W., Schiffer, L. M., Braunschweiger, P. G., and Glickson, J. D. (1984). *Biochim. Biophys. Acta* **805**, 104.
Fujiwara, N., Shimoji, K., Yuasa, T., Igarashi, H., and Miyatake, T. (1989). *NMR Biomed.* **2**, 104.
Fukuoka, M., Tanaka, A., Yamaha, T., Naito, K., Takada, K., Kobayashi, K., and Tobe, M. (1988). *J. Appl. Toxicol.* **8**, 411.
Gallis, J.-L., and Canioni, P. (1992). *In* "Magnetic Resonance Spectroscopy in Biology and Medicine" (J. D. de Certaines, W. M. M. J. Bovée, and F. Podo, eds.), pp. 345–368. Pergamon, Oxford.
Garfinkel, D. (1966). *J. Biol. Chem.* **241**, 3918.
Garlick, P. B., and Pritchard, R. D. (1993). *NMR Biomed.* **6**, 84.
Geoffrion, Y., Butler, K., Pass, M., Smith, I. C. P., and Deslauriers, R. (1985). *Exp. Parasitol.* **59**, 364.
González, G. R., Barnett, P. Aguayo, J., Cheng, H.-M., and Chylack, L. T., Jr. (1984). *Diabetes* **33**, 196.
Gopher, A., and Lapidot, A. (1991). *Proc. Soc. Magn. Reson. Med.* **10**, 1050.
Gopher, A., Segal, M., and Lapidot, A. (1990). *Proc. Soc. Magn. Reson. Med.* **9**, 1110.
Gruetter, R., Prolla, T. A., and Shulman, R. G. (1991). *Magn. Reson. Med.* **20**, 327.
Hall, J. E., Mackenzie, N. E., Massfield, J. M., McCloskey, D. E., and Scott, A. I. (1988). *Comp. Biochem. Physiol. B* **89B**, 679.

Hammer, B. E., Sacks, W., Bigler, R. E., Hennessy, M. J., Sacks, S., Fleischer, A., and Zanzonico, P. B. (1990). *Magn. Reson. Med.* **13**, 1.
Hoekenga, D. E., Brainard, J. R., and Hutson, J. Y. (1988). *Circ. Res.* **62**, 1065.
Hwang, Y. C., Kim, S.-G., Evelhoch, J. L., and Ackerman, J. J. H. (1992). *Cancer Res.* **52**, 1259.
Jans, A. W. H., and Kinne, R. K. H. (1991). *Kidney Int.* **39**, 430.
Jans, A. W. H., and Leibfritz, D. (1989). *NMR Biomed.* **1**, 171.
Jans, A. W. H., and Willem, R. (1988). *Eur. J. Biochem.* **174**, 67.
Jans, A. W. H., and Willem, R. (1989). *Biochem. J.* **263**, 231.
Jans, A. W. H., and Willem, R. (1990). *Magn. Reson. Med.* **14**, 148.
Jans, A. W. H., and Willem, R. (1991). *Eur. J. Biochem.* **195**, 97.
Jans, A. W. H., Grunewald, R. W., and Kinne, R. K. H. (1988). *Biochim. Biophys. Acta* **971**, 157.
Jans, A. W. H., Grunewald, R.-W., and Kinne, R. K. H. (1989a). *Magn. Reson. Med.* **9**, 419.
Jans, A. W. H., Winkel, C., Buitenhuis, L., and Lugtenburg, J. (1989b). *Biochem. J.* **257**, 425.
Jeffrey, F. M. H., Rajagopal, A., Malloy, C. R., and Sherry, A. D. (1991). *Trends Biochem. Sci.* **16**, 5.
Jehenson, P., Canioni, P., Hantraye, P., and Syrota, A. (1992). *Biochem. Biophys. Res. Commun.* **182**, 900.
Johnston, D. L., and Lewandowski, E. D. (1991). *Circ. Res.* **68**, 714.
Kalderon, B., Gopher, A., and Lapidot, A. (1986). *FEBS Lett.* **204**, 29.
Kalderon, B., Gopher, A., and Lapidot, A. (1987). *FEBS Lett.* **213**, 209.
Kalil-Filho, R., Gerstenblith, G., Hansford, R. G., Chacko, V. P., Vandegaer, K., and Weiss, R. G. (1991). *J. Mol. Cell. Cardiol.* **23**, 1467.
Katz, J. (1989). *Biochem. J.* **263**, 997.
Katz, J., and Grunnet, N. (1979). In "Techniques in Metabolic Research" (H. L. Kornberg, ed.), pp. 1–18. Elsevier/North-Holland Pub., Amsterdam.
Kawanaka, M., Matsushita, K., Kato, K., and Ohsaka, A. (1989). *Physiol. Chem. Phys. Med. NMR* **21**, 5.
Kotyk, J. J., Rust, R. S., Ackerman, J. J. H., and Deuel, R. K. (1989). *J. Neurochem.* **53**, 1620.
Künnecke, B., and Cerdan, S. (1989). *NMR Biomed.* **2**, 274.
Künnecke, B., and Seelig, J. (1991). *Biochim. Biophys. Acta* **1095**, 103.
Künnecke, B., Cerdan, S., and Seelig, J. (1993). *NMR Biomed.* **6**, 264.
Lajtha, A., Berl, S., and Waelsch, H. (1959). *J. Neurochem.* **3**, 322.
Laughlin, M. R., Petit, W. A., Jr., Dizon, J. M., Shulman, R. G., and Barrett, E. J. (1988). *J. Biol. Chem.* **263**, 2285.
Laughlin, M. R., Petit, W. A., Jr., Shulman, R. G., and Barrett, E. J. (1990). *Am. J. Physiol.* **258**, E184.
Laughlin, M. R., Morgan, C., and Barrett, E. J. (1991). *Diabetes* **40**, 385.
Lavanchy, N., Martin, J., and Rossi, A. (1984). *FEBS Lett.* **178**, 34.
Lawry, T. J., Weiner, M. W., and Matson, G. B. (1990). *Magn. Reson. Med.* **16**, 294.
Le Moyec, L. and Akoka, S. (1992). In "Magnetic Resonance Spectroscopy in Biology and Medicine" (J. D. de Certaines, W. M. M. J. Bovée, and F. Podo, eds.), pp. 289–294. Pergamon, Oxford.
Lerman, S., and Moran, M. (1988). *Ophthalmic Res.* **20**, 348.
Levitt, M. H., and Freeman, R. (1981). *J. Magn. Reson.* **43**, 502.
Lewandowski, E. D. (1991). *Biochemistry* **31**, 8916.
Lewandowski, E. D. (1992). *Circ. Res.* **70**, 576.
Lewandowski, E. D., and Hulbert, C. (1991). *Magn. Reson. Med.* **19**, 186.
Lewandowski, E. D., and Johnston, D. L. (1990). *Am. J. Physiol.* **258**, H1357.

Lewandowski, E. D., Chari, M. V., Roberts, R., and Johnston, D. L. (1991a). *Am. J. Physiol.* **261**, H354.
Lewandowski, E. D., Johnston, D. L., and Roberts, R. (1991b). *Circ. Res.* **68**, 578.
London, R. E., Matwiyoff, N. A., and Mueller, D. D. (1975). *J. Chem. Phys.* **63**, 4443.
Liu, K. J. M., Henderson, T. O., Kleps, R. A., Reyes, M. C., and Nyhus, L. M. (1990). *J. Surg. Res.* **49**, 179.
Liu, K. J. M., Kleps, R., Henderson, T., and Nyhus, L. (1991). *Biochem. Biophys. Res. Commun.* **179**, 366.
Lyon, R. C., Tschudin, R. G., Daly, P. F., and Cohen, J. S. (1988). *Magn. Reson. Med.* **6**, 1.
Mackenzie, N. E., Van De Waa, E. A., Gooley, P. R., Williams, J. F., Bennett, J. L., Bjorge, S. M., Baille, T. A., and Geary, T. G. (1989). *Parasitology* **99**, 427.
Malloy, C. R., Sherry, A. D., and Jeffrey, F. M. H. (1987). *FEBS Lett.* **212**, 58.
Malloy, C. R., Sherry, A. D., and Jeffrey, F. M. H. (1988). *J. Biol. Chem.* **263**, 6964.
Malloy, C. R., Sherry, A. D., and Jeffrey, F. M. H. (1990a). *Am. J. Physiol.* **259**, H987.
Malloy, C. R., Thompson, J. R., Jeffrey, F. M. H., and Sherry, A. D. (1990b). *Biochemistry* **29**, 6756.
Mason, G. F., Rothman, D. L., Behar, K. L., and Shulman, R. G. (1992). *J. Cereb. Blood Flow Metab.* **12**, 434.
Matthews, P. M., Foxall, D., Shen, L., and Mansour, T. E. (1985). *Nucl. Pharmacol.* **29**, 65.
Mégnin, F., Nedelec, J. F., Dimicoli, J. L., and Lhoste, J. M. (1989). *NMR Biomed.* **2**, 27.
Moore, M. C., Cherrington, A. D., Cline, G., Pagliassotti, M. J., Jones, E. M., Neal, D. W., Badet, C., and Shulman, G. I. (1991). *J. Clin. Invest.* **88**, 578.
Moreland, C. G., and Carroll, F. I. (1974). *J. Magn. Reson.* **15**, 596.
Navon, G., Lyon, R. C., Kaplan, O., and Cohen, J. S. (1989). *FEBS Lett.* **247**, 86.
Nedelec, J.-F., Capron-Laudereau M., Adam, R., Dimicoli, J., Gugenheim, J., Patry, J., Pin, M. L., Fredi, G., Bismuth, H., and Lhoste, J. M. (1989). *Transplant. Proc.* **21**, 1327.
Nedelec, J.-F., Capron-Laudereau, M., Adam, R., Patry, J., Dimicoli, J.-L., Bismuth, H., and Lhoste, J.-M. (1990a). *Transplant. Proc.* **22**, 492.
Nedelec, J.-F., Mégnin, F., Patry, J., Jullien, P., Lhoste, J.-M., and Dimicoli, J.-L. (1990b). *Comp. Biochem. Physiol. B* **95B**, 505.
Neurohr, K. J., Barrett, E. J., and Shulman, R. G. (1983). *Biochemistry* **80**, 1603.
Neurohr, K. J., Gollin, G., Neurohr, J. M., Rothman, D. L., and Shulman, R. G. (1984). *Biochemistry* **23**, 5029.
Nieto, R., Cruz, F., Tejedor, J. M., Barroso, G., and Cerdán, S. (1992). *Biochimie* **74**, 903.
Nishina, M., Hori, E., Matsushita, K., Takahashi, M., Kato, K., and Ohsaka, A. (1988). *Mol. Biochem. Parasitol.* **28**, 249.
Nissim, I., Yudkoff, M., and Segal, S. (1987). *Biochem. J.* **241**, 361.
Nissim, I., Yudkoff, M., and Segal, S. (1988). *Contr. Nephrol.* **63**, 60.
Nissim, I., Nissim, I., and Yudkoff, M. (1990). *Biochim. Biophys. Acta* **1033**, 194.
Novak, M., and Blackburn, B. J. (1988). *Int. J. Parasitol.* **18**, 1029.
Pahl-Wostl, C., and Seelig, J. (1986). *Biochemistry* **25**, 6799.
Pahl-Wostl, C., and Seelig, J. (1987). *Biol. Chem. Hoppe-Seyler* **368**, 205.
Petroff, O. A. C., Burlina, A. P., Black, J., and Prichard, J. W. (1991). *Neurochem. Res.* **16**, 1245.
Petroff, O. A. C., Pleban, L., and Prichard, J. W. (1993). *Magn. Reson. Med.* **30**, 559.
Portais, J. C., Pianet, I., Allard, M., Merle, M., Raffard, G., Kien, P., Biran, M., Labouesse, J., Caille, J. M., and Canioni, P. (1991). *Biochimie* **73**, 93.
Powell, J. W., Stables, J. N., and Watt, R. A. (1986a). *Mol. Biochem. Parasitol.* **18**, 171.
Powell, J. W., Stables, J. N., and Watt, R. (1986b). *Mol. Biochem. Parasitol.* **19**, 265.
Reo, N. V., Siegfried, B. A., and Ackerman, J. J. H. (1984). *J. Biol. Chem.* **259**, 13664.

Ross, B. D., Higgins, R. J., Boggan, J. E., Willis, J. A., Knittel, B., and Unger, S. W. (1988). *NMR Biomed.* **1**, 20.
Sacks, W., Cowburn, D., Bigler, R. E., Sacks, S., and Fleischer, A. (1985). *Neurochem. Res.* **10**, 201.
Scott, A. I., and Baxter, R. L. (1981). *Annu. Rev. Biophys. Bioeng.* **10**, 151.
Seelig, J., and Burlina, A. P. (1992). *Clin. Chim. Acta* **206**, 125.
Seguin, F., Le Pape, A., and Williams S. R. (1992). *In* "Magnetic Resonance Spectroscopy in Biology and Medicine" (J. D. de Certaines, W. M. M. J. Bovée, and F. Podo, F., eds.), pp. 479–489. Pergamon, Oxford.
Shalwitz, R. A., and Becker, N. N. (1990). *Semin. Perinatol.* **14**, 224.
Shalwitz, R. A., Reo, N. V., Becker, N. N., and Ackerman, J. J. H. (1987). *Magn. Reson. Med.* **5**, 462.
Shalwitz, R. A., Reo, N. V., Becker, N. N., Hill, A. C., Ewy, C. S., and Ackerman, J. J. H. (1989). *J. Biol. Chem.* **264**, 3930.
Sherry, A. D., Nunnally, R. L., and Peshock, R. M. (1985). *J. Biol. Chem.* **260**, 9272.
Sherry, A. D., Malloy, C. R., Roby, R. E., Rajagopal, A., and Jeffrey, F. M. H. (1988). *Biochem. J.* **254**, 593.
Sherry, A. D., Malloy, C. R., Zhao, P., and Thompson, J. R. (1992). *Biochemistry* **31**, 4833.
Shulman, G. I., and Landau, B. R. (1992). *Physiol. Rev.* **72**, 1019.
Shulman, G. I., Rothman, D. L., Smith, D., Johnson, C. M., Blair, J. B., Shulman, R. G., and DeFronzo, R. A. (1985). *J. Clin. Invest.* **76**, 1229.
Shulman, G. I., Rossetti, L., Rothman, D. L., Blair, J. B., and Smith, D. (1987). *J. Clin. Invest.* **80**, 387.
Shulman, G. I., Rothman, D. L., Chung, Y., Rossetti, L., Petit, W. A., Jr., Barrett, E. J., and Shulman, R. G. (1988). *J. Biol. Chem.* **263**, 5027.
Shulman, G. I., DeFronzo, R. A., and Rossetti, L. (1991). *Am. J. Physiol.* **260**, E731.
Siegfried, B. A., Reo, N. V., Ewy, C. S., Shalwitz, R. A., Ackerman, J. J. H., and McDonald, J. M. (1985). *J. Biol. Chem.* **260**, 16137.
Sillerud, L. O., and Shulman, R. G. (1983). *Biochemistry* **22**, 1087.
Sillerud, L. O., Han, C. H., Bitensky, M. W., and Francendese, A. A. (1986). *J. Biol. Chem.* **261**, 4380.
Silver, M. S., Joseph, R. I., Chen, C.-N, Sank V. J., and Hoult, D. I. (1984). *Nature (London)* **310**, 681.
Sonnewald, U., Westergaard, N., Krane, J., Unsgard, G., Petersen, S. B., and Schousboe, A. (1991). *Neurosci. Lett.* **128**, 235.
Sonnewald, U., Westergaard, N., Schousboe, A., Svendsen, J. S., Unsgard, G., and Petersen, S. B. (1993). *Neurochem. Int.* **22**, 19.
Strysower, E. H., Kohler, G. D., and Chaikoff, I. L. (1952). *J. Biol. Chem.* **198**, 115.
Szwergold, B. S. (1992). *Annu. Rev. Physiol.* **54**, 774.
Tallan, H. H., Moore, S., and Stein, W. H. (1956). *J. Biol. Chem.* **219**, 257.
Thompson, S. N. (1991). *J. Parasitol.* **77**, 1.
Tunggal, B., Hofmann, K., and Stoffel, W. (1990). *Magn. Reson. Med.* **13**, 90.
Van Cauteren, M., Miot, F., Segebarth, C. M., Eisendrath, H., Osteaux, M., and Willem, R. (1992). *Phys. Med. Biol.* **37**, 1055.
Wasylishen, R. E., and Novak, M. (1983). *Comp. Biochem. Physiol. B* **74B**, 303.
Weiss, R. G., Chacko, V. P., and Gerstenblith, G. (1989a). *J. Mol. Cell. Cardiol.* **21**, 469.
Weiss, R. G., Chacko, V. P., Glickson, J. D., and Gerstenblith, G. (1989b). *Proc. Natl. Acad. Sci. U.S.A.* **86**, 6426.
Weiss, R. G., Gloth, S. T., Kalil-Filho, R., Chacko, V. P., Stern, M. D., and Gerstenblith, G. (1992). *Circ. Res.* **70**, 392.

Weiss, R. G., Kalil-Filho, R., Herschkovitz, A., Chacko, V. P., Litt, M., Stern, M. D., and Gerstenblith, G. (1993). *Circulation* **87**, 270.
Williams, W. F., and Odom, J. D. (1986). *Science* **233**, 223.
Williams, W. F., and Odom, J. D. (1987). *Exp. Eye Res.* **44**, 717.
Willis, J. A., and Schleich, T. (1986). *Exp. Eye Res.* **43**, 329.
Willis, J. A., and Schleich, T. (1992). *Biochem. Biophys. Res. Commun.* **186**, 931.
Winkel, C., and Jans, A. W. H. (1990). *Toxicol. Lett.* **53**, 173.
Xue, M., Ng, T. C., and Majors, A. (1990). *Magn. Reson. Med.* **14**, 530.
Zang, L.-H., Laughlin, M. R., Rothman, D. L., and Shulman, R. G. (1990). *Biochemistry* **29**, 6815.

CHAPTER 6

^{13}C Magnetic Resonance Spectroscopy as a Noninvasive Tool for Metabolic Studies on Humans

Nicolau Beckmann
Biophysics Unit, Preclinical Research, Sandoz Pharma,
CH-4002 Basel, Switzerland

6.1 Introduction

A brief look at the specialized literature reveals that the number of applications of ^{13}C magnetic resonance spectroscopy (MRS) to humans is much more reduced than those of its ^1H and ^{31}P counterparts. The main reason for this situation is the fact that whole-body systems equipped with two radiofrequency (rf) channels became available only by the end of the 1980s. Another reason is the elevated price of ^{13}C-enriched substrates, which dampens the clinical application of ^{13}C MRS on a larger scale. Consequently, the systematic application of this technology to humans has until now been restricted to just a few centers worldwide.

In vivo ^{13}C MRS is especially interesting for the noninvasive study of some aspects of carbohydrate metabolism in humans. Carbohydrates constitute about 1.5% of the normal composition of the human body, fulfilling both structural and metabolic roles. Knowledge about carbohydrate metabolism in healthy indviduals, including its variations and adaptation to periods of, e.g., starvation, diet, and exercise, is a prerequisite to a good understanding of many diseases. Conversely, the study of diseased states provides important clues to the understanding of basic aspects concerning the physiology and biochemistry of organs and tissues.

The first sections of this chapter are devoted to the discussion of methodological aspects of ^{13}C MRS for applications to humans. Measurements on humans are complicated by the inherent characteristics of ^{13}C nuclei and spectra (natural abundance of 1.1%, splitting of the resonances due to coupling of the ^{13}C with the ^{1}H nuclei, large chemical-shift dispersion) and by the heterogeneous composition and anatomical distribution of the tissues. Some important facts must be kept in mind when considering applications of ^{13}C MRS to humans:

- The total measurement time should generally be limited to 1 h, at maximum.
- Power deposition by decoupling should not exceed the safety specifications for humans.
- Signals from superficial adipose tissue are much more intense than those generated by the metabolites of interest.
- Spectral localization methods based on B_0 gradients in combination with frequency-selective rf pulses currently used for ^{1}H and ^{31}P MRS are only applicable if attention is restricted to metabolite resonances occurring in a small frequency range.

Approaches that enable the acquisition of ^{13}C spectra despite these limiting conditions constitute an important subject addressed in this chapter. Of particular interest is the correct choice of surface coils, because they provide a first and decisive step toward spectral localization. Decoupling plays a decisive role in *in vivo* ^{13}C MRS not only due to the simplification of the spectral analysis but also because metabolite resonances usually remain hidden beneath the intense methylene signals of fatty acids in coupled spectra. Decoupling is considered primarily from the point of view of safety. Signal enhancement and spatial localization of spectra based on ^{13}C Chemical-Shift Imaging (CSI), polarization transfer, gradient-enhanced heteronuclear coherence, and proton-observed carbon-edited schemes are discussed next. The methodological part of the chapter concludes with a discussion of the absolute quantification of ^{13}C spectra acquired *in vivo*.

The next sections are devoted to the noninvasive assessment of carbohydrate metabolism in humans using ^{13}C MRS. Because glucose is the most important carbohydrate, constituting a major fuel for the tissues, the discussion focuses mainly on the analysis of glucose metabolism in the human brain, liver, and muscle by ^{13}C MRS. A subsection about the hepatic metabolism of galactose is also included. The remaining sections are devoted to the *in situ* assessment of the monosaturated and polyunsaturated degree of fatty acids in adipose tissue, to the *ex-vivo* analysis of body fluids by high-resolution ^{13}C nuclear magnetic resonance (NMR), and to the characterization of diseased states by ^{13}C MRS.

6.2 Coils

For tissues close to the periphery of the subject, e.g., liver or brain tissue, local coils have a distinctive advantage over larger body or head coils concerning the sensitivity of the measurement. Therefore, most applications of ^{13}C spectroscopy to humans rely on the use of circular surface coils positioned parallel to the boundary of the body. First described by Ackerman et al. (1980), circular surface coils have some well-known properties:

- Their high filling factor leads to improved sensitivity.
- Because of their inhomogeneous B_1 field distribution, the sensitive region is restricted to a half-sphere contained within a coil radius, providing some spatial localization of the signals.
- When also used for transmitting rf, the contribution of superficial signals from the adipose tissue can be reduced by proper adjustment of the rf power.

Circular surface coils, doubly tuned to $^1H/^{13}C$ or not, have thus been generally used for the excitation and detection of ^{13}C signals.

The proton coil, on the other hand, has to guarantee efficient decoupling with minimum power deposition. Several geometrical arrangements have been suggested for accomplishing this task. For instance, Heerschap et al. (1989) used a butterfly ^1H coil, and broadband decoupling was possible without violating the FDA guidelines (Young, 1988; see also Section 6.3). Bottomley et al. (1989) reported the use of a figure-eight-shaped ^1H surface coil with a crossing point on the axis of the parallel ^{13}C circular coil. For volume-selective ^{13}C spectroscopy applications of the Vosing technique, Knüttel et al. (1990a) used a bird-cage coil (Chen et al., 1983) for excitation, and a surface coil doubly tuned at the ^1H and ^{13}C frequencies for receiving.

Another important arrangement is the coplanar and concentric system of $^1H/^{13}C$ circular surface coils, with the proton coil having the larger diameter, which allows broadband decoupled natural abundance ^{13}C spectra of humans to be obtained within a 15-min measurement time or less (Beckmann et al., 1990, 1991, 1993; Jue et al., 1989a). The inherent rf field inhomogeneity of this configuration leads in principle to an inefficient use of the rf power for decoupling because the decoupling field is maximum close to the ^1H coil, and decreases by about 60% in the sensitive volume of the ^{13}C coil (Mispelter et al., 1989). In addition, the decoupling field of the outer proton coil is perturbed by the rf of the inner ^{13}C coil. However, since the perturbing ^{13}C field decreases faster with depth than the decoupling field, in some sample space the magnitude of the decoupling field presents a flat maximum

(Mispelter et al., 1989). By properly selecting the coil dimensions, an optimized ^{13}C–{^{1}H} surface coil system can be projected, for which the maximum homogeneous decoupling field region is matched with the volume of maximum ^{13}C sensitivity. The geometric constraints are:

$$\frac{d_w}{d_C} < 0.1 \tag{6.1}$$

and

$$1.4 \leq \frac{d_H}{d_C} \leq 2.0, \tag{6.2}$$

where d_w is the diameter of the wire, and d_C and d_H are the diameters of the ^{13}C and ^{1}H coils, respectively (Mispelter et al., 1989). Some signal loss is expected due to the electrical coupling between the coils. However, usually no active electrical decoupling scheme is required; it is sufficient to introduce a low-pass filter in front of the carbon receiver in order to suppress the noise introduced by broadband proton decoupling. Thus, the use of a system of coplanar and concentric ^{13}C–{^{1}H} surface coils seems to represent an adequate compromise between optimal decoupling efficiency and minimum power deposition. This arrangement is also adequate for the indirect detection of carbons via protons (Rothman et al., 1985, 1992). For this application, however, the ^{13}C coil is larger than the proton coil.

6.3 Decoupling

The sensitivity and spectral resolution of ^{13}C spectra can be improved by proton decoupling due to the collapse of both short- and long-range ^{1}H–^{13}C couplings. However, applications to humans require special precautions so that the power deposition does not lead to tissue overheating. The present U.S. Food and Drug Administration (FDA) guidelines recommend two levels of concern for local specific absorption rate (SAR) of extremities and trunk: (1) SAR $\leq$ 8 W/kg or (2) localized rf heating not greater than 40°C at extremities and 39°C in the trunk (Young, 1988). Using the magnetic vector potential method, Bottomley et al. (1989) determined the average power deposited in an M-kg subject when a proton-decoupling field of power W_i is applied to the decoupler to be, excluding all input cable losses,

$$P = W_i (1 - Q_{loaded}/Q_{unloaded})/M, \tag{6.3}$$

where Q_i is the quality factor of the coil, loaded and unloaded.

Studies performed at 1.5 and 2.0 T indicate that the SAR levels achieved by decoupling either with cw or WALTZ (Shaka *et al.*, 1983) applied with surface coils are within the guidelines recommended by the FDA (Beckmann *et al.*, 1990; Beckmann, 1992; Heerschap *et al.*, 1989; Jue *et al.*, 1989a,b). Nevertheless, it is advisable to reduce the influence of the high-flux regions close to the decoupling coil by using spacers between the coil arrangement and the human body (Beckmann *et al.*, 1990).

With ^{13}C CSI it has been demonstrated both in phantom studies (Müller and Beckmann, 1989) and under *in vivo* conditions (Beckmann and Müller, 1991a) that WALTZ leads to homogeneous decoupling over a large volume, despite the use of a coplanar and concentric system of surface coils (see Figures 3 and 7 later in this chapter). As already pointed out in Section 6.2, this can be attributed to the fact that for this arrangement of surface coils the maximum homogeneous region of the decoupling field overlaps to a large extent with the volume of maximum ^{13}C sensitivity. The strength of the decoupling rf field is usually calibrated with a simple procedure introduced by Bax (1983).

6.4 Schemes for the Acquisition and Enhancement of ^{13}C Signals and for ^{13}C Editing

6.4.1 Pulse-Acquire Scheme

The acquisition of ^{13}C spectra has been accomplished by means of different approaches, depending, e.g., on the localization of the organ, the metabolite(s) to be detected, the relaxation time(s) of the metabolite(s), and the measurement time available. The simplest and most robust approach is the pulse-acquire scheme in combination with surface coils, in which excitation is performed by a rectangular rf pulse, followed by broadband proton decoupling (Beckmann *et al.*, 1990; Heerschap *et al.*, 1989). Metabolites of interest are detected by applying an rf pulse of θ degrees in the plane of the surface coil, $120° < \theta < 200°$ depending on the depth of the organ or tissue. Usually θ is adjusted to $180°$ at the center of the surface coil, thereby reducing considerably the contribution of superficial signals (nulling approach).

Figure 1 shows a proton broadband-decoupled, natural abundance ^{13}C spectrum of the human abdominal region with $\theta \approx 200°$. Although the spectrum is dominated by fatty acid signals, additional resonances arising from liver metabolites can also be seen, e.g., C1 of glycogen at 100.5 ppm.

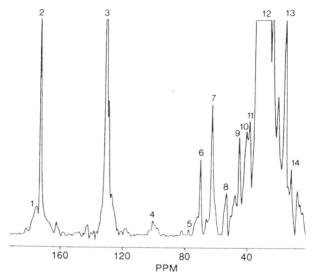

Figure 1. Natural abundance ^{13}C spectrum from the abdominal region of a healthy volunteer acquired at 1.5 T using the pulse-acquire scheme applied with a system of coplanar and concentric ^{1}H/^{13}C surface coils; 3072 acquisitions with a repetition time of 0.51 s (26.0-min measurement time). Decoupling with WALTZ-8. Resonances {2}, {3}, {6}, {7}, {12}, and {13} are from carbonyl carbons, olefinic carbons, C2 of glycerol backbone, C1 and C3 of glycerol backbone, methylene, and methyl groups of fatty acids, respectively. Other tentative assignments are carbonyl carbons in the peptide bonds of proteins {1}; C1 {4} and C4 {5} of glycogen; –N$^+$(CH$_3$)$_3$ of phosphatidylcholine {8}; –CH$_2$–NH$_2$ group from phosphatidylethanolamine {9}; and β-carbons of amino acid sidechains {10} and {11}. (Reproduced with permission from Beckmann *et al.*, 1990. Copyright © 1990 Academic Press, Orlando.)

6.4.2 Frequency-Selective Pulses

A 1-*t*-1 sequence was used by Jue *et al.* (1989a) to suppress the intense lipid signals at 30.0 ppm and to optimally excite the C1 resonance of glycogen. Each pulse was 75 μs long, and the interpulse delay was 320 μs. A total of 11,000 acquisitions and a repetition time of 80 ms, optimized for the C1 of glycogen whose T_1 and T_2 relaxation times at 2.0 T are of the order of 240 and 30 ms (Alger *et al.*, 1984), respectively, were chosen to improve SNR. With a short decoupling time, only the C1 of glycogen was decoupled well; other resonances remained partially coupled. Gaussian pulses of 3.0-ms duration were also used to excite selectively the glycogen resonance (Rothman *et al.*, 1991a).

6.4.3 Adiabatic Pulses

Adiabatic pulses, which provide a homogeneous excitation profile even when applied with surface coils, have also been used to excite ^{13}C nuclei. Heerschap *et al.* (1989) used an adiabatic rapid half-passage pulse (Bendall and Pegg, 1986; Luyten and den Hollander, 1986), 3.0 ms long, for the acquisition of ^{13}C spectra of muscle and liver tissue. Werner *et al.* (1990) applied numerically optimized sin/cos pulses (Bendall and Pegg, 1986; Ugurbil *et al.*, 1988) to increase the signal-to-noise ratio (SNR) of the C1 signal of glycogen.

6.4.4 Polarization Transfer and Spectral Editing

Polarization transfer from protons to ^{13}C (Beckmann and Müller, 1991b; Saner *et al.*, 1989, 1992), cyclic polarization transfer (Knüttel *et al.*, 1991), and heteronuclear editing techniques (Rothman *et al.*, 1985, 1992) have been applied to human studies with the purpose of signal enhancement and spatial localization (for a more detailed description, see Section 6.5). Whereas in the first approach ^{13}C signals are detected, the other two approaches rely on the detection of protons bound to ^{13}C nuclei. Commonly used polarization transfer schemes from ^{1}H to ^{13}C were DEPT (Doddrell *et al.*, 1982) and SINEPT (Jakobsen *et al.*, 1983). Cyclic polarization transfer has been achieved with gradient-enhanced heteronuclear coherence spectroscopy, i.e., the use of techniques based on the selection of coherence pathways by gradients (Knüttel *et al.*, 1991). Heteronuclear editing has been effected with the proton-observed carbon-edited technique (Rothman *et al.*, 1985, 1992).

To avoid signal losses in spectral editing due to the use of conventional rf pulses applied by surface coils, Garwood and Merkle (1991) introduced the B_1-*insensitive spectral editing pulse* (BISEP) for the selective excitation of ^{1}H spins engaged in heteronuclear J modulation. BISEP is based on the adiabatic BIR-4 pulse, which can induce any desired flip angle with compensation for B_1 inhomogeneity and resonance offset (Garwood and Ke, 1991).

Although the approaches just mentioned should provide significant signal enhancements (see Table 1 of Chapter 2), their application *in vivo* is complicated by factors such as the rather short T_2 relaxation times of the metabolites, the inhomogeneous B_1 field of surface coils, motion, and the heterogeneous composition and distribution of tissues. For instance, Saner *et al.* (1992) demonstrated that the C1 resonance of glycogen presents the same SNR in spectra acquired with polarization transfer (from ^{1}H to ^{13}C) or pulse-acquire schemes. On the other hand, polarization transfer sequences are not as robust as the simple pulse-acquire scheme, since they have to be optimized according to the J-coupling value of the resonance of interest. Nevertheless, when applied in conjunction with the administration of ^{13}C-enriched substrates, gradient-enhanced and heteronuclear editing techniques have proven to be very useful. Novotny *et al.* (1990) determined a 14-fold improvement in the sensitivity of proton heteronuclear editing over direct ^{13}C MRS for the detection of [2-^{13}C]ethanol in the rabbit brain *in vivo*. Therefore, using these techniques, we should be able to follow the flux through specific metabolic pathways with a time resolution of a few minutes.

6.4.5 Nuclear Overhauser Effect

Enhancement of *in vivo* ^{13}C spectra through the nuclear Overhauser effect (NOE) was systematically studied by Ende and Bachert (1993). In particular, the effects of transient and truncated driven NOE on the ^{13}C resonances from methylene carbons were observed in human calf tissue. These techniques are based on the application of, respectively, one and a series of rf pulses of flip-angle α to the proton spins, prior to the acquisition of the ^{13}C signal. The SNR of fatty acid signals was significantly higher in NOE-enhanced spectra. Nonetheless, typical enhancement factors η_I for the methyl groups of fatty acids were of the order of 0.52, i.e., only a fraction of the theoretical maximum $\eta_{max} = \gamma_H/2\gamma_C = 1.984$, indicating the presence of other competitive relaxation mechanisms beside the dipolar, e.g., via chemical-shift anisotropy, and/or the violation of the extreme narrowing condition for tumbling acyl chains (Ende and Bachert, 1993).

A systematic study of NOE-enhancement of resonances generated by metabolites of interest, e.g., in the liver and brain, is lacking. Although some resonances might be enhanced by NOE, their absolute quantification becomes less straightforward (see also Section 6.6).

6.5 Spatially Localized ^{13}C Spectroscopy

An essential prerequisite for *in vivo* MRS of any nucleus is the spatial localization of the spectra. Image-guided localization schemes such as PRESS (Bottomley, 1984; Ordidge *et al.*, 1985), STEAM (Frahm *et al.*, 1985), and ISIS (Ordidge *et al.*, 1986), based on the application of frequency-selective pulses in combination with B_0 magnetic field gradients, play a key role in accomplishing this requirement. Although these techniques are successful in ^{1}H and ^{31}P MRS, the rather short T_2 relaxation times of the metabolites, the low natural abundance, and the large chemical-shift range of the ^{13}C nuclei make their direct application in ^{13}C MRS difficult.

Therefore, in studies on animals (see Chapter 5) as well as on humans the surface coil itself has been the most important device for spatially localizing the ^{13}C signals. In most cases the B_1 inhomogeneity provided by the surface coil is sufficient to achieve a reasonable spatial localization by simply using rectangular rf pulses and applying the nulling approach, i.e., by adjusting the pulse to 180° in the center of the coil. An inversion of the magnetization is thus achieved for sample regions close to the surface coil, resulting in a partial suppression of the corresponding signals; at a distance of approximately 0.6 times the coil radius the flip angle is about 90°. Complex DEPTH pulses (Bendall and Gordon, 1983), relying on extensive phase cycling of nonselective rf pulses, further restrict the sensitive volume of the rf coil to regions characterized by specific B_1 field intensities.

A drawback of these approaches is that the spatial region from which the ^{13}C signals arise is difficult to define. Furthermore, the *in vivo* ^{13}C spectra are still dominated by intense fatty acid signals, which may overlap with resonances of interest. Therefore, many applications require an additional step for the spatial localization of the ^{13}C signals than that provided by the surface coil, in particular when an absolute quantification of a metabolite is required. Some approaches used to achieve this goal under certain conditions are discussed in the next subsections.

6.5.1 Suppression of Surface Signals

The suppression of surface signals may be important for reducing the contamination of the ^{13}C spectra not only by fatty acid resonances, but also by signals from muscle tissue, e.g., in the quantification of liver glycogen in more obese subjects. Beside the nulling approach just described, basically three approaches have been used to accomplish this suppression, two of them based on the application of B_0 gradients, and the third of an electrically generated gradient:

1. Preceding the acquisition of ^{13}C signals, Heerschap *et al.* (1989) introduced a rectangular pulse with a flip angle of 50° in the center of the coil followed by a dephasing B_0 gradient.

2. Rothman *et al.* (1991b) applied a one-dimensional ISIS (Ordidge *et al.*, 1986) scheme before the acquisition of ^{13}C signals. A 4-ms phase-swept hyperbolic secant inversion pulse (Silver *et al.*, 1984) was applied in conjunction with a gradient parallel to the axis of the surface coil, resulting in an inversion of the signal from tissue within 2.5 cm of the boundary of the body. Superficial signal was reduced by adding scans obtained with an inversion pulse to scans obtained without the inversion. Subsequent excitation of the ^{13}C nuclei was performed with a rectangular rf pulse. On the basis of proton

density profiles and assuming equal glycogen concentration throughout the volume of observation, it was shown that less than 15% of the glycogen signal could arise from tissues within 2.5 cm of the surface of the subject. Localization errors for the glycogen were avoided by setting the glycogen signal on resonance.

3. A method based on an electrically generated phase-alternating surface gradient applied during the two T periods of a spin-echo pulse sequence (θ-T-2θ-T-AQ) was used to remove the signal contribution from superficial tissues, thus allowing observation of deeper lying tissues with a surface coil without contamination (Chen and Ackerman, 1989; Crowley and Ackerman, 1985). The method relied on the application of an inhomogeneous B_z field gradient localized to the superficial region between the coil and the deeper lying tissue of interest. The gradient was created by passing a low current through a planar array of wires. The usefulness of this approach was first demonstrated *in vivo* for the localization of ^{31}P and ^{13}C signals from rat liver (Chen and Ackerman, 1989, 1990). It was further improved by Jehenson and Bloch (1991), who developed a technique based on the pulse-acquire scheme with a delay T between the rf pulse and the acquisition for surface-gradient application. This method enables the theoretical determination of the optimal coil parameters (size and current) for a given application and avoids the testing of different gradient-coil sizes and current values on phantoms and subjects (Jehenson and Bloch, 1991). As an example, a time delay of $T = 200$ μs and a current of 35 A were used during the detection of hepatic glycogen in humans by natural abundance ^{13}C MRS (Jehenson and Bloch, 1991).

6.5.2 ^{13}C Chemical-Shift Imaging

In humans CSI (Brown *et al.*, 1982) has been applied primarily to ^{1}H and ^{31}P nuclei. Of great interest have been studies concerning the heterogeneity of brain tumors (Luyten *et al.*, 1990) and of ischemic regions (Bottomley *et al.*, 1988). Extension of the technique to ^{13}C nuclei showed some advantages, in particular regarding the spectral localization (Beckmann and Müller, 1991a; Müller and Beckmann, 1989):

- Phase encoding with pulsed B_0 gradients does not introduce any chemical-shift dispersion in the spatial dimension, i.e., spatial and spectroscopic information are completely separated. This is a consequence of the fact that the gradients are not applied during the excitation of the ^{13}C nuclei and during signal acquisition.
- Since spatial resolution is only introduced via B_0 gradients, the B_1 field does not interfere with the localization mechanism, so that surface coils can be used.

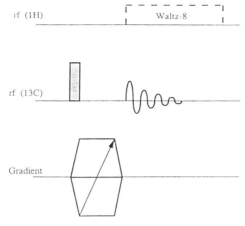

Figure 2. Pulse sequence for ^{13}C spectroscopic imaging. For one-dimensional maps acquired with a surface coil, the gradient is normal to the plane of the coil. (Reproduced with permission from Beckmann and Müller, 1991a. Copyright © 1991 Academic Press, Orlando.)

- The low sensitivity of ^{13}C usually requires extensive signal averaging. Thus the phase encoding period does not prolong significantly the measurement time, and spatial mapping of the metabolites is obtained without compromising substantially the time resolution of the ^{13}C experiment.

Although CSI is a multidimensional experiment providing spatial resolution in up to three dimensions, for sensitivity reasons applications *in vivo* of natural abundance ^{13}C CSI have been limited to one spatial and one spectroscopic dimension (Beckmann and Müller, 1991a). In this case, spectra are acquired along a single phase-encoding gradient, i.e., each spectrum has its origin in a slice orthogonal to the direction of the phase-encoding gradient (Figure 2). When a surface coil is used for excitation and/or detection, the gradient is chosen to be parallel to the coil's axis. With the surface coil positioned at the periphery of the human body, the first sections of the acquired spectroscopic image contain mainly resonances from subcutaneous adipose tissue. Signals from deeper lying organs can be found in the remaining sections of the image. To enable the detection of metabolites of shorter T_2 relaxation times, the delay between excitation and acquisition has to be kept as small as possible. Typical gradient lengths for measurements *in vivo* have been 500 to 700 μs with additional ramp times of 200 μs, for strengths up to 5 mT/m, resulting in a maximum spatial resolution of 2.5 mm (Beckmann and Müller, 1991a).

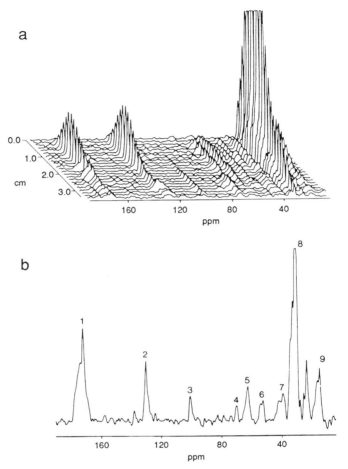

Figure 3. (a) ^{13}C spectroscopic image from the human abdominal region acquired in 25.6 min at 1.5 T using the sequence of Figure 2 applied with a concentric and coplanar system of ^{1}H/^{13}C surface coils. Repetition time 1.2 s, 20 averages, 64 phase-encoding steps (−2.0 to 2.0 mT/m). Decoupling with WALTZ-8. Real part map of dimensions 1024 × 64. (b) A spectrum extracted from the spectroscopic map. Resonances {1}, {2}, {4}, {5}, {8}, and {9} correspond to carbonyl carbons, olefinic carbons, C2 of glycerol backbone, C1 and C3 of glycerol backbone, methylene groups, and methyl groups of fatty acids, respectively. Other tentative assignments are C1 of glycogen {3}; methyl carbons of choline {6}; and β-carbons of amino acid sidechains {7}. (Reproduced with permission from Beckmann and Müller, 1991a. Copyright © 1991 Academic Press, Orlando.)

Figure 3 presents a natural abundance ^{13}C chemical-shift image of the abdominal region generated in 25.6 min. A spectrum extracted from the map (Figure 3b) shows several metabolites contained in the liver with an SNR comparable to that achieved with the pulse-acquire scheme (Figure 1). A metabolite of particular interest is glycogen, readily seen at 100.5 ppm in the map. From ^{13}C chemical-shift images of this kind, it has been estimated that in ^{13}C spectra of the abdominal region the contribution of muscle glycogen to the overall glycogen resonance is less than 10% for nonobese subjects (Beckmann et al., 1993).

6.5.3 Localization and Signal Enhancement Based on Polarization Transfer, Heteronuclear Editing, and Gradient-Enhanced Heteronuclear Multiple-Quantum Coherence

Polarization transfer and heteronuclear proton editing were also applied to *in vivo* ^{13}C MRS with the aim of improving the localization of the spectra and at the same time enhancing the sensitivity of the measurements. In the next subsections acquisition schemes based on direct and cyclic polarization transfer as well as on ^{13}C editing, designed for spatial localization purposes, are discussed. These techniques share the important property that volume selection is guided by proton imaging.

6.5.3.1 *Direct Polarization Transfer from ^{1}H to ^{13}C*

An approach that allows the image-guided spatial localization of ^{13}C signals was described by Aue et al. (1985). It consists of performing the volume selection with B_0 gradients in combination with frequency-selective pulses on protons, where localization is excellent due to the small chemical-shift range, and then transferring the localized proton magnetization to the ^{13}C nuclei in the same volume (Figure 4). Localization errors due to chemical-shift dispersion are therefore minimized because the ^{13}C signals are localized via the volume selected in the proton space. It was demonstrated experimentally and theoretically that an excellent spatial localization of the ^{13}C signals can be achieved by this approach even when using ^{1}H and ^{13}C surface coils for excitation and detection (Beckmann and Müller, 1991b). The main advantages of the present technique follow:

- Selection is guided by a proton image.
- Since the resulting ^{13}C magnetization is directly proportional to the ^{1}H magnetization preceding the polarization transfer step (Beckmann and Müller, 1991b), selectivity may be verified with the more sensitive proton signals, for example, by looking at proton projections or images of the selected region.

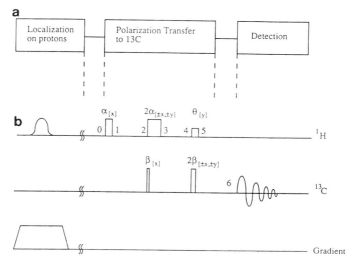

Figure 4. Pulse scheme for localized ^{13}C spectroscopy using proton volume selection followed by polarization transfer to ^{13}C. The general scheme (a) may involve various localization and polarization transfer techniques. In scheme (b) localization on protons is performed with 1D ISIS and polarization transfer with DEPT. Flip angles on the nonselective pulses should ideally be multiples of 45°; however, in the inhomogeneous B_1 field of the surface coil they are spatially dependent, and therefore indicated generally as α, β, and θ. The ISIS inversion pulse is of the hypersecant type (Silver et al., 1984). (Reproduced with permission from Beckmann and Müller, 1991b. Copyright © 1991 Academic Press, Orlando.)

- As the selection is performed on protons, different image-guided spatially selective sequences can precede the polarization transfer step. Selection may be performed in one, two, or three dimensions, and multifrequency excitation (Müller et al., 1988, 1989; Reddy et al., 1993) is equally possible.
- Any polarization transfer scheme may be used.
- High-flux signals originating from regions close to the surface coils are substantially reduced or even eliminated from the spectra.

A disadvantage of this method, common to all approaches based on direct and cyclic polarization transfer, is that signals from nuclei that relax very fast are difficult to detect because the sequence is rather long. Nevertheless, spectra from the human muscle, obtained with one-dimensional ISIS (Ordidge et al., 1986) for signal localization and DEPT (Doddrell et al., 1982) for polarization transfer, showed that the contribution of fat signals was substantially reduced and some metabolite resonances, especially in the frequency interval between 40 and 80 ppm, were characterized better in the spectra acquired with this approach than with the pulse-acquire scheme (Beckmann and Müller, 1991b).

6.5.3.2 Proton-Observed ^{13}C-Edited Spectroscopy

The Proton-Observed Carbon-Edited (POCE) approach, particularly useful in combination with the administration of ^{13}C-labeled substrates, is based on the multinuclear multipulse difference sequence described by Bendall *et al.* (1981), which allows the observation of protons coupled to ^{13}C while eliminating signals of protons bound to ^{12}C nuclei. The sequence, which is basically a spin-echo for protons, includes an inversion pulse for carbons applied at $t = 1/(2J)$ after the ^{1}H excitation pulse on alternate scans only. The inversion pulse has no effect on protons attached to molecules containing ^{12}C nuclei, and signals from such protons are canceled by subtraction of alternate scans. Magnetization of protons directly attached to ^{13}C nuclei refocuses along the $\pm y$ axes of the rotating frame on consecutive scans. Subtraction of the signals from consecutive acquisitions thus causes a summation of the corresponding magnetization components.

This technique was adapted to generate POCE spectra from the human brain with surface coils for measuring the glutamate turnover in the visual cortex during the intravenous infusion of 99%-enriched [1-^{13}C]glucose (Rothman *et al.*, 1992; see also Section 6.7.1.1). Image-guided volume selection was achieved by a three-dimensional ISIS sequence (Ordidge *et al.*, 1986) in proton space. Additional three-dimensional outer volume suppression was also applied. The acquisition sequence consisted of a sinc pulse in an x gradient for excitation and a semiselective 2$\overline{2}$ pulse (Hore, 1983) followed by dephasing gradients for water suppression and refocusing. A 1-ms ^{13}C inversion pulse centered on a time interval $(2J_{^1H-^{13}C})^{-1}$ prior to acquisition was applied to the [4-^{13}C]glutamate resonance on alternate scans. Free induction decays (FIDs) acquired with the inversion pulse were subtracted from FIDs acquired with a 90_x90_{-x} ^{13}C pulse to obtain the edited spectrum. Additional water suppression was achieved by applying a 30-ms frequency-selective pulse to the H$_2$O resonance frequency, and the spectra were ^{13}C decoupled (Rothman *et al.*, 1992).

The POCE approach has the disadvantage that protons resonating close to water are difficult to detect. In addition, the method is motion sensitive, since multiple scans are necessary to attain coherence selection.

6.5.3.3 Gradient-Enhanced Heteronuclear Multiple-Quantum Coherence

Heteronuclear Multiple-Quantum Vosing Technique In contrast to the two-scan subtraction sequence originally proposed for line editing (Bendall *et al.*, 1981; see also 6.5.3.2), the Vosing technique (Knüttel *et al.*, 1990a) for image-guided ^{13}C volume-selective spectroscopy is a single-scan editing

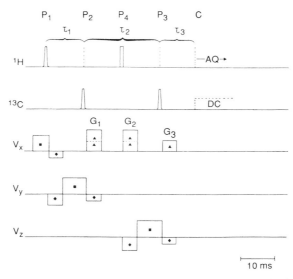

Figure 5. RF and field-gradient pulse sequence of the heteronuclear multiple-quantum filter Vosing technique. Field-gradient pulses serve three different purposes: spoiling of undesired coherences (▲), balancing of pairs of dephasing/rephasing intervals (♦), and selection of slices (■). RF pulses P_1 (protons) and P_2, P_3 (carbons) are frequency selective, whereas P_4 (protons) is a hard pulse. P_1: $\{90°\}^H_{\pm x}$, P_2: $\{90°\}^C_{\pm x}$, P_3: $\{90°\}^C_{\phi}$, and P_4: $\{180°\}^H_{\phi}$. Phase ϕ is arbitrary, and G_1, G_2 and G_3 are the gradients that determine the pathway selection. (Reproduced with permission from Knüttel et al., 1990a. Copyright © 1990 Academic Press, Orlando.)

procedure (Figure 5). This technique is based on cyclic polarization transfer (Cyclpot) and multiple-quantum filtering (MQF); B_0 field gradients are applied for slice selection, for spoiling undesired coherences, and for balancing pairs of dephasing/rephasing intervals. In the Cyclpot procedure (Knüttel et al., 1990a), which consists of applying successively forward and backward INEPT-like polarization transfer sequences, coherences resulting from a cyclic polarization transfer pathway are detected while all other resonances are suppressed. The proton magnetization is first transferred to the ^{13}C coherences for all ^1H nuclei bound and properly coupled to the ^{13}C nuclei. At this point, spoiler gradients are applied. Coherences of coupled protons are selectively refocused at a later stage of the sequence after having transferred their polarization back from the ^{13}C to the ^1H nuclei. MQF line editing on the other hand is based on the transfer of single-quantum ^1H coherences to heteronuclear multiple-quantum coherences, which are then transferred back to single-quantum ^1H coherences. Echoes from all other coherence pathways, in particular from uncoupled protons, are suppressed. Volume selection is provided by three slice-selective 90° pulses (P_1, P_2 and P_3 in Figure 5) applied in different channels.

The phase ϕ_i induced by a gradient $\mathbf{G}_i$ during a period τ_i on a given coherence is (see also Section 2.4.1.5)

$$\phi_i = \left(\sum_i p_i \gamma_i\right)_i \int_0^{\tau_i} \mathbf{G}_i(t) \cdot \mathbf{r}_i(t) \, dt, \qquad (6.4)$$

where p_i and γ_i are the coherence order and the gyrogmagnetic ratio of the coupled nuclei, respectively, and $\mathbf{r}_i$ their position in the sample. Complete rephasing occurs only when the sum of the phases ϕ_i induced by all the gradients in a sequence is zero, hence

$$\sum_i \phi_i = 0. \qquad (6.5)$$

Let us apply these conditions for a CH group submitted to the Vosing sequence (Figure 5). Considering the position of the nuclei to be equal to r during the whole evolution period and the gradients to have rectangular shapes, Eq. (6.4) reduces to

$$\phi_i = (p_H \gamma_H + p_C \gamma_C)_i \, r \cdot G_i \cdot \tau_i. \qquad (6.6)$$

During the application of the gradient G_1 double-quantum I^+S^+ coherence will accumulate a phase $(\gamma_H + \gamma_C)rG_1\tau_1$. The 180° P_4 pulse creates zero-quantum I^-S^+ coherence, which acquires a phase $(-\gamma_H + \gamma_C)rG_2\tau_2$ during the application of G_2. The 90° P_5 pulse creates I^-S_z antiphase magnetization, which accumulates a phase of $-\gamma_H G_3 \tau_3$ during the application of G_3. Since $\gamma_H \approx 4\gamma_C$, refocusing will occur if the gradients have the same duration and an intensity ratio of, e.g., $G_1:G_2:G_3 = 2:2:1$, as depicted in Figure 5. For this combination of gradients, the pathway selected is double-quantum, zero-quantum, and single-quantum, respectively. Another possibility would be to choose a gradient ratio of $G_1:G_2:G_3 = 3:5:0$. In this case double- and zero-quantum coherences are selected during the application of G_1 and G_2, respectively. The sensitivity of the method to residual gradients for different gradient combinations is discussed by Ruiz-Cabello et al. (1992).

The present approach has the advantages that (1) it is a one-scan technique and (2) gradient-based coherence selection does not depend on phase coherence between scans (when multiple scans are necessary). It is, therefore, rather insensitive to motion. In addition, because only protons attached to ^{13}C nuclei are selected, an inherent frequency-independent water suppression is achieved, while lipids are suppressed to the natural abundance ^{13}C level. Thus, metabolites close to the water resonance are also detectable.

An interesting *in vivo* application of the Vosing technique was the detection of glycogen in the human liver within a 22-min measurement time (Knüttel *et al.*, 1991). The Vosing technique has also been extended to image selectively protons coupled to ^{13}C nuclei (Knüttel *et al.*, 1990b).

Again attention must be given to relaxation times. Since the acquired echoes are strongly affected by T_2 relaxation, signals resulting from metabolites with short T_2 relaxation times are difficult to be detected *in vivo*.

With the advent of actively shielded gradient systems, the gradient-enhanced heteronuclear coherence experiments as described by Knüttel *et al.* (1990a) and improved by Ruiz-Cabello *et al.* (1992) are becoming increasingly important, in particular for dynamic measurements concerning the metabolism of ^{13}C-enriched substrates. An elegant experiment using this approach and dealing with the metabolism of [1–^{13}C]glucose in the cat brain was presented recently by van Zijl *et al.* (1993a). The resonances of C1 of α- and β-glucose, C2, C3, and C4 of glutamate, C4 of glutamine, and C3 of lactate could be detected with a time resolution of 3 min. In addition, heteronuclear (^{1}H–^{13}C) 2D spectra were acquired in 4.5 min, enabling the separation of the resonances of C4 of glutamate and glutamine (van Zijl *et al.*, 1993a). Applications to humans are expected soon because hardware modifications that enable us to perform such experiments with a whole-body system have already been reported (van Zijl *et al.*, 1993b).

6.5.3.4 Polarization Transfer in the Rotating Frame

The recently described Cyclrop-Losy technique (Kunze *et al.*, 1993) is based on cyclic and spatially resolved coherent rotating-frame polarization transfer. Like the Cyclpot-Vosing technique, it consists of a forward and a backward polarization transfer segment, each of which is composed of a spin-locking pulse on the primary and a contact pulse on the secondary side (Figure 6). These rf pulse pairs are matched according to the Hartmann-Hahn condition (Hartmann and Hahn, 1962). After having transferred the polarization from the I (^{1}H) to the S (^{13}C) spins, the in-phase coherences of the S spins are stored as z magnetization by the aid of a 90_{-x} pulse. The antiphase coherences are dephased together with the residual I-spin coherences by a subsequent homospoil gradient. The S-spin magnetization is then spin-locked again and the polarization is transferred back to the I spins, producing I-spin coherences of the coupled spins only. In the presence of gradients the Hartmann-Hahn matching condition is spatially dependent; the cross-polarization process becomes slice selective. Spatial selectivity may also be achieved by different inhomogeneities of the superimposed double-resonance radiofrequency fields, for example, by combining a ^{13}C surface coil with a ^{1}H bird-cage resonator (Köstler and Kimmich, 1993a). J cross-polarization is thus restricted to the region where the Hartmann-Hahn con-

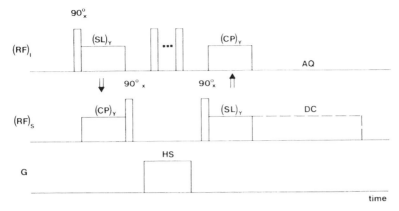

Figure 6. Pulse sequence for heteronuclear editing by cyclic J cross-polarization (Cyclrop). Abbreviations: SL, spin-lock pulse; CP, contact pulse; HS, homospoil pulse using a field gradient G; DC, decoupling pulse; AQ, acquisition of the induction signals. The contact pulses are matched to the spin-locking pulses according to the Hartmann-Hahn condition. The cross-polarization intervals are optimized for maximum polarization transfer. This depends on the spin-spin coupling constant J, on the size of the spin system, and on deviations from the Hartmann-Hahn matching. After the polarization is transferred from the I to the S spins, the magnetization is stored in the z direction. Coherences of the abundant I spins can be presaturated by a comb of 90° pulses, and are also dephased by the homospoil pulse. The S-spin magnetization is then spin-locked again and its polarization transferred back to the I spins, where the corresponding coherences are detected. Application of an S-spin decoupling pulse is optional.

dition is approached. A version of the experiment based on adiabatic J cross-polarization may be found in Köstler and Kimmich (1993b).

As for gradient-enhanced spectroscopy, inherent frequency-independent water suppression is attained because only coherences from protons attached to ^{13}C nuclei are detected. For the moment, the Cyclrop-Losy technique has been applied only to phantom studies. When applied *in vivo*, care should be taken to ensure that the power necessary for the Hartmann-Hahn matching does not exceed the safety recommendations.

6.5.4 Frequency-Selective Pulses in Combination with B_0 Gradients Applied Directly to ^{13}C Nuclei

If attention is limited to metabolite signals on or close to resonance, localization errors introduced by the direct application of frequency-selective pulses in combination with B_0 gradients are avoided or considerably reduced.

Under these conditions, some of the image-guided localization schemes commonly used in ^{1}H and ^{31}P MRS may also be applied directly to ^{13}C nuclei.

6.5.4.1 Depth-Resolved Surface Coil Spectroscopy (DRESS)

The DRESS technique (Bottomley *et al.*, 1984), consisting of the application of a frequency-selective pulse in the presence of a gradient directed normal to a surface detection coil, allows for the acquisition of signals from a slice parallel to the coil as soon as the gradient is refocused. Cardiac-gated ^{13}C DRESS spectra, acquired in 7 min at 1.5 T from a 2-cm-thick section positioned in the anterior myocardium of the human heart, were reported by Bottomley (1989). The spectra presented resonances in the region about 100 ppm, which were attributed to glycogen and glucose. Signal enhancement was achieved through NOE.

The main difficulty of the DRESS technique is the delay between excitation and acquisition; this delay is necessary for refocusing the magnetization. Signal decay is affected by T_2^* relaxation; therefore, a dramatic signal reduction for species of small T_2 relaxation times is expected. Furthermore, the acquisition delay may also cause rolling baseline artifacts.

6.5.4.2 Image-Selected in Vivo Spectroscopy (ISIS)

A 3D ISIS sequence (Ordidge *et al.*, 1986) was applied to ^{13}C nuclei for measuring the glucose content in 144-ml volumes (6 × 4 × 6 cm^3) localized within the occipitoparietal region of the human brain, excluding major blood vessels and the ventricles, after the intravenous infusion of [1–^{13}C]glucose (Gruetter *et al.*, 1992a; see also Section 6.7.1.1). While 8-ms-long hypersecant pulses (Silver *et al.*, 1984) were used for spin inversion, excitation was performed by a 5-ms adiabatic half-passage sin/cos pulse (Bendall and Pegg, 1986) with a numerically optimized modulation function. To minimize the z magnetization and to ensure proper localization, the pulse used for excitation was applied again at the end of the acquisition. The time resolution for each spectrum was 30 min.

The main advantage of the ISIS technique is that it is not affected by T_2 relaxation. However, since ISIS relies on the addition and subtraction of signals (for the 3D version of the experiment, eight signals), it is extremely susceptible to movement during acquisition.

6.6 Absolute Quantification of Metabolites

Since the intensity of an NMR signal is a complicated function of many parameters, e.g., coil loading, relaxation times, and spatial distribution of the rf field, usually only relative changes of peak areas are followed in the *in vivo* spectra. If a resonance remains fairly constant during the measurements, it can serve as an internal standard. Furthermore, if the concentration of the substrate acting as standard can be established by another means, then the concentration of other metabolites in the spectra can be inferred from the relative areas under the peaks. However, in most situations such an internal standard is lacking, or the concentration of the standard varies with the physiological and/or pathological conditions. A direct absolute quantification of metabolites is therefore desirable in most circumstances.

The absolute quantification of ^{13}C spectra from humans was generally based on a comparison between the signal generated *in vivo* and the signal intensity *in vitro* from a phantom containing a solution of the metabolite to be quantified. An important condition was that the distance between the region of interest, containing the substrate in question, and the coil was the same in the *in vivo* and *in vitro* experiments. This method was applied, for instance, to the determination of the glycogen concentration in the human liver (Beckmann *et al.*, 1993; Rothman *et al.*, 1991b; see also Section 6.7.2.1). To simulate liver depth, which ranged between 2.0 and 2.4 cm, the phantom containing glycogen was raised 2.2 cm above the surface coil. Ill-defined tissue geometry was overcome by comparing the coil sensitivity profile *in vivo* and *in vitro* using proton imaging. Differences in the spectrometer sensitivity and coil loading for the *in vivo* and the *in vitro* measurements were corrected by integrating the signal from a reference sample having the same relative position to the coil in both experiments. This procedure allowed a glycogen quantification even when the volunteers were removed from the magnet between the measurements, compensating also for artifacts and differences in coil loading due to respiration and other voluntary movements during the examinations. Liver glycogen concentrations determined noninvasively with ^{13}C MRS (Beckmann *et al.*, 1993; Fried *et al.*, 1994; Rothman *et al.*, 1991b), about 255 mM after a carbohydrate-rich diet and an overnight fasting period, are in good agreement with the value of 270 mM obtained invasively with needle biopsies (Nilsson and Hultman, 1973).

6.7 *In Vivo* Metabolic Studies with ^{13}C MRS

Until now attention has been concentrated on the technical aspects of the acquisition, localization, and quantification of ^{13}C signals acquired under *in vivo* conditions. The next sections are devoted to the applications of ^{13}C MRS to metabolic studies in healthy humans. The organs of major interest have been the brain and liver, followed by the muscle. The following step will be diseased states, the final goal of clinical studies.

6.7.1 Human Brain Function

Recently, powerful methods have been developed for the noninvasive investigation of human brain function. They aim at, for example, visualizing within the brain local physiological changes associated with visual or motor activation. The basis for these methods is the association between discrete anatomical regions in the brain and sensory, cognitive, and motor functions, a fact that was recognized long ago. The technique that has so far contributed most to functional brain mapping is positron emission tomography (PET), which depends on the administration of radioactive-labeled substrates (for a review, see Chadwick and Whelan, 1991). More recently functional imaging of the human brain has also been achieved using proton magnetic resonance imaging without the administration of any contrast agent (Belliveau et al., 1991; Kwong et al., 1992; Ogawa et al., 1992), thus opening another avenue for the noninvasive examination of certain aspects of brain function in the normal and diseased states.

The energy necessary for these functional processes is derived from glucose, which constitutes the primary carbon source for energy production and amino acid synthesis in the brain. The rate of glucose transport into the brain must therefore be high enough to cover the energy demand of the cells under basal conditions. However, it may become rate-limiting during hypoglycemia and in certain diseases. For instance, alterations in brain glucose transport may be involved in the central nervous morbidity associated with diabetes mellitus, seizures, and hypoxic-ischemic encephalopathy.

Regional glucose uptake in the human brain has been determined with PET by measuring the distribution of radioactivity after the infusion of isotopically labeled glucose. The main drawback of this approach, beside the fact that radioactive substances are used, is that the label cannot be observed on specific positions of the molecules, i.e., signals from glucose and glucose metabolites cannot be distinguished.

As pointed out in Chapter 5, a distinctive advantage of ^{13}C MRS is the possibility of following noninvasively and simultaneously the kinetics of ^{13}C labeling in individual carbons of several metabolites after the administration of stable ^{13}C-labeled substrates. During the metabolism of the externally supplied compounds, the ^{13}C label is transferred to various intermediates, which may be detected by ^{13}C MRS. In this case, metabolic pathways and fluxes become accessible to an *in vivo* analysis. Advantage was taken of these properties for assessing noninvasively the glucose transport and metabolism in the human brain with ^{13}C MRS, the subject of the next subsection.

Before discussing this topic, let us examine which brain metabolites can be detected *in vivo* with natural abundance ^{13}C MRS in a reasonable measurement time of about 30 min or less. Although a coupled ^{13}C spectrum of the human head was published as early as 1983 (Bottomley et al., 1983), the first broadband proton-decoupled natural abundance ^{13}C spectra were reported only seven years later (Beckmann *et al.*, 1990). The reason for this delay has certainly been the fact that whole-body systems equipped with two rf channels became available only toward the end of the 1980s. Figure 7a shows a 1D ^{13}C chemical-shift image of the human head, acquired within 27 min (Beckmann and Müller, 1991a). The map is dominated by fatty acid signals, but some metabolite resonances can also be recognized, as demonstrated in an individual spectrum extracted from the map (Figure 7b). A tentative assignment of these resonances, based on ^{13}C spectra from isolated tissues (Bárány *et al.*, 1984, 1985), reveals the metabolites (phospho)creatine, phosphorylcholine, phosphoethanolamine, γ-aminobutyrate, and N-acetyl-aspartate, which have concentrations between 1.1 μmol/g-brain and 9.6 μmol/g-brain (Clarke *et al.*, 1989; Petroff *et al.*, 1989). Unfortunately there is a strong overlap in the *in vivo* spectra, and not all resonances can be properly resolved. By increasing the measurement time to 90 min, Gruetter *et al.* (1992b) were able to detect *myo*-inositol in the human brain with natural abundance ^{13}C MRS in a 144-cm^3 volume selected with ISIS (Ordidge *et al.*, 1986). Quantification of the signals revealed a concentration of 7.2 ± 0.5 μmol/g-brain for *myo*-inositol (C5 at 75 ppm, C1–C3 around 73 ppm, C4 and C6 around 72 ppm), which is in agreement with concentration measurements performed by chemical methods (Petroff *et al.*, 1989; Stokes *et al.*, 1989).

6.7.1.1 Glucose Transport and Metabolism

The concentration and metabolism of glucose, the primary brain fuel, have been assessed in the human brain by ^{1}H MRS (Bruhn *et al.*, 1991; Michaelis *et al.*, 1991; Gruetter *et al.*, 1992c), by ^{13}C MRS (Beckmann *et al.*, 1991; Gruetter *et al.*, 1992a), and by POCE spectroscopy (Rothman *et al.*, 1992). Proton MRS is particularly attractive for this purpose since the sensitivity of the measurements allows the use of natural abundance glucose in the experiments. However, direct observation of glucose signals in the proton spectra is made much more difficult by the proximity of the water signal and by the overlap with resonances from taurine, glutamate, glutamine, and inositol. The quantification of the glucose signal in ^{13}C spectra is much easier since it appears in a spectral region that is free from other resonances. For this reason ^{13}C MRS remains an important approach to follow the concentration and metabolism of glucose in the human brain, although it requires the administration of ^{13}C-enriched glucose.

One study reported qualitatively the dynamics of incorporation of the ^{13}C label into metabolites of the brain during the intravenous infusion of 99%-enriched [1–^{13}C]glucose in a hyperglycemic clamp (Beckmann *et al.*, 1991). Figure 8 shows spectra resulting from the difference between signals acquired during the referred time intervals and the reference signal, acquired

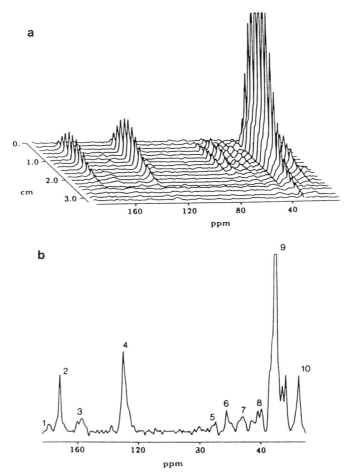

Figure 7. (a) ^{13}C spectroscopic image from the human head acquired in 28.2 min at 1.5 T using the sequence of Figure 2 applied with a concentric and coplanar system of ^{1}H/^{13}C surface coils. Repetition time 1.1 s, 24 averages, 64 phase-encoding steps (−5.0 to 5.0 mT/m). Decoupling with WALTZ-8. Real part map of dimensions 1024 × 64. (b) A spectrum extracted from the spectroscopic map. Resonances {2}, {4}, {5}, {6}, {9}, and {10} correspond to carboxyl carbons, olefinic carbons, C2 of glycerol backbone, C1 and C3 of glycerol backbone, methylene groups, and methyl groups of fatty acids, respectively. Other tentative assignments are C1, C4 of N-acetylaspartate (NAA) {1}; C3 of creatine {3}; C2 of NAA, C4 of creatine, C5 of sn-glycerol-3-phosphorylcholine {7}; C2 of glycine and ethanolamine {8, left}; C3 of NAA, C4 of γ-aminobutyrate, and C2 of creatine {8, right}. (Reproduced with permission from Beckmann and Müller, 1991a. Copyright © 1991 Academic Press, Orlando.)

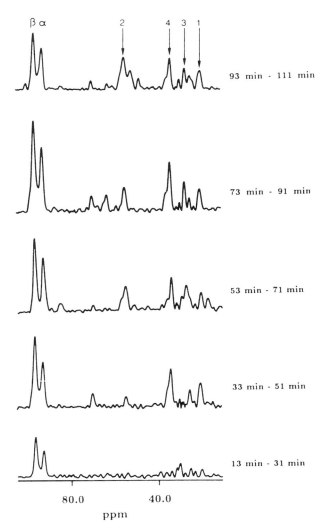

Figure 8. Spectra obtained from the difference between the signals acquired during an infusion of 99% [1-^{13}C]glucose and the control spectrum. The timescale is relative to the beginning of the infusion. The lettering indicated in the spectra correspond to the following substrates: β, C1 of β-glucose (96.6 ppm); α, C1 of α-glucose (92.4 ppm); 2, C2 of glutamate/glutamine (55.0 ppm); 4, C4 of glutamate/glutamine (34.4 ppm); 3, C3 of glutamate/glutamine (27.8 ppm); 1, C3 of lactate (21.0 ppm). Individual spectra were acquired at 1.5 T applying the pulse-acquire scheme with a system of concentric and coplanar ^{1}H/^{13}C surface coils. (Reproduced from Beckmann et al., 1991. Copyright © 1991 The American Chemical Society.)

prior to the beginning of the infusion. The C1 resonances of α- and β-glucose, resulting from glucose in the brain and in the circulation, were detected after only 10 min of infusion time. The difference spectra show additionally the appearance of resonances at 55.0, 27.8, and 34.4 ppm, attributed to C2, C3, and C4 of glutamate/glutamine, respectively (Bárány et al., 1985; see also Figure 14 of Chapter 5). Of note is that the ^{13}C label reached the C4 position earlier ($\approx$30 to 40 min) than the C2 and C3 positions of glutamate/glutamine ($\approx$50 to 70 min). As mentioned in Section 5.7.1, this labeling pattern is characteristic for the generation of [3-^{13}C]pyruvate in glycolysis and its subsequent incorporation into the tricarboxylic acid cycle via pyruvate dehydrogenase. There are also indications of an additional resonance at 21.0 ppm (C3 of lactate).

To quantify the brain glucose by ^{13}C MRS, a further localization step is necessary. This was described by Gruetter et al. (1992a), who used the ISIS sequence (Ordidge et al., 1986) to determine directly brain glucose concentrations at euglycemia and hyperglycemia in healthy children (13 to 16 years old) with ^{13}C MRS. The selected 144-ml volumes were positioned within the occipitoparietal region to exclude the major blood vessels and the ventricles. Brain glucose concentrations averaged 1.0 ± 0.1 mM at euglycemia (4.7 ± 0.3 mM in plasma) and 1.8 to 2.7 mM at hyperglycemia (7.3 to 12.1 mM in plasma). From the relationship between plasma and brain glucose concentrations the Michaelis-Menten parameters of transport were calculated to be K_t = 4.9 ± 0.9 mM and T_{max} = 1.1 ± 0.1 μmol/g^{-1} min^{-1}. The authors concluded that brain glucose concentrations and transport constants were consistent with transport not being rate-limiting for resting brain metabolism at plasma levels larger than 3 mM. By applying a similar procedure, Gruetter et al. (1992d) were also able to detect some amino acids in localized ^{13}C spectra of the human brain after 2 to 3 h of infusion of [1-^{13}C]glucose.

Instead of acquiring ^{13}C spectra, Rothman et al. (1992) used the POCE approach (Rothman et al., 1985; see also Section 6.5.3.2) to follow the metabolism of glucose in the human brain during an intravenous infusion of [1-^{13}C]glucose. The time course of the ^{13}C label incorporation into C4 of glutamate was followed in 24-ml volumes placed in the occipital cortex. An unedited ^{1}H-[^{13}C] spectrum of human brain, acquired in 3 min, is shown in Figure 9a. In vivo ^{13}C-edited proton spectra of glutamate obtained before (spectrum A) and after (spectrum B) [1-^{13}C]glucose infusion are presented in Figure 9b. The resonance that appeared at 2.29 ppm during the infusion was assigned to [4-^{13}C]glutamate, with possible contributions from [3-^{13}C]glutamate and [4-^{13}C]glutamine. From these data the isotopic turnover time for the C4 glutamate pool was estimated to be $\approx$21 min (Rothman et al., 1992), which is consistent with O_2 consumption measurements performed with PET (Fox et al., 1988).

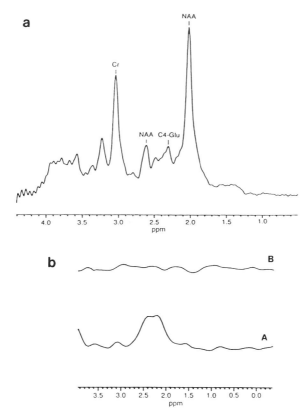

Figure 9. (a) Unedited ^{1}H–[^{13}C] spectrum of human brain. Assigned resonances are the acetyl and aspartyl groups of NAA, the C4 resonance of glutamate (C4–Glu), and the methyl resonance of creatine (Cr). The spectrum was obtained from 3 min of data acquisition. (b) *In vivo* ^{13}C-edited glutamate spectra obtained with the POCE sequence. Twelve-minute ^{1}H–[^{13}C] spectra obtained before (spectrum A) and after (time = 60 min, spectrum B) [1–^{13}C]glucose infusion are shown. The resonance at 2.29 ppm in the 60-min spectrum corresponds to [4–^{13}C]glutamate. Spectra were obtained at 2.1 T with a system of concentric and coplanar ^{1}H/^{13}C surface coils. (Reproduced with permission from Rothman *et al.*, 1992.)

In the edited ^{1}H–[^{13}C] spectra the C4 signal of glutamate at 2.29 ppm overlaps to some extent with the C4 resonances of glutamine and C3 of glutamate. This overlap could be avoided with 2D heteronuclear ^{1}H–^{13}C methods based on gradient-enhanced heteronuclear techniques, recently applied to similar experiments on the cat brain (van Zijl *et al.*, 1993a; see also Section 6.5.3.3).

6.7.2 Carbohydrate Metabolism in the Liver

Amino acids, glucose, and other water-soluble products of digestion are initially directed to the liver since they share a common route of absorption via the hepatic portal vein. The liver has the primary function of regulating the blood concentration of most metabolites, especially of glucose and amino acids. In the case of glucose, this is achieved by taking up excess glucose and converting it to glycogen (glycogenesis) or to fat (lipogenesis). Between meals, the liver can use its glycogen stores to replenish glucose in the blood (glycogenolysis) or, in conjunction with the kidneys, to convert non-carbohydrate metabolites such as lactate, glycerol, and amino acids to glucose (gluconeogenesis). The maintenance of an adequate level of blood glucose is vital for certain tissues where glucose is an obligatory fuel, e.g., brain and erythrocytes.

The next subsections are devoted to the noninvasive analysis of glucose and galactose metabolism in the human liver by ^{13}C MRS.

6.7.2.1 Glucose Metabolism

Quantification of Glycogen The publication of the first ^{13}C spectrum from human liver glycogen *in vivo* (Jue *et al.*, 1987) has opened the door to its noninvasive quantification at multiple time points based on the C1 resonance at 100.5 ppm. Previously hepatic glycogen could only be assessed by biochemical analysis of samples obtained from needle biopsies. It is evident that such an analysis was hardly requested, and repetitive sampling was impossible.

However, quantification of glycogen by ^{13}C MRS requires some caution. It is very important to standardize glycogen determinations because of the simultaneous and continuous formation and breakdown of this macromolecule. Thus, a standardized diet before the measurements may be helpful to reduce the interindividual differences in glycogen concentration. Another point of concern is the MR visibility of glycogen, a matter of extensive debate in the literature (for a discussion of this subject, see Section 5.5.2). To clarify this point, it would be of interest to compare the hepatic glycogen determinations by ^{13}C MRS with biochemical assessments on individuals who are going to undergo biopsy anyway.

Quantification of glycogen was usually carried out in the morning following a 12-h fasting period since under these conditions glycogen was estimated to be $\approx$100% MR visible (Künnecke and Seelig, 1991). Typical average hepatic glycogen concentrations determined after a carbohydrate-rich diet and a 12-h fast were of 251 $\pm$ 30 (Rothman *et al.*, 1991b) and 254.3 $\pm$ 38.8 mM (Fried *et al.*, 1994), which are consistent with the value of 270 mM obtained with needle biopsies (Nilsson and Hultman, 1973). Without any special dietary preparation, average liver glycogen concentration in 12-h overnight fasted volunteers was assessed to be 229 ± 34 mM (Beckmann et al., 1993). For a brief description of the quantification procedure, see Section 6.6.

Influence of the Route of Glucose Administration on Hepatic Glycogen Repletion The formation of hepatic glycogen in normal volunteers was followed noninvasively with ^{13}C MRS in two different situations: (1) intravenous infusion of [1-^{13}C]glucose under hyperglycemic and hyperinsulinemic conditions, and (2) oral intake of glucose in the form of a bolus (Beckmann et al., 1993). For the intravenous infusion, [1-^{13}C]glucose with an enrichment level of 99% was used. The C1 signals of α- and β-glucose were detected after an infusion period of only 8 min. However, a significant increase in the glycogen signal was observed only after a prolonged infusion of about 60 min.

Formation of hepatic glycogen was also followed after the ingestion of nonlabeled glucose or [1-^{13}C]glucose with a low level of enrichment (Beckmann et al., 1993). The use of nonlabeled glucose, besides lowering significantly the costs of the experiments, simplifies the quantification of net liver glycogen synthesis since it can be based directly on changes in the ^{13}C MRS glycogen signal, avoiding label dilution through the various metabolic pathways of glucose (see also Section 6.9.1). Figure 10 contains spectra of the abdominal region obtained before and after the ingestion of 250 g of natural abundance glucose. The intense glucose signals in the spectra, detected 20 min after ingestion, originated from glucose in the stomach, the gut, the circulation, and in part only from the liver.

From the initial glycogen concentrations and the changes in glycogen signals after the administration of glucose, the authors estimated that a maximum of 10% of the intravenously infused glucose was stored as liver glycogen (Beckmann et al., 1993). On the other hand, the glucose uptake for the oral route was assessed to be between 24% and 30%. These estimates are consistent with findings from more invasive studies of glucose uptake in the liver (Katz et al., 1983).

Several factors regulate the amount of glucose uptake in the human liver, such as the hormonal milieu (insulin, glucagon), the glucose load, the hepatic glycogen content itself, and the route of glucose administration. DeFronzo et al. (1983) demonstrated that only combined hyperglycemia and hyperinsulinemia can increase the splanchnic glucose uptake. In humans rendered hyperglycemic by intravenous glucose in the presence of hyperinsulinemia, addition of oral glucose resulted in a sixfold increase in net splanchnic uptake (DeFronzo et al., 1983). Therefore, it has been initially suggested that the gut produces a factor that increases the hepatic glucose uptake (DeFronzo et al.,

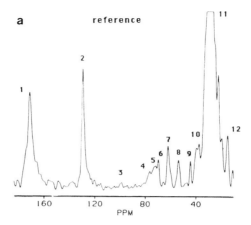
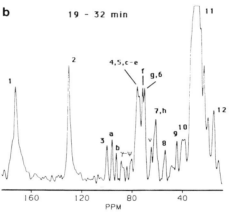
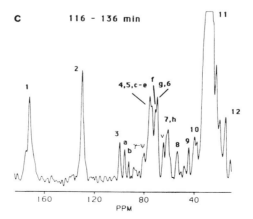

1983). However, more recent animal studies that examined the hepatic glucose balance after a glucose load suggested that not a gut factor, but a portal-arterial glucose concentration gradient may serve as a signal for net glucose uptake by the liver (Adkins *et al.*, 1987; Gardemann *et al.*, 1986). These studies demonstrated an enhanced uptake of glucose when it was infused intraportally versus a peripheral infusion, despite identical loads of insulin and glucose presented to the liver.

Significant glycogen signal changes were also reported after the ingestion of meals. Hwang *et al.* (1993) examined the time course of hepatic glycogen concentration following the ingestion of three isocaloric mixed meals during the course of a day. The mean fasting hepatic glycogen concentration was 242 ± 10 mM and increased progressively following each meal (296 ± 19 mM, 3 to 4 h after breakfast; 342 ± 32 mM, 3 to 4 h after lunch; 358 ± 29 mM, 3 to 4 h after dinner). No decrease in liver glycogen concentration was detected up to 5 h following any of the meals, suggesting that net hepatic glycogenolysis contributes minimally to postprandial hepatic glucose production (Hwang *et al.*, 1993).

Assessment of Glycogenolysis and Gluconeogenesis by Combining ^{13}C MRS, ^1H Imaging, and Radioactive Isotope Dilution Hepatic glycogenolysis and gluconeogenesis are difficult to quantify on humans. Methods based on the incorporation of isotopically labeled gluconeogenic precursors into plasma glucose or on the determination of the net splanchnic uptake of gluconeogenic substrates have been used to assess the rate of gluconeogenesis (for reviews, see Cherrington and Vranic, 1986; Wolfe, 1992). However, they have the drawback of relying on a series of assumptions of uncertain validity (Katz, 1985; Krebs *et al.*, 1966).

The net rates of glycogenolysis and gluconeogenesis were quantified directly by combining ^{13}C MRS, proton imaging of the liver, and constant infusions of a radioactive tracer ([6–^3H]glucose) for 2.5 h (Rothman *et al.*, 1991b; see also Section 6.7.2.2). The rate of total glucose production R_{tgp} was determined from activity measurements of [6–^3H]glucose by ion-exchange chromatography, using the equation:

Figure 10. Natural abundance ^{13}C spectra of the abdominal region of a healthy volunteer, (a) before and (b,c) after the ingestion of 250 g of glucose. Reference spectrum: 1024 acquisitions, 9.6-min measurement time. Spectra after the ingestion: 768 acquisitions, 7.2-min measurement time. In all acquisitions the repetition time was 0.56 s. Measurements were performed at 1.5 T with a system of concentric and coplanar $^1H/^{13}C$ surface coils and applying the pulse-acquire scheme. Resonances {3}, {a}, and {b} are from C1 of glycogen, and C1 of β- and α-glucose, respectively. The other resonances, {4}–{7} and {c}–{h}, overlapped with fatty acid signals, are from, respectively, liver glycogen and glucose in the gut, the stomach, and the liver. (Reproduced with permission from Beckmann *et al.*, 1993. Copyright © 1993 Williams & Wilkins, Baltimore.)

$$R_{\text{tgp}} = \text{infusion rate of tracer/[6-}^3\text{H]glucose activity.} \quad (6.7)$$

The net rate of gluconeogenesis was assessed by combining the rate of hepatic glycogen production R_{hgp}, determined from glycogen measurements by ^{13}C MRS and liver volumes by ^1H imaging, and the rate of total glucose production R_{tgp} according to the following expressions (Rothman et al., 1991b):

$$R_{\text{hgp}} = \text{net change in liver glycogen over time} \times \text{liver volume} \quad (6.8)$$

$$\text{net rate of gluconeogenesis} = R_{\text{tgp}} - R_{\text{hgp}}. \quad (6.9)$$

Glycogenolysis decreased during a 64-h fasting period, from 4.3 ± 0.6 μmol/(kg-body-weight)$^{-1}$ min^{-1} ($\pm$SEM) in the first 22 h to 0.3 ± 0.6 μmol/(kg-body-weight)$^{-1}$ min^{-1} in the interval from 46 to 64 h (Rothman et al., 1991b). Hepatic gluconeogenesis did not vary significantly in the same time interval, being 7.9 ± 1.0 μmol/(kg-body-weight)$^{-1}$ min^{-1} in the first 22 h, and 8.3 ± 0.5 μmol/(kg-body-weight)$^{-1}$ min^{-1} from 46 to 64 h of fasting. Glucose production from gluconeogenesis increased from $64 \pm 5\%$ to 96% in the same time interval (Rothman et al., 1991b).

Hepatic Glycogen Turnover   Animal studies suggested that liver glycogen turnover may be an important mechanism for regulating the net rate of glycogen synthesis (David et al., 1990; see also 5.5.3). Rothman et al. (1991a) determined the liver glycogen turnover in humans by measuring the incorporation of infused [1-^{13}C]glucose into liver glycogen and the loss of label during a subsequent chase of unlabeled glucose. The rate of hepatic glycogen synthase was determined from the rate of increase in [1-^{13}C]-labeled glycogen concentration divided by the plasma glucose fractional enrichment. As the infusate was switched to unlabeled glucose, the C1 resonance declined during the period when glucose fractional enrichment decreased. The rate of glycogenolysis relative to net glycogen synthesis was calculated by comparing the actual time course of the [1-^{13}C]glycogen concentration measured during the chase period and the time course predicted from the synthase rate and the residual plasma glucose fractional enrichment. The rate of glycogenolysis relative to the glycogen synthase rate (%turnover) was $29 \pm 7\%$ in the fasted state, increasing significantly to $59 \pm 5\%$ ($P < 0.05$) in the fed state (Rothman et al., 1991a).

6.7.2.2 Galactose Metabolism

Recently Fried et al. (1993) combined ^{13}C MRS and a stable isotope dilution technique based on the administration of [6,6-^{2}H$_2$]glucose to determine simultaneously the hepatic glycogen turnover and the hepatic glucose output (HGO) during an acute carbohydrate challenge by galactose. For many years radioactive tracers were used to study glucose turnover in humans. However, for ethical reasons the radioactive isotopes have more recently been exchanged for stable ones, such as ^{13}C and ^{2}H. Using ^{2}H- or ^{13}C-labeled substrates, stable isotope methods were established that are free of risk for volunteers and patients, and are in accordance with the legal requirements for radiation protection. (For a review on this topic, see Wolfe, 1992.)

The galactose elimination capacity has been used as a measure of hepatic function for almost 30 years (Tygstrup, 1966). It correlates well with clinical scores and may be predictive of liver failure. Galactose is predominantly metabolized in the liver, being incorporated into glycogen by a different metabolic pathway than glucose or fructose (Cohn and Segal, 1973). Despite the widespread use of the galactose elimination test, no data were available on the interaction of galactose with hepatic glycogen in humans. A few studies concerning the hepatic metabolism of galactose were conducted on animals. Niewoehner *et al.* (1990) have shown in fasted rats that oral galactose leads to an increase in liver glycogen. They also found an initial decrease in glycogen concentration during their experiments (Niewoehner *et al.*, 1990; Niewoehner and Neil, 1992).

Measurements were conducted in the morning after a 12-h fasting period following a carbohydrate-rich diet. For the quantification of the HGO, volunteers received a primed (5 mg/kg-body-wt) constant infusion (50 μg/(kg-body-weight)$^{-1}$ min^{-1}) of [6,6-^{2}H$_{2}$]glucose beginning 120 min prior to the injection of galactose and lasting until the end of the experiment. Position and volume of the liver were determined by proton imaging, and the initial glycogen content was assessed by ^{13}C MRS. A bolus of galactose (0.5 g/kg-body-weight) was administered intravenously during 5 min, and ^{13}C measurements were then performed during the following 180 min. Blood samples were drawn throughout the study. The fractional enrichment of the deuterated glucose in the blood plasma was determined by gas chromatography/mass spectrometry (GC/MS). Calculation of the HGO as a function of time after the administration of galactose was based on Steele's equations for non-steady-state conditions (Cobelli and Toffolo, 1984).

The temporal evolution of the average hepatic glycogen concentration after the galactose bolus is depicted in Figure 11. This characteristic behavior of the glycogen signal, i.e., a decrease during approximately 60 min after the bolus followed by an increase of the signal, was observed in all the studies. The average HGO increased by a factor of 3 in the first 20 min following the bolus; an increase of the plasma glucose in the same time interval was also observed. Later, both the HGO and plasma glucose returned to their basal levels (Fried *et al.*, 1993).

The characteristic initial decrease of glycogen concentration following the administration of galactose (Figure 11), also described for invasive experiments performed on rats (Niewoehner *et al.*, 1990; Niewoehner and

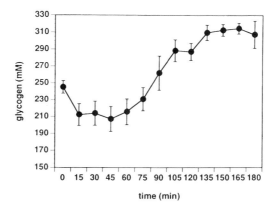

Figure 11. Time course of the average hepatic glycogen concentration for 12 male volunteers after an intravenous galactose bolus of 0.5 g/kg-body-weight. Glycogen concentration was assessed using the procedure outlined in Section 6.6. Galactose was administered at time point $t = 0$.

Neil, 1992), can be explained by the pathway for incorporation of galactose. The first step involves a phosphorylation of galactose with the aid of galactokinase. Galactose-1-phosphate and UDP-glucose are subsequently transglycosylated by uridyltransferase to glucose-1-phosphate and UDP-galactose. The net result is the conversion of galactose-1-P to glucose-1-P, and the conversion of UDP-glucose to UDP-galactose. Glucose-1-phosphate is then available both for the formation of glycogen via UDP-glucose and for metabolization via glucose-6-P. UDP-galactose is epimerized to UDP-glucose by UDP-gal-4-epimerase. The transglycosylation reaction between galactose-1-P and UDP-glucose depletes the UDP-glucose pool from which glycogen is synthesized. Because glycogen synthesis and glycogenolysis occur simultaneously, even in the fasted state, a net decrease in hepatic glycogen occurs. However, the initial decrease in liver glycogen could not be explained if epimerization of UDP-galactose to UDP-glucose were instantaneous, thus offsetting the effect of conversion of UDP-glucose to glucose-1-P. These results suggest therefore that the epimerization reaction is a rate-limiting step during the metabolism of galactose.

^{13}C MRS and infusion of deuterated glucose as the only tracer allow simultaneous assessment of the dynamics of liver glycogen and hepatic glucose output in humans. The use of deuterated glucose avoids all of the difficulties associated with the radioactive tracers. Therefore, the combination of ^{13}C MRS with the administration of nonenriched substrates and stable isotope tracers is a noninvasive, safe, and comparatively cheap method for the assessment of carbohydrate metabolism in humans. Furthermore, in studies of carbohydrate metabolism of the liver, net gluconeogenesis may be calculated applying Eqs. (6.7) through (6.9), or a modification thereof in the case of a carbohydrate challenge.

6.7.3 Glucose Metabolism in the Muscle

Skeletal muscle utilizes glucose as a fuel, forming both lactate and CO_2. It stores glycogen as a fuel for use in muscular contraction and synthesizes muscle protein from plasma amino acids. Muscle accounts for approximately 50% of body mass, representing therefore a considerable store of protein that can be drawn on to supply plasma amino acids, especially during dietary shortage.

Due to its favorable anatomical location, the gastrocnemius muscle of the calf has also been a tissue of interest to study under *in vivo* conditions with natural abundance ^{13}C MRS. Broadband proton-decoupled ^{13}C spectra of the muscle (Beckmann *et al.*, 1990; Heerschap *et al.*, 1989) have shown a number of metabolites, e.g., carnosine, (P)creatine, carnitine, glycogen, and phosphoethanolamine. Assignment of the resonances was based on ^{13}C NMR studies of muscle tissue extracts from animals (Doyle *et al.*, 1981; Lundberg *et al.*, 1986) and humans (Bárány *et al.*, 1984). Most spectra were acquired with the surface coil as the sole localization device. Further steps of spatial localization based on ^{13}C CSI (Beckmann and Müller, 1991a) and polarization transfer (Beckmann and Müller, 1991b) enabled the improvement of the detection of some of these metabolites, especially in the spectral region from 40 to 80 ppm, where partial overlapping with fatty acid signals occurs (see also Section 6.5).

In the muscle, glycogen has been the metabolite of major interest, in particular for studies involving patients suffering from diabetes mellitus (Shulman *et al.*, 1990b; see also Section 6.10.1.1). The average glycogen concentration in the muscle, 70 to 110 mmol of glucosyl units per kilogram of wet tissue in normal individuals (Karlsson *et al.*, 1971; Nilsson and Hultman, 1974), is lower than in the liver. Muscle glycogen quantification by ^{13}C MRS and direct biochemical assay of needle biopsies were compared by Taylor *et al.* (1992). In this study individuals underwent six NMR scans and three biopsies of the gastrocnemius muscle on the same day. The average muscle glycogen concentration was 87.4 ± 28.1 mM ($\pm$SD) by ^{13}C MRS and 88.3 ± 28.8 mM by biopsy. There was a close correlation between the pairs of observations on each subject ($R = 0.95$; $P < 0.0001$), demonstrating that *in vivo* ^{13}C MRS measurements of human muscle glycogen are accurate.

Avison *et al.* (1988) and Heerschap *et al.* (1989) presented data concerning glycogen depletion in the gastrocnemius muscle as a result of long-distance runs by well-trained volunteers. These authors showed that after the runs, the muscle glycogen level fell to approximately 30% of the pre-exercise level. Glycogen recovered to about 80% of the pre-exercise level 20 h after the runs. The total accumulation time for attaining a reasonable SNR of the C1 resonance in natural abundance ^{13}C spectra was of the order of 25 min.

^{13}C MRS was used to measure directly the rate of muscle glycogen formation from intravenously infused, isotopically labeled [1-^{13}C]glucose (Jue et al., 1989b). The data showed that under hyperglycemic and hyperinsulinemic conditions the major part (70% to 90%) of the infused glucose was converted to muscle glycogen. Previously, more invasive measurements had also furnished evidence that the majority of a glucose load, whether given orally or intravenously, is disposed of by the muscle (Ferrannini et al., 1985; see also Section 6.7.2.1).

6.8 Noninvasive Determination of the Degree of Unsaturation of Fatty Acids from Adipose Tissue

For the measurements reported until now, suppression of the intense signals from mobile fatty acids constituted an essential step during the acquisition of *in vivo* spectra. However, information about the degree of unsaturation of fatty acids can be easily derived from these signals. Conventional methods for the determination of the composition of adipose tissue fatty acids involve either tissue removal during a surgical procedure or aspiration of subcutaneous tissue under local anesthesia. It is evident that these invasive procedures are only reluctantly requested, especially if repeated sampling is desired.

The relative amount of mono- and polyunsaturation in adipose tissue was assessed by determining the ratio between the olefinic resonances and the carbonyl signal at 180 ppm, after correcting the signals for relaxation (Beckmann et al., 1992; Moonen et al., 1988; see also Section 5.5.1). Distinct signals around 129 ppm are generated by the olefinic carbons of unsaturated fatty acids, for which two or more double bounds are separated by a single methylene group. Olefinic carbons adjacent to the intermediate methylene group, known as "inner" olefinic carbons, give rise to a resonance at approximately 128.5 ppm. Another resonance at 130 ppm is produced by the two "outer" olefinic carbons in polyunsaturated fatty acids, and by the olefinic carbons in monounsaturated fatty acids.

Average values of 44.7 ± 4.2% and 15.4 ± 4.3% have been obtained for the degree of mono- and polyunsaturation, respectively, of adipose tissue in individuals following a normal diet (Beckmann et al., 1992). These data are in good agreement with determinations based on invasive procedures

(Malcolm *et al.*, 1989; Riemersma *et al.*, 1986). For subjects who followed a fat-reduced diet for at least half a year before the ^{13}C measurements, the degree of mono- and polyunsaturation were of 52.6 ± 5.5% ($P < 0.0002$ with respect to controls) and 17.5 ± 3.3% (NS), respectively (Beckmann *et al.*, 1992). Moonen *et al.* (1988) compared the ^{13}C MRS results with gas chromatography determinations of biopsy samples from the same volunteers, showing good agreement between both methods.

These studies were motivated by the fact that low levels of linoleic acid ($18:2n - 6$) in the circulation and in adipose tissue may be linked to an increased risk of coronary heart disease (Riemersma *et al.*, 1986; Wood *et al.*, 1984). These and additional studies suggested furthermore that:

- The relative amount of polyunsaturated fatty acids in adipose tissue reflects differences in dietary intake (Riemersma *et al.*, 1986).
- A determination of the concentration of fatty acids in adipose tissue is a valid tool to assess the dietary fat intake for the preceding one to three years (Katan *et al.*, 1986). Beynen *et al.* (1980) even showed a direct relationship between the fatty acid composition of the habitual diet and of the subcutaneous adipose tissue in humans.

Thus, for experimental and epidemiological nutritional surveys, the average fat composition of the adipose tissue is a useful parameter for long-term dietary habits (Beynen *et al.*, 1980). This association is not only of pathophysiological interest but also of practical importance since an objective assessment of long-term dietary compliance is generally recognized to be an unresolved issue.

6.9 Body Fluids and Isolated Tissues

The characterization of body fluids and isolated tissues with ^{13}C NMR is not only important in animal studies, as extensively shown in Chapter 5, but also in applications to humans, in particular because a large number of resonances can be observed with a high spectral resolution and sensitivity under natural abundance conditions without the spatial localization problems common to all *in vivo* measurements. In addition, ^{13}C-enriched substrates may be detected even when present in small quantities. Therefore, the strategy of analyzing body fluids, e.g., blood and urine, with NMR after the administration of ^{13}C-enriched substrates, and of acquiring natural abundance ^{13}C spectra from tissue extracts constitutes an important complement to *in vivo* studies (see Section 6.10). Keep in mind that body fluids are of great biochemical value since their composition reflects the body's functional state as a whole.

An example is the use of multiple ^{13}C labeling and ^{13}C NMR to detect low levels of exogenous metabolites in the presence of large endogenous pools for the determination of glucose turnover. A significant problem that could be encountered in ^{13}C NMR studies of metabolism is the contribution that background levels of ^{13}C may make to the observed spectra when low levels of the ^{13}C label are present. Brainard et al. (1989) showed that the introduction of two or more labeled sites in the same tracer molecule is an effective means for reducing this difficulty.

The glucose production in a human subject was determined by a continuous infusion of tracer levels of [U–^{13}C]glucose over a 4-h period following a 12-h overnight fast and subsequent analysis of plasma levels of the tracer by *in vitro* NMR. Assessment of the relative percentages of the multiply labeled (exogenous) and 1.1% labeled (endogenous) pools of glucose was straightforward as a result of analyzing the resonances of the C1α and C1β anomers of glucose, both characterized by a singlet and a doublet (see Figure 1 of chapter 5). The intensity of the center resonances, arising from singly labeled [1–^{13}C]glucose, was considered to be proportional to the concentration of the endogenous pool. The two doublet resonances surrounding these peaks arise from the multiply labeled exogenous glucose pool, with the splittings reflecting the ^{13}C–^{13}C coupling. A glucose production rate of 9.3 $\pm$ 0.72 μmol/(kg-body-weight)$^{-1}$ min^{-1} and a gluconeogenic rate of 0.89 μmol/(kg-body-weight)$^{-1}$ min^{-1} were determined by ^{13}C NMR. These results were identical within the estimated experimental errors to the values determined with mass spectrometry (Brainard et al., 1989). The gluconeogenic rate was about 10% of the glucose production rate, a normal postabsorptive value for gluconeogenesis from three-carbon intermediates.

6.9.1 Mechanisms of Hepatic Glycogen Repletion

It has been shown that, in addition to the widely accepted mechanism of glycogen repletion by direct incorporation of glucose into glycogen (glucose → glucose-6-phosphate → glucose-1-phosphate → UDP-glucose → glycogen), hepatic glycogen is also formed by an indirect pathway possibly involving this sequence: glucose → lactate → oxaloacetate → phosphoenolpyruvate → UDP-glucose → glycogen (McGarry et al., 1987; Newgard et al., 1984; see also Section 5.5.4). By using radioisotope techniques Magnusson et al. (1989) estimated that after an overnight fast a maximum of 65% of glycogen is formed from a glucose load via the direct pathway, and that this contribution rises to a maximum of 77% when a glucose load is ingested at lunchtime.

To determine the effect of fasting versus refeeding on hepatic glycogen repletion by the direct pathway, Shulman et al. (1990a) used high-resolution ^{13}C NMR and GC/MS to determine the isotope scrambling and enrichment of blood plasma and urine, respectively, from samples taken during infusions of [1-^{13}C]- and [6-^{14}C]glucose. Acetaminophen was also administered during the infusions of labeled glucose, which lasted for 270 min. Studies were conducted after an overnight fast and 4 h after breakfast. Glycogen formation by the direct pathway was estimated in the following ways:

- High-resolution ^{13}C NMR, comparing the ^{13}C enrichments in C1 and C6 of glucose formed from urinary acetaminophen glucuronide with enrichments in C1 and C6 of plasma glucose. This approach had the advantage of furnishing estimates that were independent of label dilution through the indirect pathway (Landau and Wahren, 1988).
- GC/MS, comparing the specific activity of glucose from glucuronide and the specific activity of plasma glucose, along with the percentages of ^{14}C in C1 and C6 of the glucose from the glucuronide.

Both approaches furnished similar results. After an overnight fast, 49 ± 3% of the glycogen was estimated to be formed via the direct pathway. This estimate rose to 69 ± 7% after breakfast (Shulman et al., 1990a).

6.10 Diseased States

6.10.1 Genetic Diseases

The initial observations made by the English physician Archibald Garrod on a small group of inborn errors of metabolism in the early 1900s stimulated the investigation of the biochemical pathways affected by genetic diseases. It is estimated that more than 3000 diseases have a genetic basis. Genetic diseases account for approximately 10% of the hospitalized children in a number of centers, and many of the chronic diseases that afflict adults (e.g., diabetes mellitus, atherosclerosis) possess a significant genetic component.

6.10.1.1 *Non-Insulin-Dependent Diabetes Mellitus*

The cardinal manifestation of diabetes mellitus is elevated blood glucose levels (hyperglycemia), which result from (1) a decreased entry of glucose into the cells, (2) a decreased utilization of glucose by various tissues, and (3) an increased production of glucose (gluconeogenesis) by the liver. About 90% of persons with diabetes have non-insulin-dependent (type II) diabetes mellitus (NIDDM). Such patients are usually obese, have elevated plasma insulin levels, and have down-regulated insulin receptors. The other 10% have insulin-dependent (type I) diabetes mellitus (IDDM), which is additionally characterized by glycosuria (increased glucose levels in the urine) and may be accompanied by changes in fat metabolism.

Using natural abundance ^{13}C MRS, Shulman et al. (1990b) determined significantly lower basal levels of muscle glycogen in NIDDM (39 ± 6 mM, mean $\pm$ SEM) than in normal subjects (73 ± 11 mM) ($P < 0.01$). A similar reduction had been observed previously by Roch-Norland et al. (1972) using muscle biopsies. In addition, during the infusion of [1-^{13}C]glucose the mean rate of muscle glycogen synthesis was 180 ± 39 μmol/(kg-muscle-tissue)$^{-1}$ min^{-1} in normal subjects and reduced to 78 ± 28 μmol/(kg-muscle-tissue)$^{-1}$ min^{-1} in diabetic subjects ($P < 0.05$) (Shulman et al., 1990b)—direct evidence that glycogen synthesis is impaired in NIDDM. The rate of nonoxidative glucose metabolism in the same period was determined to be 42 ± 4 μmol/(kg-body-weight)$^{-1}$ min^{-1} in controls and 22 ± 4 μmol/(kg-body-weight)$^{-1}$ min^{-1} in diabetic individuals. From these data the authors concluded the following:

- Under hyperglycemic and hyperinsulinemic conditions, muscle glycogen synthesis is the major pathway of glucose disposal in both normal and diabetic individuals.
- Defective muscle glycogen synthesis has a predominant role in impairing glucose metabolism in NIDDM patients (Shulman et al., 1990b).

In another study, Magnusson et al. (1992) showed that the hepatic glycogen concentration 4 h after a meal was significantly lower in diabetes patients than in matched controls, 131 ± 20 mM versus 282 ± 60 mM ($P < 0.05$). For three days before the study all subjects received a standardized, high-carbohydrate diet. Net hepatic glycogenolysis, determined from ^{13}C spectra acquired at different time points up to a fasting period of 23 h, was decreased in the diabetics, 1.3 ± 0.2 as compared to 2.8 ± 0.7 μmol/(kg-body-weight)$^{-1}$ min^{-1} in the controls ($P < 0.05$). Whole-body glucose production, assessed from dilution studies using an infusion of [6-^{3}H]glucose and applying Eq. (6.7), was increased in the diabetics, 11.1 ± 0.6 versus 8.9 ± 0.5 μmol/(kg-body-weight)$^{-1}$ min^{-1} ($P < 0.05$). Gluconeogenesis, determined by applying Eq. (6.9), was consequently increased in the diabetics, 9.8 ± 0.7 as compared to 6.1 ± 0.5 μmol/(kg-body-weight)$^{-1}$ min^{-1} in the controls ($P < 0.01$). It accounted for $88 \pm 2\%$ of total glucose production in the diabetics, compared with $70 \pm 6\%$ in the controls ($P < 0.05$). The authors concluded that increased gluconeogenesis is responsible for the increased whole-body glucose production in type II diabetes mellitus after an overnight fast (Magnusson et al., 1992).

6.10.1.2 Glycogen Storage Disease

Glycogen storage disease or glycogenosis designates a group of inherited disorders characterized by deficient mobilization of glycogen and deposition of abnormal forms or quantities of glycogen in the tissues, leading to muscular weakness or even death. Eight types of glycogenosis have been reported

(Hers *et al.*, 1989), each being caused by a deficiency or inactivation of a certain enzyme. In type I glycogenosis (von Gierke's disease), for instance, both the liver cells and the cells of the renal convoluted tubules are characteristically loaded with glycogen. However, these glycogen stores are unavailable. In liver, kidney, and intestinal tissue the activity of glucose-6-phosphatase is either extremely low or entirely absent. Patients with type III glycogen storage disease are characterized by an increased glycogen concentration in the liver and muscle due to inactivation of the amylo-1,6-glucosidase debrancher enzyme (Hers *et al.*, 1989; Howell, 1972). The accumulated glycogen has also an abnormal structure, with short outer branches. Early in life patients exhibit hepatomegaly, and growth retardation may be striking (Hers *et al.*, 1989).

In Vivo Studies   Natural abundance ^{13}C spectra from the calf and the abdominal region of a patient with type IIIA glycogen storage disease are presented in Figure 12. Measurements showed that the patients had a two- to threefold glycogen content in the muscle and liver compared to well-trained volunteers and individuals prepared with a carbohydrate-rich diet (Beckmann *et al.*, 1990). Additional resonances in the spectral region between 71 and 85 ppm were also detected in ^{13}C spectra of the heads of these patients (Beckmann *et al.*, 1990). The molecular nature of these resonances remains unclear at present, although resonances of glycogen (C2–C5), signals of α- and β-glucose, and signals of inositol may be found in this frequency interval in the spectra of a variety of tissues (Bárány *et al.*, 1984, 1985). It is interesting to note that an increase of the glycogen content in glial cells of the brain, which perform several energy-dependent functions that may aid neuronal survival under pathological conditions, has been reported for patients with type II glycogen storage disease (Howell, 1972). However, ^{13}C MRS would hardly be sensitive enough to allow the detection of such small levels of glycogen under *in vivo* conditions.

In Vitro Studies   *In vitro* ^{13}C NMR spectra were acquired from blood plasma samples taken from patients with glycogen storage disease type I (GSD-I) and type III (GSD-III) to elucidate the mechanism by which glucose is produced (Kalderon *et al.*, 1989). A primed, dose-constant nasogastric infusion of D-[U–^{13}C]glucose (99% ^{13}C-enriched) or an infusion diluted with nonlabeled glucose was administered following different periods of fasting. Recycling parameters were derived from the plasma β-glucose C1 splitting pattern, i.e., from the doublet/singlet (d/s) values of plasma glucose C1 in comparison to the d/s values of known mixtures of [U–^{13}C]glucose and unlabeled glucose as a function of ^{13}C enrichment of glucose C1. The fractional glucose C1 enrichment of plasma glucose was determined by ^{1}H spectroscopy and confirmed by GC/MS.

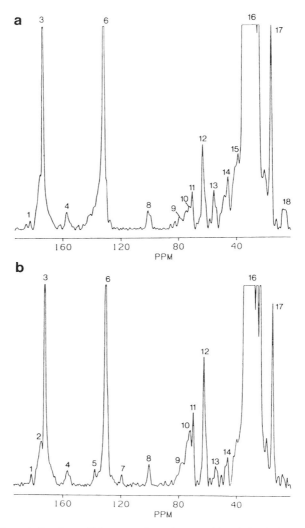

Figure 12. Natural abundance ^{13}C spectra from the (a) calf and (b) abdominal region of a patient with type IIIA glycogen storage disease. We had 1536 acquisitions for the (a) calf and 1664 acquisitions for the (b) abdomen with a repetition time of 1.1 s (28.6- and 31.0-min measurement times, respectively) performed at 1.5 T with a system of coplanar and concentric $^1H/^{13}C$ surface coils using the pulse-acquire scheme. Resonances {3}, {6}, {11}, {12}, {16}, and {17} are from carbonyl carbons, olefinic carbons, C2 of glycerol backbone, C1 and C3 of glycerol backbone, methylene groups, and methyl groups of fatty acids, respectively. Other tentative assignments are carbonyl carbons in the peptide bonds of proteins {1} and from carnosine, creatine, carnitine {2}; guanidino carbons of (P)creatine and arginine {4}; rings C1, C2, and C8 from phenylalanine, histidine, and tryptophan {5}, respectively; rings C4 of histidine, C3,5 of tyrosine, and C4,6 of tryptophan {7}; C1 {8}, C4 {9}, and C2,3,5 {10} of glycogen; methyl groups of (P)creatine, choline, carnitine {13}; $-CH_2-NH_2$ group from phosphatidylethanolamine {14}; and β-carbons of amino acid sidechains {15}. (Reproduced with permission from Beckmann *et al.*, 1990. Copyright © 1990 Academic Press, Orlando.)

Recycling parameters derived from the spectra indicated that the plasma glucose of GSD-I patients comprised only a mixture of 99% ^{13}C-enriched D-[U-^{13}C]glucose and unlabeled glucose but lacked any recycled glucose, eliminating therefore a mechanism involving gluconeogenesis for glucose production in these subjects.

Significantly different results were obtained for GSD-III patients. The d/s ratio as a function of ^{13}C enrichments of glucose C1 showed a greatly reduced degree of C1–C2 coupling. The singlet C1 resonance of glucose no longer arose only from the endogenous nonlabeled glucose. This fact indicated that a great portion of glucose molecules derived from the plasma of GSD-III patients did not have the C1 and C2 positions intact, thereby explaining the reduced d/s ratio. It was therefore suggested that gluconeogenesis is the major route for endogenous glucose synthesis in GSD-III patients (Kalderon *et al.*, 1989).

6.10.1.3 *Fructose Intolerance*

An inborn deficiency in the ability of aldolase B to split fructose-1-phosphate is found in humans with hereditary fructose intolerance (HFI) (Gitzelmann *et al.*, 1985). A continuous exposure of these subjects to parental fructose during infancy may result in liver cirrhosis, mental retardation, and death (Gitzelmann *et al.*, 1985). The final diagnosis of aldolase B deficiency is usually performed using liver biopsy specimens.

To elucidate the mechanism by which fructose is converted to glucose in normal children and in HFI children, blood plasma samples, drawn during the nasogastrical infusion of [U-^{13}C]fructose, were analyzed by high-resolution ^{13}C NMR (Gopher *et al.*, 1990). Spectra obtained from a normal control and from an HFI patient are shown in Figure 13. The multiplet structure of the glucose carbon resonances are due to ^{13}C–^{13}C coupling, either with one or two adjacent ^{13}C atoms. The ^{13}C enrichment of plasma glucose C1 of the control subject was 3.2%, whereas that of the patient was 1.0%. Both individuals were subjected to similar rates of [U-^{13}C]fructose infusion. Lactate carbons were observed as multiplet structures, indicating that their resonances were of two adjacent ^{13}C. From the triplet structure of lactate (expanded scale) the relative ^{13}C enrichment in comparison to glucose C1 was deduced. Glycerol, a fructose metabolite, was identified in HFI plasma. The glycerol might be due to the activity of aldolase A on Fru(1,6)P$_2$ producing glyceraldehyde phosphate and dihydroxyacetone phosphate, and to the subsequent activities of the respective dehydrogenases, resulting in *sn*-glycerol-phosphate and glycerol.

Significantly lower values ($\approx$threefold reduction) for fructose conversion to glucose were obtained in HFI children in comparison to matched controls. A quantitative determination of the metabolic pathways of fructose conver-

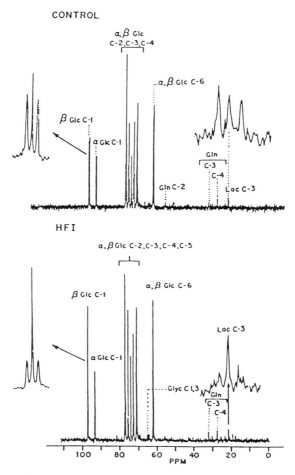

Figure 13. Proton-decoupled ^{13}C NMR spectra (at 125.76 MHz) of deproteinized plasma ($\approx$3 ml) derived from a control subject (3600 accumulations) and from HFI patient (12,000 accumulations). Abbreviations: glucose (Glc), glutamine (Gln), glycerol (Glyc), and lactate (Lac). (Reproduced with permission from Gopher et al., 1990.)

sion to glucose was derived from ^{13}C measurements of plasma [^{13}C]glucose isotopomer populations. The results of Gopher et al. (1990) suggest that splitting of fructose-1-phosphate by aldolase B accounts for only $\approx$50% of the total amount of hepatic fructose conversion to glucose in normal subjects. The finding of isotopomer populations consisting of three adjacent ^{13}C atoms at glucose C4 (^{13}C3–^{13}C4–^{13}C5) suggests additionally that there is a direct pathway from fructose, bypassing fructose-1-phosphate aldolase, to fructose-1,6-biphosphate, accounting for 47 and 27% of the conversion in control and HFI children, respectively (Gopher et al., 1990).

6.10.1.4 Methylmalonic Acidemia

A pioneering study concerning the application of ^{13}C spectroscopy in the clinical environment was performed by Tanaka et al. (1975), who analyzed urine samples, drawn from a patient having methylmalonic acidemia after the sequential ingestion of $[1-^{13}C]$- and $[1,2-^{13}C_2]$valine, with high-resolution ^{13}C NMR. Methylmalonic acidemia, first described by Oberholzer et al. (1967), is an inborn error of propionate metabolism caused by four different genetic defects either at the methylmalonyl CoA (MMCoA) racemase or mutase steps (Rosenberg and Mahoney, 1973). The results obtained by Tanaka et al. (1975) suggest that in the metabolism of valine, which enters the MMCoA pool through methylmalonic acid semialdehyde (MMS), MMS is first decarboxylated to propionate and only then carboxylated to MMCoA. Prior to this study it was believed that there was a direct oxidative route from MMS to MMCoA.

6.10.1.5 Cystic Fibrosis

Cystic fibrosis is a common genetic disease of the exocrine glands and of the eccrine sweat glands. It is characterized by abnormally viscous secretions that plug up the secretory ducts of the pancreas and the bronchioles. Patients with cystic fibrosis also exhibit elevated amounts of chloride in their sweat. Victims often die at an early age from lung infections.

A number of studies demonstrated abnormalities of essential fatty acids (EFAs) in circulating triglycerides and phospholipids in patients with cystic fibrosis (Chase and Dupont, 1978; Rogers et al., 1984). Although most investigations focused on these circulating lipid classes, some studies also demonstrated abnormal levels of EFAs stored in the adipose tissue of cystic fibrosis patients with malabsorption (Farrell et al., 1985).

The poly- and monounsaturated fatty acid content of adipose tissue of the lower leg was determined noninvasively with ^{13}C MRS by Dimand et al. (1988) in cystic fibrosis patients and in normal volunteers (see also Section 6.8). On controls the average poly- and monounsaturated fatty acid content was 17.8 ± 2.1% and 44.8 ± 3.8%, respectively. On patients, these values were 15.0 ± 2.0% ($P < 0.005$ versus normal volunteers) and 47.8 ± 6.5% (NS). One cystic fibrosis patient without fat malabsorption had decreased adipose polyunsaturateds, whereas another patient on high caloric gastrostomy feeds had normal levels.

6.10.2 Tumors

Cancer is the second most common cause of death in many countries, being surpassed only by cardiovascular disease. Humans of all ages can develop cancer, and a wide variety of organs are affected. Cancer cells are characterized by (1) diminished or unrestrained control of growth, (2) invasion of local tissues, and (3) spread, or metastasis, to other parts of the body. Cells of benign tumors also show diminished control of growth but do not invade local tissue or spread to other parts of the body.

6.10.2.1 *Differentiation of Human Tumors from Nonmalignant Tissue by Natural Abundance ^{13}C NMR*

High-resolution natural abundance ^{13}C spectra were obtained *in vitro* at 100.6 MHz from human pathology specimens of tumors and adjacent nonneoplastic control tissues from colon, lung, and prostate (Halliday *et al.*, 1988). The aim of this study was to identify in the spectra specific molecular parameters that could distinguish neoplastic from nonneoplastic tissue. The amount of tissue analyzed varied between 0.4 and 3.5 g, with a corresponding acquisition time between 3 and 6 h. The data suggest that high-resolution ^{13}C spectroscopy of unprocessed human pathological tissues might be useful in finding qualitative and quantitative biochemical markers that complement histological methods for separating tumors into biologically more similar groups, thus improving the tumor prognosis and the design of effective treatment modalities (Drewinko *et al.*, 1984; Halliday *et al.*, 1988; Pretlow *et al.*, 1985; see also Section 6.9).

Colon Tissue In comparison to signals from adjacent normal colon, ^{13}C spectra of colonic adenocarcinoma revealed that this type of tumor contained smaller signals from triacylglycerols, larger signals from phospholipids and lactate, and decreased lipid fatty acyl chain saturation.

Colloid carcinoma, a variant of colonic adenocarcinoma, demonstrated abundant mucin production, in contrast to the common tumor type, which generally has less mucin production than does normal colon tissue. ^{13}C NMR spectra of the colloid tumors were dominated by resonances from glycoproteins, rather than by resonances from lipids as found for the more common colon tumors. Significant signals from the sugar portions of the mucins were also present.

Lung Tissue Resonance assignments showed that the components of lung tissue that were mobile on the NMR timescale (correlation times of less than 2 ns at 9.4 T) were predominantly glycoproteins. Virtually all of the signals arose from amino acids and saccharides. ^{13}C NMR spectra of poorly differentiated human lung carcinoma showed a general replacement of the glycoprotein component of nonneoplastic lung tissue with lipid during tumor formation. Signals from proline were decreased in the tumor. There was also a striking lack of signals from the anomeric carbons of mucin's oligosaccharide chains in the tumor.

Prostate Tissue A comparison of the ^{13}C NMR spectra of prostatic adenocarcinoma with that of adjacent hyperplastic tissue revealed that the tumors contained smaller amounts of triacylglycerols, citrate, and acidic mucins. These deviations from the normally elevated amounts of citrate (Franklin and Costello, 1984) and low amounts of lipids in the prostate are consistent with an alteration in either the concentration or the activity of ATP-citrate lyase in the tumors. Indeed, the normal metabolism of the human prostate supports and maintains extraordinarily high (up to 60 mM) tissue concentrations of citrate (Franklin and Costello, 1984), whereas prostatic adenocarcinoma appears to be characterized by a dramatic, differentiation-dependent drop in the concentration of this important metabolite (Costello *et al.*, 1978; Pretlow *et al.*, 1985).

6.10.3 Steatosis

The liver carries out major functions in lipid metabolism. For a variety of reasons, lipids can accumulate in the liver, mainly as triacylglycerols. However, an extensive accumulation is regarded as a pathological condition. When accumulation of lipids in the liver becomes chronic, fibrotic changes occur in the cells that progress to cirrhosis and impair liver function.

The diagnosis of steatosis ("fatty liver") is usually made by histological grading of the triglycerides in liver tissue obtained from needle biopsies or during abdominal surgery. Recently the hepatic triglyceride content determined with ^{13}C MRS was compared with liver biopsy determinations on 13 patients having liver disease and a triglyceride content ranging from grade 0 to grade 4 (Petersen *et al.*, 1993). Localized ^{13}C spectra of the liver were obtained with 2D ISIS (Ordidge *et al.*, 1986), with the $-(CH_2)_n-$ signal at 30 ppm set on resonance. Signal intensity was computed by integrating over a bandwidth of 200 Hz centered at 30 ppm. Spectra were also acquired *in vitro* from a phantom containing a solution of a corn oil, by selecting a volume at the same coordinates relative to the coil as in the *in vivo* measurements. The hepatic triglyceride content was determined by comparing the signal intensities obtained *in vivo* and *in vitro*. The area of the NMR signal from a patient with fatty liver was in general several times larger than that from a normal fasted person, showing the large dynamic range for measuring hepatic triglyceride content. The coefficient of correlation between the ^{13}C MRS and the biopsy measurements was 0.76 (Petersen et al., 1993).

6.10.4 Rheumatic Diseases

Albert et al. (1993) used ^{13}C NMR to characterize the structural changes of hyaluronic acid of synovial fluids from knee punctates of patients suffering from rheumatic diseases. It is known that pathological changes in both the structure and concentration of synovial hyaluronic acid, in particular its depolymerization, affect the viscoelasticity of the synovial fluid with negative biomechanical implications and potential joint damage (Dahl et al., 1985).

Although normal synovial fluid does not contain collagen, the authors demonstrated that collagen breakdown products could be detected by ^{13}C NMR in the samples from patients. In addition, it was shown that the half-width of the C1' resonance of hyaluronic acid was significantly lower in synovial fluids from patients suffering from rheumatoid and reactive arthritis than in controls (Albert et al., 1993). Depolymerization of synovial hyaluronic acid in patients with arthritides could be monitored by assessing the half-width of the C1' resonance of hyaluronic acid in ^{13}C spectra of synovial fluid.

6.11 Concluding Remarks

The subjects addressed in this chapter show that ^{13}C MRS is a useful tool for noninvasive metabolic investigations in humans. The studies of carbohydrate metabolism in the liver and the brain have certainly been the most important contributions made by the technique. The demonstration that hepatic and muscle glycogen can be detected and quantified using ^{13}C MRS opened the door for the serial quantification of this macromolecule under several hormonal conditions and in diseased states. A series of elegant approaches have been devised, which combine the noninvasive determination of glycogen by ^{13}C MRS with the determination of hepatic glucose production by isotope dilution studies. These approaches enable the assessment of otherwise only indirectly accessible parameters such as glycogen turnover and gluconeogenesis. Furthermore, determination of the hepatic glycogen content from needle biopsies is largely used as a diagnostic procedure for the characterization of a number of liver diseases. The noninvasive assessment of the glycogen content by ^{13}C MRS not only eliminates the risks for the patients, but can be made repetitively.

From the methodological point of view, heteronuclear editing procedures, in particular gradient-enhanced heteronuclear multiple-quantum coherence (ge-HMQC), are very promising for improving the temporal resolution of experiments exploring the dynamics of incorporation of enriched substrates into a given organ. The time resolution of 3 to 4.5 min achieved for 1D and 2D ge-HMQC spectra of the cat brain (van Zijl et al., 1993a) is sufficient to measure ^{13}C glucose uptake and metabolic rates, such as glycolysis and TCA-cycle turnover. The potential of imaging metabolic rates in the human brain by means of these techniques is fascinating. This would be an important advantage over PET studies with fluorodeoxyglucose, where perfusion and uptake of a glucose analogue cannot be separated from the actual specific metabolism. The spatial resolution provided by NMR is also superior (the temporal resolution is, however, poorer than in PET). In addition, the combination of metabolic imaging with functional imaging of the brain might open the door to a series of noninvasive studies aimed at investigating in situ biochemical aspects of diseases such as stroke and cancer. One important aspect is, of course, that the price of the ^{13}C-enriched substrates diminishes.

Operating at higher magnetic fields should be, in principle, an advantage. However, since the required total power increases with the square of the frequency range irradiated and because of the frequency dependence of the rf absorption, broadband decoupling requires much higher power levels at higher magnetic fields (Bomsdorf et al., 1988). Thus, the field strength at which decoupling is still feasible for applications to humans must be verified carefully.

The recently reported studies concerning the metabolism of natural abundance substrates in the liver (Beckmann et al., 1993; Fried et al., 1994; Rothman et al., 1991b) show that a cheaper version of ^{13}C MRS for applications to humans is feasible. Spectra obtained during the hepatic metabolism of natural abundance carbohydrates are even easier to interpret, since no distinction between the direct and indirect pathways of glycogen repletion must be made. These pathways affect differently the fate of the label, and must be quantified separately when using ^{13}C-enriched glucose. Appropriate adaptations of the protocols might allow the performance of analogue studies on patients.

The isotopomer analysis addressed in Chapter 5, based on the investigation of ^{13}C–^{13}C homonuclear spin-coupling patterns after the administration of multilabeled substrates, constitutes an interesting approach for metabolic studies on humans as well. Due to the elevated price of these substances, ^{13}C NMR of body fluids rather than in vivo ^{13}C MRS would be the method of choice in this particular case since small, still affordable amounts of tracer are needed. The studies concerning the analysis by high-resolution ^{13}C NMR of plasma from blood samples drawn during the infusion of D-[U–^{13}C]glucose and [U–^{13}C]fructose, respectively, to patients suffering from glycogen storage disease (Kalderon *et al.*,1989) and fructose intolerance (Gopher *et al.*, 1990) serve as good examples. It is my personal feeling that ^{13}C NMR of body fluids will be increasingly important both as a complement to *in vivo* measurements and as an auxiliary tool for the characterization of metabolic diseases.

Acknowledgments

I am indebted to Dr. Ronald Fried from the Department of Gastroenterology, University Hospital, Basle, for enjoyable discussions about carbohydrate metabolism in the liver. Some of the results presented in this review were obtained during my stay at the MR Center and Biocenter of the University of Basle. I acknowledge the support of Prof. Joachim Seelig and from the Swiss National Science Foundation during that period.

References

Ackerman, J. J. H., Grove, T. H., Wong, G. C., Gadian, D. G., and Radda, G. K. (1980). *Nature (London)* **283**, 167.
Adkins, B. A., Myers, S. R., Hendrick, G. K., Stevenson, R. W., Williams, P. E., and Cherrington, A. D. (1987). *J. Clin. Invest.* **79**, 557.
Albert, K., Michele, S., Günther, U., Fial, M., Gall, H., and Saal, J. (1993). *Magn. Reson. Med.* **30**, 236.
Alger, J. R., Behar, K. L., Rothman, D. L., and Shulman, R. G. (1984). *J. Magn. Reson.* **56**, 334.
Aue, W. P., Müller, S., and Seelig, J. (1985). *J. Magn. Reson.* **61**, 392.
Avison, M. J., Rothman, D. L., Nadel, E., and Shulman, R. G. (1988). *Proc. Natl. Acad. Sci. U.S.A.* **85**, 1634.
Bárány, M., Doyle, D. D., Graff, G., Westler, W. M., and Markley, J. L. (1984). *Magn. Reson. Med.* **1**, 30.
Bárány, M., Arús, C., and Chang, Y. C. (1985). *Magn. Reson. Med.* **2**, 289.
Bax, A. (1983). *J. Magn. Reson.* **52**, 76.
Beckmann, N. (1992). In "NMR: Basic Principles and Progress," (P. Diehl, E. Fluck, H. Günther, R. Kosfeld, J. Seelig, eds.), Vol 28, p. 75. Springer, Berlin.
Beckmann, N., and Müller, S. (1991a). *J. Magn. Reson.* **93**, 186.
Beckmann, N., and Müller, S. (1991b). *J. Magn. Reson.* **93**, 299.
Beckmann, N., Wick, H., and Seelig, J. (1990). *Magn. Reson. Med.* **16**, 150.
Beckmann, N., Turkalj, I., Seelig, J., and Keller, U. (1991). *Biochemistry* **30**, 6362.
Beckmann, N., Brocard, J.-J., Keller, U., and Seelig, J. (1992). *Magn. Reson. Med.* **27**, 97; erratum: **29**, 581 (1993).
Beckmann, N., Fried, R., Turkalj, I., Seelig, J., Keller, U., and Stalder, G. (1993). *Magn. Reson. Med.* **29**, 583.
Belliveau, J. W., Kennedy, D. N., McKinstry, R. C., Buchbinder, B. R., Weisskoff, R. M., Cohen, M. S., Vevea, J. M., Brady, T. J., and Rosen, B. R. (1991). *Science* **254**, 716.
Bendall, M. R., and Gordon, R. E. (1983). *J. Magn. Reson.* **53**, 365.
Bendall, M. R., and Pegg, D. T. (1986). *J. Magn. Reson.* **67**, 376.
Bendall, M. R., Pegg, D. T., Doddrell, D. M., and Field, J. (1981). *J. Am. Chem. Soc.* **103**, 934.

Beynen, A. C., Hermus, R. J. J., and Hautvast, J. G. A. J. (1980). *Am. J. Clin. Nutr.* **33**, 81.
Bomsdorf, H., Helzel, T., Knuz, D., Roeschmann, P., Tschendel, O., and Wieland, J. (1988). *NMR Biomed.* **1**, 151.
Bottomley, P. A. (1984). U.S. Pat. 4,480,228.
Bottomley, P. A. (1989). *Radiology* **170**, 1.
Bottomley, P. A., Hart, H. R., Edelstein, W. A., Schenck, J. F., Smith, L. S., Leue, W. M., Mueller, O. M., and Redington, R. W. (1983). *Lancet 1*, 273.
Bottomley, P. A., Foster, T. H., and Darrow, R. D. (1984). *J. Magn. Reson.* **59**, 338.
Bottomley, P. A., Charles, H. C., Roemer, P. B., Flamig, D., Engeseth, H., Edelstein, W. A., and Mueller, O. M. (1988). *Magn. Reson. Med.* **7**, 319.
Bottomley, P. A., Hardy, C. J., Roemer, P. B., and Mueller, O. M. (1989). *Magn. Reson. Med.* **12**, 348.
Brainard, J. R., Downey, R. S., Bier, D. M., and London, R. E. (1989). *Anal. Biochem.* **176**, 307.
Brown, T. R., Kincaid, B. M., and Ugurbil, K. (1982). *Proc. Natl. Acad. Sci. U.S.A.* **79**, 3523.
Bruhn, H., Michaelis, T., Merboldt, K. D., Hänicke, W., Gyngell, M. L., and Frahm, J. (1991). *Lancet* **337**, 745.
Chadwick, D. J. and Whelan, J. (1991). *Ciba Found. Symp.* **163**, 1.
Chase, E. P., and Dupont, J. (1978). **2**, 236.
Chen, C.-N., Hoult, D. I., and Sank, V. J. (1983). *J. Magn. Reson.* **54**, 324.
Chen, W., and Ackerman, J. J. H. (1989). *J. Magn. Reson.* **82**, 655.
Chen, W., and Ackerman, J. J. H. (1990). *NMR Biomed.* **3**, 158.
Cherrington, A. D., and Vranic, M. (1986). In "Hormonal Control of Gluconeogenesis" (N. Kraus-Freidman, ed.), p. 204. CRC Press, Boca Raton, FL.
Clarke, D. D., Lajtha, A. L., and Maker, H. S. (1989). In "Basic Neurochemistry: Molecular, Cellular, and Medical Aspects" (J. Siegel, B. W. Agranoff, R. W. Albers, and P. B. Molinoff, eds.), 4th ed., p. 541. Raven Press, New York.
Cobelli, C., and Toffolo, G. (1984). *Math. Biosci.* **71**, 237.
Cohn, R. M., and Segal, S. (1973). *Metab., Clin. Exp.* **22**, 627.
Costello, L. C., Kittleton, G. K., and Franklin, R. B. (1978). In "Endocrine Control in Neoplasia" (R. K. Sharma and W. E. Criss, eds.), p. 303. Raven Press, New York.
Crowley, M. G., and Ackerman, J. J. H. (1985). *J. Magn. Reson.* **65**, 522.
Dahl, L. B., Dahl, I. M. S., Engström-Laurent, A., and Granath, K. (1985). *Ann. Rheum. Dis.* **44**, 817.
David, M., Petit, W. A., Laughlin, M. R., Shulman, R. G., King, J. E., and Barrett, E. J. (1990). *J. Clin. Invest.* **86**, 612.
DeFronzo, R. A., Ferrannini, R., Hendler, R. *et al.* (1983). *Diabetes* **32**, 35.
Dimand, R. J., Moonen, C. T. W., Chu, S. C., Bradbury, E. M., Kurland, G., and Cox, K. L. (1988). *Pediatr. Res.* **24**, 243.
Doddrell, D. M., Pegg, D. T., and Bendall, M. R. (1982). *J. Magn. Reson.* **48**, 323.
Doyle, D. D., Chalovich, J. M., and Bárány, M. (1981). *FEBS Lett.* **131**, 147.
Drewinko, B., Yang, L. Y., Leibovitz, A., Barlogie, B., Lutz, D., Jansson, B., Stragand, J. J., and Trujillo, J. M. (1984). *Cancer Res.* **44**, 4241.
Ende, G., and Bachert, P. (1993). *Magn. Reson. Med.* **30**, 415 (1993).
Farrell, P. M., Mischler, E. H., Engle, M. J., Brown, D. J., and Lou, S. M. (1985). *Pediatr. Res.* **19**, 104.
Ferrannini, E., Reichard, G. A., Bjorkman, O., Wahren, J., Pilo, A., Olsson, M., and DeFronzo, R. A. (1985). *Diabetes* **34**, 580.
Fox, P. T., Raichle, M. E., Mintun, M. A., and Dence, C. (1988). *Science* **241**, 462.
Frahm, J., Merboldt, K. D., Hänicke, W., and Haase, A. (1985). *J. Magn. Reson.* **64**, 81.
Franklin, R. B., and Costello, L. C. (1984). *J. Urol.* **132**, 1239.

Fried, R., Beckmann, N., Ninnis, R., Keller, U., Stalder, G., and Seelig, J. (1993). *Soc. Magn. Reson. Med., 12th Annu. Meet.*, New York, Book of Abstracts, p. 85.
Gardemann, A., Strulik, H., and Jungermann, K. (1986). *FEBS Lett.* **202**, 255.
Garwood, M., and Ke, Y. (1991). *J. Magn. Reson.* **94**, 511.
Garwood, M., and Merkle, H. (1991). *J. Magn. Reson.* **94**, 180.
Gitzelmann, R., Steinmann, B., and van den Berghe, G. (1985). In "The Metabolic Basis of Inherited Disease" (C. R. Scriver, A. L. Beaudet, W. L. Sly, and D. Valle, eds.), 5th ed., p. 118. McGraw-Hill, New York.
Gopher, A., Vaisman, N., Mandel, H., and Lapidot, A. (1990). *Proc. Natl. Acad. Sci. U.S.A.* **87**, 5449.
Gruetter, R., Novotny, E. J., Boulware, S. D., Rothman, D. L., Mason, G. F., Shulman, G. I., Shulman, R. G., and Tamborlane, W. V. (1992a). *Proc. Natl. Acad. Sci. U.S.A.* **89**, 1109; erratum: p. 12208.
Gruetter, R., Rothman, D. L., Novotny, E. J., and Shulman, R. G. (1992b). *Magn. Reson. Med.* **25**, 204.
Gruetter, R., Rothman, D. L., Novotny, E. J., Shulman, G. I., Prichard, J. W., and Shulman, R. G. (1992c). *Magn. Reson. Med.* **27**, 183.
Gruetter, R., Novotny, E. J., Boulware, S. D., Rothman, D. L., Tamborlane, W. V., and Shulman, R. G. (1992d). *Soc. Magn. Reson. Med., 11th Annu. Meet.*, Berlin, Book of Abstracts, p. 1921.
Halliday, K. R., Fenoglio-Preiser, C., and Sillerud, L. O. (1988). *Magn. Reson. Med.* **7**, 384.
Hartmann, S. R., and Hahn, E. L. (1962). *Phys. Rev.* **128**, 2042.
Heerschap, A., Luyten, P. R., van der Heyden, J. I., Oosterwaal, L. J. M. P., and den Hollander, J. A. (1989). *NMR Biomed.* **2**, 124.
Hers, H.-G., Hoof, F. V., and de Barsy, T. (1989). In "The Metabolic Basis of Inherited Disease" (C. R. Scriver, A. L. Beaudet, W. L. Sly, and D. Valle, eds.), 6th ed., Vol. I, p. 425. McGraw-Hill, New York.
Hore, P. J. (1983). *J. Magn. Reson.* **55**, 283.
Howell, R. R. (1972). In "The Metabolic Basis of Inherited Disease" (J. B. Standbury, J. B. Wyngaarden, and D. S. Fredrickson, eds.), 3rd ed., p. 149. McGraw-Hill, New York.
Hwang, J.-H., Rothman, D. L., Cline, G., Magnusson, I., Gerard, D., Petersen, K., and Shulman, G. I. (1993). *Soc. Magn. Reson. Med., 12th Annu. Meet.*, New York, Book of Abstracts, p. 86.
Jakobsen, H. J., Sørensen, O. W., and Bildsoe, H. (1983). *J. Magn. Reson.* **51**, 157.
Jehenson, P., and Bloch, G. (1991). *J. Magn. Reson.* **94**, 59.
Jue, T., Lohman, J. A. B., Ordidge, R. J., and Shulman, R. G. (1987). *Magn. Reson. Med.* **5**, 377.
Jue, T., Rothman, D. L., Tavitian, D. A., and Shulman, R. G. (1989a). *Proc. Natl. Acad. Sci. U.S.A.* **86**, 1439.
Jue, T., Rothman, D. L., Shulman, G. I., Tavitian, B. A., DeFronzo, R. A., and Shulman, R. G. (1989b). *Proc. Natl. Acad. Sci. U.S.A.* **86**, 4489.
Kalderon, B., Korman, S. H., Gutman, A., and Lapidot, A. (1989). *Proc. Natl. Acad. Sci. U.S.A.* **86**, 4690.
Karlsson, J., Diamant, B., and Saltin, B. (1971). *Scand. J. Clin. Lab. Invest.* **26**, 385.
Katan, M. B., van Staveren, W. A., Deurenberg, P., Bardendeuse-van Leeuwen, J., Germing-Nouwen, C., Soffers, A., Berkel, J., and Beynen, A. C. (1986). *Prog. Lipid Res.* **25**, 193.
Katz, J. (1985). *Am. J. Physiol.* **248**, R391.
Katz, L. D., Glickman, M. G., Rapoport, S., Ferrannini, E., and DeFronzo, R. A. (1983). *Diabetes* **32**, 675.
Knüttel, A., Kimmich, R., and Spohn, K.-H. (1990a). *J. Magn. Reson.* **86**, 526.
Knüttel, A., Spohn, K.-H., and Kimmich, R. (1990b). *J. Magn. Reson.* **86**, 542.

Knüttel, A., Kimmich, R., and Spohn, K.-H. (1991). *Magn. Reson. Med.* **17**, 470.
Köstler, H., and Kimmich, R. (1993a). *J. Magn. Reson. B* **102**, 177.
Köstler, H. and Kimmich, R. (1993b). *J. Magn. Reson. B* **102**, 285.
Krebs, H. A., Hems, R., Weidemann, M. J., and Speake, R. N. (1966). *Biochem. J.* **101**, 242.
Künnecke, B. and Seelig, J. (1991). *Biochim. Biophys. Acta* **1095**, 103.
Kunze, C., Kimmich, R., and Demco, D. E. (1993). *J. Magn. Reson.* **A 101**, 277.
Kwong, K. K., Belliveau, J. W., Chesler, D. A., Goldberg, I. E., Weisskoff, R. M., Poncelet, B. P., Kennedy, D. N., Hoppel, B. E., Cohen, M. S., Turner, R., Cheng, H.-M., Brady, T. J., and Rosen, B. R. (1992). *Proc. Natl. Acad. Sci. U.S.A.* **89**, 5675.
Landau, B. R., and Wahren, J. (1988). *FASEB J.* **2**, 2368.
Lundberg, P., Vogel, H. J., and Ruderus, H. (1986). *Meat Sci.* **18**, 133.
Luyten, P. R., and den Hollander, J. A. (1986). *Radiology* **161**, 795.
Luyten, P. R., Marien, A. J. H., Heindel, W., van Gerwen, P. H. J., Herholz, K., den Hollander, J. A., Friedmann, G., and Heiss, W.-D. (1990). *Radiology* **176**, 791.
Magnusson, I., Chandramouli, V., Schumann, W. C., Kumaran, K., Wahren, J., and Landau, B. R. (1989). *Metab. Clin. Exp.* **38**, 583.
Magnusson, I., Rothman, D. L., Katz, L. D., Shulman, R. G., and Shulman, G. I. (1992). *J. Clin. Invest.* **90**, 1323.
Malcolm, G. T., Bhattacharyya, A. K., Velez-Duran, M., Guzman, M. A., Oalmann, M. C., and Strong, J. P. (1989). *Am. J. Clin. Nutr.* **50**, 288.
McGarry, J. D., Kuwajima, M., Newgard, C. B., Foster, D. W., and Katz, J. (1987). *Annu. Rev. Nutr.* **7**, 51.
Michaelis, T., Merboldt, K. D., Hänicke, W., Gyngell, M. L., and Frahm, J. (1991). *NMR Biomed.* **5**, 90.
Mispelter, J., Tiffon, B., Quiniou, E., and Lhoste, J. M. (1989). *J. Magn. Reson.* **82**, 622.
Moonen, C. T. W., Dimand, R. J., and Cox, K. L. (1988). *Magn. Reson. Med.* **6**, 140.
Müller, S., and Beckmann, N. (1989). *Magn. Reson. Med.* **12**, 400.
Müller, S., Sauter, R., Weber, H., and Seelig, J. (1988). *J. Magn. Reson.* **76**, 155.
Müller, S., Hafner, H.-P., and Beckmann, N. (1989). *NMR Biomed.* **2**, 209.
Newgard, C. B., Moore, S. V., Foster, D. W., and McGarry, J. D. (1984). *J. Biol. Chem.* **259**, 6958.
Niewoehner, C. B., and Neil, B. (1992). *Am. J. Physiol.* **263** (Endocrinol. Metab. **26**), E42.
Niewoehner, C. B., Neil, B., and Martin, T. (1990). *Am. J. Physiol.* **259** (Endocrinol. Metab. **22**), E804.
Nilsson, L. H., and Hultman, E. (1973). *Scand. J. Clin. Lab. Invest.* **32**, 325.
Nilsson, L. H., and Hultman, E. (1974). *Scand. J. Clin. Lab. Invest.* **33**, 5.
Novotny, E. J., Ogino, T., Rothman, D. L., Petroff, O. A. C., Prichard, J. W., and Shulman, R. G. (1990). *Magn. Reson. Med.* **16**, 431.
Oberholzer, V. G., Levin, B., Burgess, E. A., and Young, W. F. (1967). *Arch. Dis. Child.* **42**, 492.
Ogawa, S., Tank, D. W., Menon, R., Ellermann, J. M., Kim, S.-G., Merkle, H., and Ugurbil, K. (1992). *Proc. Natl. Acad. Sci. U.S.A.* **89**, 5951.
Ordidge, R. J., Bendall, M. R., Gordon, R. E., and Connelly, A. (1985). In "Magnetic Resonance in Biology and Medicine" (Govil, Khetrapal, and Saran, eds.), p. 387. McGraw-Hill, New Delhi.
Ordidge, R. J., Connelly, A., and Lohman, J. A. B. (1986). *J. Magn. Reson.* **66**, 283.
Petersen, K., Rothman, D. L., Reuben, A., West, A. B., and Shulman, G. I. (1993). *Soc. Magn. Reson. Med., 12th Annu. Meet.*, New York, Book of Abstracts, p. 88.
Petroff, O. A. C., Spencer, D. D., Alger, J. R., and Prichard, J. W. (1989). *Neurology* **39**, 1197.
Pretlow, T. G., II, Harris, B. E., Bradley, E. L., Jr., Bueschen, A. J., Loyd, K. L., and Pretlow, T. P. (1985). *Cancer Res.* **45**, 442.
Reddy, R., Leigh, J. S., and Goelman, G. (1993). *J. Magn. Reson. B* **101**, 139.

Riemersma, R. A., Wood, D. A., Buttler, S., Elton, R. A., Oliver, M., Salo, M., Nikkari, T., Vartiainen, E., Puska, P., Gey, F., Rubba, P., Mancini, M., and Fidanza, F. (1986). *Br. Med. J.* **292**, 1423.
Roch-Norland, A. E., Bergström, J., and Hultman, E. (1972). *Scand. J. Clin. Invest.* **30**, 77.
Rogers, V., Dab, I., Michotte, Y., Vercruysse, A., Crokaert, R., and Vis, H. (1984). *Pediatr. Res.* **18**, 704.
Rosenberg, L. E., and Mahoney, M. J. (1973). In "Inborn Errors of Metabolism" (F. A. Hommes and C. J. van den Berg, eds.), p. 303. Academic Press, London.
Rothman, D. L., Behar, K. L., Hetherington, H. P., and Shulman, R. G. (1985). *Proc. Natl. Acad. Sci. U.S.A.* **82**, 1633.
Rothman, D. L., Magnusson, I., Jucker, B., Shulman, R. G., and Shulman, G. I. (1991a). *Soc. Magn. Reson. Med., 10th Annu. Meet.*, San Francisco, Book of Abstracts, p. 656.
Rothman, D. L., Magnusson, I., Katz, L. D., Shulman, R. G., and Shulman, G. I. (1991b). *Science* **254**, 573.
Rothman, D. L., Novotny, E. J., Shulman, G. I., Howseman, A. M., Petroff, O. A. C., Mason, G. F., Nixon, T., Hanstock, C. C., Prichard, J. W., and Shulman, R. G. (1992). *Proc. Natl. Acad. Sci. U.S.A.* **89**, 9603.
Ruiz-Cabello, J., Vuister, G. W., Moonen, C. T. W., van Gelderen, P., Cohen, J. S., and van Zijl, P. C. M. (1992). *J. Magn. Reson.* **100**, 282.
Saner, M., McKinnon, G., and Boesiger, P. (1989). *Soc. Magn. Reson. Med., 8th Annu. Meet.*, Amsterdam, Book of Abstracts, p. 602.
Saner, M., McKinnon, G., and Boesiger, P. (1992). *Magn. Reson. Med.* **28**, 65.
Shaka, A. J., Keeler, J., Frenkiel, T., and Freeman, R. (1983). *J. Magn. Reson.* **52**, 335.
Shulman, G. I., Cline, G., Schumann, W. C., Chandramouli, V., Kumaran, K., and Landau, B. R. (1990a). *Am. J. Physiol.* **259**, E335.
Shulman, G. I., Rothman, D. L., Jue, T., Stein, P., DeFronzo, R. A., and Shulman, R. G. (1990b). *N. Engl. J. Med.* **322**, 223.
Silver, M. S., Joseph, R. I., Chen, C. N., Sarky, V. J., and Hoult, D. I. (1984). *Nature (London)* **310**, 681.
Stokes, C. E., Gillon, K. R. W., and Hawthorne, J. N. (1989). *Biochim. Biophys. Acta* **753**, 136.
Tanaka, K., Armitage, I. M., Ramsdell, H. S., Hsia, Y. E., Lipsky, S. R., and Rosenberg, L. E. (1975). *Proc. Natl. Acad. Sci. U.S.A.* **72**, 3692.
Taylor, R., Price, T. B., Rothman, D. L., Shulman, R. G., and Shulman, G. I. (1992). *Magn. Reson. Med.* **27**, 13.
Tygstrup, N. (1966). *Scand. J. Lab. Clin. Invest.* **18**, Suppl. 92, 118.
Ugurbil, K., Garwood, M., and Rath, A. R. (1988). *J. Magn. Reson.* **80**, 448.
van Hoof, F., and Hers, H. G. (1967). *Eur. J. Biochem.* **2**, 265.
van Zijl, P. C. M., Chesnick, A. S., DesPres, D., Moonen, C. T. W., Ruiz-Cabello, J., and van Gelderen, P. (1993a). *Magn. Reson. Med.* **30**, 544.
van Zijl, P. C. M., Barker, P. B., Soher, B. J., Gillen, J., Bottomley, P. A., Duyn, J., Moonen, C. T. W., and Weiss, R. G. (1993b). *Soc. Magn. Reson. Med., 12th Annu. Meet.*, New York, Book of Abstracts, p. 373.
Werner, B., Boesch, C., Gruetter, R., and Martin, E. (1990). *Soc. Magn. Reson. Med., 9th Annu. Meet.*, New York, Book of Abstracts, p. 868.
Wolfe, R. R. (1992). In "Radioactive and Stable Isotope Tracers in Biomedicine." Wiley-Liss, New York.
Wood, D. A., Butler, S., Riemersma, S. R. A., Thomson, M., Oliver, M. F., Fulton, M., Birthwhistle, A., and Elton, R. (1984). *Lancet* **2**, 117.
Young, F. E. (1988). *Fed. Regist.* **53**, 7575.

Index

Acetaldehyde, 207
Acetate, 171–172, 189, 208–210, 215–219, 221–222, 224–225, 227–228, 231–233, 234, 236–237, 240, 252, 254–257
Acetoacetate, 171, 210–213, 248
Acetoacetyl-CoA, 213
N-acetyl-aspartate, see NAA
Acetylcarnitine, 216, 219
Acetylcholine, 240
Acetyl-CoA, 170–171, 173–174, 187, 189, 207–208, 211–213, 215–218, 225, 228, 233, 234, 237, 248, 257
Acid
  adipic, 213–214
  amino, 119, 123–124, 138, 212, 218, 250, 258, 290, 296, 303, 310, 315
  γ-amynobutyric, see GABA
  bile, 119
  decanoic, 118, 130, 132–134, 149, 154
  dicarboxylic, 213–214
  dodecanoic (lauric), 130–132, 149, 171, 213–214
  fatty, see Fatty acid
  formic, 186
  glutamic, 121
  hexacosanoic, 118, 143, 149–150
  hexanoic, 219
  hyaluronic, 316
  linoleic, 118, 305
  malonic, 101
  myristic (tetradecanoic), 124–125, 128–129, 140
  nucleic, see also DNA, 41
  octadecanoic, 121
  octanoic, 118, 130–134, 142–143, 148–149, 154–155
  oleic, 118, 120–124, 126–128, 130, 132, 135–142, 144–149, 151, 153
  palmitic, 118, 122, 135–136, 138–139, 146, 149, 151
  phosphatidic, 140
  retinoic, 109
  sebacic, 213
  sialic, 249
  stearic, 118, 122, 130, 140, 149, 151
  suberic, 213
Acidosis, 245, 247–248, 260
cis-aconitate, 247
Acyl
  chain, 143, 153
  CoA, 139, 155
Adenocarcinoma
  colloid, 314
  colonic, 314
  prostatic, 315
Adenosine-5'-triphosphate, see ATP
Adipocyte, 189
Adipose tissue, see Tissue
Adrenoleukodystrophy, 118
Adriamycin, 225, 226
Alanine, 75–76, 79, 100–101, 108, 170–171, 173, 196–198, 202–210, 213, 215, 221, 225, 234, 238, 245–247, 249, 252, 254, 257–259

Alanine aminotransferase, 204
Albumin, see also BSA, HSA, 117–121, 123–126, 128, 130–132, 134, 138–139, 143–144, 146, 148–155
Alcohol dehydrogenase, 207
Aldolase, 256, 311–312
Aldose reductase, 241, 243, 249
Amylo-1,6-glucosidase, 192–193, 309
γ-Aminobutyrate, 291–292
Ammonia, 202, 240, 248–249
ε-Ammonium group, 124
Ammonium chloride, 210
Amyloid protein, 69, 85
Anesthesia, see also Halothane, 177, 201
Angiostrongylus cantonensis, 257
Anaplerosis, 217–218, 220–221, 238
Anaplerotic pathway, see Pathway
Anoxia, 222
Arginine, 124–125, 138, 310
Arthritis, 316
Ascardiasis, 251
Asparagine, 91–92
Aspartate, 120, 170–171, 187, 202–203, 206, 209–210, 215, 219, 222, 234, 245–248
Astrocyte, 227–228, 231, 240
ATP, 203, 221–222, 224, 226, 255
ATP-citrate lyase, 315
axon, 227

B_0-field gradient, see Gradient
B_1-gradient pulse, see Pulse
Bacteriorhodopsin, 69, 108–112
Betaine, 249
Bicarbonate, 211–212, 231, 256
Biceps femoris, 251
Bilayer
 membrane, see Membrane bilayer
 phospholipid, see Phospholipid bilayer
Binding
 affinity, 117
 site, 117, 120–121, 124–126, 130, 132–134, 148–151, 153–154
Binomial excitation, 13
BIRD
 filter, see Filter
 pulse, see Pulse
BISEP, see pulse
Blindness, river, 257

Blood–brain barrier, 226
Body fluid, see Fluid
Bovine serum albumin, see BSA
Broadband Dipolar Recoupling, 99
Brugia pahangi, 256–257
BSA, 119–133, 144–151, 153
Butyrate, 171, 210–213, 224

Cadmium, 249
Caffeine, 250
Calorimetry, differential scanning, 123
Canavan's disease, 240
N-Carbamoyl aspartate, 202, 209–210
Carbonate, 210–211
Carcinoma, 259, 314, 315
Cardiomyopathy, 225
Cardioprotectant, 224
Cardiotoxin, 224
Carnitine, 216, 303, 310
Carnitine acyl transferase system, 210
Carnosine, 161, 303, 310
Cataract, 241
Cataractogenesis, 241
CBCA(CO)NH, 48
Cell suspension, 173
Cestode, 251–252, 254–256
Channel
 Ca^{2+}, 118
 ion, 139
 K^+, 118
Chelating agent, 178
Chemical Shift Anisotropy, see CSA
Chemical Shift Imaging, see CSI
Chloride, 313
Cholesterol, 134–135, 137, 140–141, 152
Cholesteryl ester, 134
Choline, 185, 188, 240–241, 250, 252, 280, 310
Chromatography
 gas, see GC/MS
 ion exchange, 299
Chylomicron, 134–135, 151
Cirrhosis, 315
Citrate, see also Isocitrate, 152, 161, 170–171, 207, 216–217, 231, 245–249, 257, 315
Citrate
 lyase, 245
 synthase, 207, 247

CO_2, 171, 254, 256, 303
Coherence
  double antiphase, 49
  double quantum, 285
  heteronuclear multiquantum, see also HMQC, 18, 33, 284
  quadruple-quantum, 40
  three-spin (^{1}H, ^{15}N, ^{13}C$^\alpha$), 49–50
  zero quantum, 285
Coherence
  order, 32, 34–35, 285
  selection
    by phase cycling, 31
    by B_0-gradients, 31–36, 283–286
  transfer, heteronuclear, 9, 13
Coil
  birdcage, 286
  body, 271
  butterfly, 271
  figure-eight-shaped surface, 271
  head, 271
  surface, 179–182, 185, 187, 270–275, 277–283, 286, 288, 292–293, 295, 299, 303, 310
Collagen, 316
COLOC, 9, 19, 27, 30, 58
Colon, 259, 314
Compartmentation, metabolic, 169, 226–228, 231, 248
Complex, supramolecular, 42
Connectivity, ^{13}C-^{13}C, 41
Cortex, visual, 283
Cory cycle, see cycle
COSY, 10, 36, 38, 47
  ^{13}C-, 10
  H,C-, 11, 19
  E.Cosy, 56, 57
Coupling
  dipolar, 67, 168
  long-range, 9, 30, 53–59, 61, 166
  passive scalar, 15, 57
  $^1J_{N,C\alpha}$, $^2J_{N,C\alpha}$, 49
  $^2J_{H\alpha,C}$, $^3J_{H\beta,C}$, $^3J_{HN,C}$, $^3J_{HN,C\beta}$, 57
CP, 3, 13, 91, 94, 100, 104, 286
CPMAS, 110
Creatine, see also Phospho-creatine, 234, 238, 250, 258, 291–292, 295, 303, 310
Cross-polarization
  adiabatic J, 4, 287
  cyclic J, 287

Crystallography, see X-ray
CSI, 270, 273, 278–281, 291–292, 303
Cycle
  Cory, 196
  futile, 203, 210, 223
  glyoxylate, 257
  tricarboxilic acid, see TCA
Cycloserine, 222
Cyclpot, 275, 281–282, 284
Cyclpot-Vosing, 283–286
Cyclrop-Losy, 286, 287
Cysteine, 9, 171, 244
Cystic fibrosis, 313
Cytosol, 138, 226

Database, 8
DANTE pulse, see Pulse
Decoupling
  bilevel, 183
  selective, 39
2-Deoxyglucose, 230–231, 259
2-Deoxyglucose-6-phosphate, 230–231, 259
DEPT, 4, 7, 11, 13, 17–20, 52–53, 180–181, 275, 282
DEPTH, see Pulse
Diabetes mellitus, see also Glycogen, Glycosuria, 153, 176, 202–203, 210, 213, 231, 241–243, 245, 290, 303, 307–308
Diabetes mellitus
  insulin dependent, 307
  non-insulin dependent, 307–308
Diacylglycerol, 155
2,3-dibromopropyl phosphate, 249
DICSY, 99
Differential equation, for modelling the TCA cycle activity, see Equation
Differential scanning calorimetry, see Calorimetry
Dihedral angle (ϕ, ψ), 9, 53–54, 84
Dihydroxyacetone-3-phosphate, 196, 201, 255, 311
2,4-dihydroxybutyrate, 204
Dipetalonema viteae, 256
Dipolar
  coupling, carbon-nitrogen, 70
  coupling, carbon-phosphorus, 70
  Hamiltonian, 96, 99
  tensor, 103

Dipolar Recovery at the Magic Angle,
    see DRAMA
DIPSI, 46, 47
DNA, 12, 27
Dopamine receptor, 79
DRAMA, 95–98
DRAMA, 2D-, 97
DRESS, 288
Duct, papillary, 249–250

Escherichia coli, 47, 138
Eddy current, 31
Editing
    heteronuclear proton, 281, 287, 317
    line, 283
    spectral, 4, 275
Electron Spin Resonance, see ESR
Emerimicin, 69
Encephalopathy, 290
5-Enolpyruvylshikimate-3-phosphate, 87–88
5-Enolpyruvylshikimate-3-phosphate synthase, 87, 89
Enrichment
    fractional, 163, 169, 183–184, 226, 232, 300–301
    positional, 163, 169, 183–185, 197, 215, 233, 244, 247
EPSPS/S3P/glyphosate complex, 63, 87–89
Equation
    differential, 169, 171–172, 215
    Henderson-Hasselbach, 136, 139
    input-output, 169, 173, 204, 233
    Steele's, 301
Erythrocyte, 296
ESR, 119, 130, 143
Ethanol, 171, 189–190, 204–205, 207–210, 257, 276
Ethanolamine, 292
Euglycemia, 294
Exchange
    magnetization, see Magnetization
    spin, see Spin
Extract
    chloroform/methanol, 178
    perchloric acid, 172–173, 178, 208, 215–217, 231–234, 244, 246, 248–249, 252, 258–260

Fasciola hepatica, 255
Fatty acid
    dicarboxylic, 213–214
    essential, 313
    long-chain, 118, 122, 124, 126, 130, 132, 134, 140–143, 146, 149–151, 154, 155, 210
    medium-chain, 118, 125, 132–134, 148, 155, 190, 210, 212
    monocarboxylic, 213
    monomeric, 124, 142
    monounsaturated, 189, 252, 270, 304, 313
    polyunsaturated, 189, 250, 270, 304–305, 313
    short-chain, 210, 212, 216
    unesterified, 117, 118
    unsaturated, 189, 252, 304
    very-long chain, 118, 143, 149–150
Fatty acid binding protein, 117–118, 138–139, 144, 151, 154
Fatty acyl chain, 188–189
FDA, 271–273
Fibrosarcoma, 258
Filter
    analog frequency, 39
    BIRD, see also Pulse, 24–25, 30, 37, 40, 52
    double, 45
    isotope, 42, 44–45
    J-, 29–30
    low-pass, 272
    multiquantum, 284
    X-half, 43–44, 57
    z, 35
Flow, coronary, 220–221, 225
Fluid
    body, 178, 261, 305–307, 318
    cerebro-spinal, 178
    synovial, 316
Fluorescence, 132, 140–141, 146, 151
Fluorodeoxyglucose, see also PET, 317
Flux
    anaplerotic, 218, 220, 224
    gluconeogenic, 245
    glycolytic, 225, 255–256
    metabolic, 163, 214, 221
    TCA, 238
Folding, 12, 39, 40
Fractional enrichment, see Enrichment

Freeman-Hill approach, 52
Fructose, 171, 241–243, 301, 311, 313, 317
    diphosphatase, 255
    intolerance, 311–313, 318
Fructose-1,6-diphosphate, 171, 256, 313
Fructose-1-phosphate, 311–312
Fructose-1-phosphate aldolase, 311, 313
Fructose-6-phosphate, 171, 256
Fumarase, 206, 254
Fumarate, 170–171, 196, 202, 204, 206–207, 222, 245, 254–255
Functional imaging, see Imagining
Futile cycle, see Cycle

GABA, 171, 187, 227–228, 231–234, 236–238, 248, 257
Galactokinase, 302
Galactose, see also UDP-galactose, UDP-glucose, 270, 296, 300–302
Galactose-1-phosphate, 302
Galactose elimination capacity, 301
Gas chromatography-mass spectrometry, 184, 257, 301, 305–306, 309
Gaussian cascade, G3, G4, see Pulse
Ge-HMQC, see also Vosing, 31–36, 270, 275, 281, 283, 286, 295, 316–317
Gland
    eccrine sweat, 313
    exocrine, 313
Glass, inorganic, 65
Glial cel, 227–228, 231–234, 237–238
Glioma, 259
Glucagon, 176, 192, 194–195, 297
Gluconeogenesis
    in the animal liver, 187, 196–204, 206, 212
    in the animal kidney, 244–245, 248–250
    in tumor cells, 259
    in the human liver, 296, 299–300, 303, 306–309, 311, 316
Glucopyranose residue, chain, 191
Glucose
    concentration gradient, portal-arterial, 299
    metabolism
        in the animal liver, 190–210
        in the animal heart, 214–226
        in the animal brain, 227–238
        in the animal kidney, 244–249

    in the human brain, 291–295
    in the human liver, 296–300
    in the human muscle, 303–304
    output, 300–302
    paradox, 195–196, 198
    production, synthesis
        in the animal liver, 199, 201–202
        in the human liver, 299–300, 306, 308, 311, 316
    transport, 290–291
    turnover, 300, 306
    uptake, 230, 242, 297, 317
Glucose-1-phosphate, 171, 195, 302, 306
Glucose-6-phosphate, 171, 195–196, 205, 230, 302, 306
Glucose-6-phosphatase, 202, 309
Glycosyl unit, 197, 222
Glucuronide, 307
Glutamate, 120–121, 125, 170–171, 173–174, 187, 198, 203–204, 206–211, 215–216, 218–234, 236–238, 240, 246–249, 256–259, 283, 286, 291, 293–295
Glutamate
    decarboxylase, 231
    dehydrogenase, 249
    pool, 227–228, 232–233, 238
    turnover, 283
Glutamate, myocardial, 173
Glutaminase, 210
Glutamine, 171, 187, 203, 207–211, 226–234, 236–238, 246, 248–249, 256–257, 259, 286, 291, 293–295, 312
Glutamine
    pool, 228
    synthase, 209–210, 227–228, 231–232, 240
Glutaminolysis, 256
Glutarate, 247
Glutathione, 171, 208–209, 243–244, 246–247
Glutathione reductase, 244
Glyceraldehyde-3-phosphate, 171, 196, 201, 205, 247, 255, 311
Glycerol, 171, 187–188, 201–202, 204–205, 244–246, 250, 252, 274, 280, 292, 296, 310–312
Glycerolester, 190
Glycerol-3-phosphate, 161, 171, 201–202

sn-Glycerol-3-phosphorylcholine, 292, 311
Glycerophosphocholine, 234, 238, 249
Glycine, 75–76, 82, 171, 234, 247, 258, 292
N-(phosphonomethyl)glycine, see Glyphosate
Glycogen
  glial cells
    in gliomas, 259
    in glycogenosis patients, 309
  hepatic
    in animals, 190, 196–201, 203, 223
    in humans, 277–278, 286, 289, 296, 297, 299–302, 306, 308–309, 316
  muscle
    in humans, 281, 303–304, 308, 316
  myocardial
    in animals, 192, 223
    in humans, 288
Glycogen
  concentration, 289, 296–297, 299, 301–303, 316
  depletion, 303
  glycogenosis, 308–311, 318
  hydrolysis, 192–193, 197
  phosphorylase, 195, 223–224
  repletion, 198–199, 297, 306, 317
  repletion pathway, see Pathway
  synthase, 195, 222, 300
  synthesis
    in the animal liver, 192, 195–199, 201, 203
    in the animal heart, 222–225
    in the human liver, 297, 300, 302
    in the human muscle, 308
  turnover
    animal heart, 222
    human liver, 300
  visibility (by ^{13}C MRS), 191, 192, 223, 251, 296
Glycogenosis, see Glycogen, storage disease
Glycogenolysis
  in animals, 195, 223–224
  in humans, 296, 299–300, 302, 308
Glycolysis, 171, 187, 196, 201, 214, 217, 222–223, 230, 240, 243, 254, 258–260, 294
Glycoprotein, 314
Glycoside, 9
Glycosuria, 307
Glycosyl residue, 190–192, 195, 200, 223
Glyoxal shunt, 254

Glyoxylate, 257
Glyoxylate cycle, see Cycle
Glyphosate, 69, 87–89
Gradient
  actively shielded, 1, 16, 31, 286
  B_0-field, 13, 276–279, 281, 284–285, 287–288
  dephasing, 283
  electrically generated surface, 181, 277–278
  glucose concentration, see Glucose concentration
  homospoil, 35, 286–287
  phase encoding, 278–280, 292
Gradient-enhanced heteronuclear multiquantum coherence, see ge-HMQC
Group
  chromophoric, 119
  fluorescent, 119
  nitroxide, 119
  reporter, 119
Guanosine-5′-triphosphate, 222

Halothane, see also Anesthesia, 196
Hartmann-Hahn, heteronuclear, 7, 13, 19
Hartmann-Hahn condition, 18, 286–287
3D HCCH-COSY, 60
HCCH-TOCSY, 48
3D HCCH-TOCSY, 48, 60
4D HCCH-TOCSY, 48
HDL, (see also Lipoprotein) 134–136, 138, 151
Heart, perfused, 172–173, 214–216, 218, 221, 223
Helminth, 251–252, 257
Henderson-Hasselbach equation, see Equation
Hepatic glycogen production, see Glycogen
Hepatocyte, periportal, perivenous, 201–204, 210, 213
Herzfeld-Berger analysis, 67, 109, 112
Heteronuclear Multi-Bond Correlation, see HMBC
Heteronuclear multiquantum coherence, see Coherence, HMQC
HETLOC, 57–60
Hexanoate, 216
Hexokinase, lenticular, 241
Hexose monophosphate shunt, 243–244

Histidine, 9, 250, 310
HMBC, inverse, 30
HMBC, 9, 19, 27–31, 55–56, 59–60
HMBCS, 28–29
HMQC
  BIRD-filtered, 22
  sensitivity-enhanced, 25
HMQC, 11–12, 18–23, 25–28, 30, 36, 54, 59–60
HMQC-COSY, 37–38, 60
HMQC-NOESY, 25
HMQC-TOCSY, 25, 36–38, 40, 60
HNCA, 48–51, 60
HN(CA)CO, 48
HNCACO, 48
HN(CA)H, 10
Homeostasis, 243
Homospoil pulse, see Pulse
HPLC, 184
HQQC, 39
HQQC-TOCSY, ^{13}C-, 40
HSA, 119, 124, 130, 132–134, 151–152, 154
HSQC, 11–12, 19–23, 25–27, 30, 33, 35–36, 38, 52, 54, 59–60
HSQC
  BIRD-filtered, 22
  gradient-selected, 33, 35
  sensitivity-enhanced, 25–26
Human serum albumin, see HSA
3-Hydroxybutyrate, 171, 209–213, 227–228, 231, 234, 236, 237, 240, 247–248
Hydroxymethylglutaryl-CoA, 213
Hydroxymethylglutaryl pathway, see Pathway
Hymenolepis
  diminuta, 251–252, 254
  microstoma, 251
  nana, 251
Hyperammonia, 238
Hyperglycemia, 294, 297, 307
Hyperinsulinemia, 297
Hyperthermia, 260
Hyperthyroidism, 203
Hypoxia, 214, 220, 222–223, 225, 229–230

Imaging
  chemical shift, see CSI
  functional, 290

INADEQUATE, 19
INADEQUATE, ^{13}C, ^{13}C-, 41
INEPT, 4, 7, 13–20, 25, 35, 46–47, 49–53, 91, 284
INEPT
  refocused, 14–16, 18
  reverse-, 15, 20, 35
Infarction, myocardial, 224–225
Infinite convergent series, see Series
Inosine, 225
Inositol, 171, 234, 249, 291
Input-output equation, see Equation
Insulin, 176, 194–195, 202, 207, 209, 216, 222, 224–225, 297, 299, 307
Inversion-recovery, 52
Ischemia, 214, 217, 220, 222–225
ISIS, 276–277, 282–283, 288, 294, 315
Isocitrate, 257
Isotope
  effect, 167–168
  filter, see Filter
Isotopic steady-state, see Steady-state
Isotopomer, 166–170, 172–175, 183, 185, 201, 215, 222, 232–233, 246, 254–255, 312
Isotopomer analysis, 166–167, 169–175, 185, 214–215, 217–218, 225, 237–238, 255–256, 317

J-filter, low pass, see Filter

K$^+$ channel, see Channel
Keeler/Neuhaus technique, 55–56
Ketogenesis, 171, 187, 210–212
α-Ketoglutarate, 170–171, 206–208, 218, 220, 222, 232, 234, 238, 248–249, 257
Ketone body, 231, 248
Kinetics, zero-order, 243

Label, selective, 160, 162
Labeling pattern, 168, 172
Laboratory animals
  baboon, 197
  cat, 286
  dog, 198
  frog, 250
  guinea pig, 221–223, 225, 232

Laboratory animals, (continued)
  mouse, 187, 203–204, 207
  rat, 189–190, 193–194, 196–200, 203–204, 207, 209, 211–213, 218, 222
  rabbit, 218, 226, 228–229, 237, 240, 243, 245, 247–249, 251, 276
Lactate, 170–171, 197–198, 203–204, 216–217, 220–221, 223, 225–226, 229–230, 234, 242–243, 245–247, 249–250, 252, 254–260, 286, 293–294, 296, 303, 306, 311–312, 314
Langendorff method, 214
LDL, see also Lipoprotein, 134–136, 138, 151
Lens
  crystalline, 241
  ocular, 187, 241, 243
  translucent, 241
Leucyl, 82–83
Leukemia, 208, 259
Line shape analysis, 55
Lipase, 190
Lipid, 117, 120, 135, 139–142, 146, 155, 230, 250–252, 258, 313–315
Lipid-membrane interaction, 117
Lipid-protein interaction, 117
Lipid
  metabolism, see Metabolism
  transport, 117
Lipogenesis, 171, 187, 189–190, 296
Lipolysis, 153
Lipoprotein, see also HDL, LDL, VLDL, 117–118, 134–135, 151–154
Lipoprotein lipase, 134
Liver glycogen, see also Steatosis, Glycogen
ϵ-Lysine, 121, 124–125

Macrofilariae, 256
Magic Angle Spinning, see also CP, MAS
Magnetic vector potential method, 272
Magnetization exchange, 105
Malaria, 251
Malate, 170–171, 196, 202, 204, 206–207, 222, 234, 246–248, 252, 254–255, 257
Malate dehydrogenase, 206, 254
Malic enzyme, 204, 218, 234, 237, 254
Malonic acid, see Acid
Malonyl-CoA, 189

MAS, see also CP, 3, 66, 72–73, 92–93, 98, 101–102, 112–113, 117, 141–143
Matrix
  mitochondrial, 210
  phospholipid, 141
Melanostatin, 69, 79–84, 87
Membrane
  bilayer, 135, 139
  lipid, 139
  mitochondrial, 210
  model, 117, 139–141, 143, 149, 151
Mesocestoides corti, 252
Metabolism
  aerobic, 251, 255–256
  anaerobic, 222, 251, 255
  anaplerotic, 207
  lipid, 117–118
  oxidative, 207, 224, 226
Methionine, 240–241
Methionine sulfoximine, 232, 238
3-O-Methyl-glucose, 260
Methylmalonic acid semialdehyde, 313
Methylmalonic acidemia, 313
Methylmalonyl CoA, 313
Methylmalonyl CoA
  mutase, 313
  racemase, 313
MEXICO, 45
Micelle, 118, 122
Michaelis-Menten model, 230, 294
Mitochondria, 206, 210, 214, 222, 248, 254
MLEV, see also Decoupling, Spin-lock, 18, 182
Modeling, metabolic, 168–170
Monte-Carlo simulation, 175
Mucin, 314–315
Multifrequency excitation, 282
Multiple Quantum Filtering, see Filter
Multiplet pattern, 55–56, 167
Multiplicity selection, 39
Muscle
  gastrocnemius, 250, 303
  heart, 214, 222
  pectoralis, 250
  skeletal, 187, 214, 250, 303
Muscle glycogen, see Glycogen
Myelin, 240
Myelogenesis, 240
Myocardium, 214, 225, 288
Myocyte, 214

Index / 331

NAA (N-acetyl-aspartate), 234, 240, 291–292, 295
NADH, 212, 248
NADPH, 205, 241, 243–244
Nematode, 256
Network, J-coupling, 46
Neuron, 227–228, 231–234, 237–238
Nitrogen metabolism, 165, 187
NOE, 3, 11, 31, 43–44, 53, 120, 129–130, 161, 168, 182–183, 185, 191, 223, 246, 276, 288
NOE
  transient, 276
  truncated driven, 276
NOESY, 31, 36, 38, 43–44, 55
Nucleic acid see also DNA, 41
Nulling, surgace, 179, 273, 277

Octanoate, 171, 190, 212
Olygodendrocyte, 227
Onchocerca volvulus, 257
Operator, product, see Product operator
Oxaloacetate, 170–171, 196, 198–199, 202–203, 206–208, 210, 217–218, 222, 227, 234, 245, 248–249, 254–255, 306
β-Oxidation, of fatty acids, 149, 171, 187, 210, 212–214, 217–218, 224
ω-Oxidation, of fatty acids, 213
Oxoacid, 222
Oxygen, 258
Oxygen consumption, 220, 294

Pake powder pattern, 77–78
Pake-spun pattern, 77–78
Palmitate, 152, 224
Parasite, neurotropic, 257
Parasite, 251–252, 255
α-Particle, β-particle, of glycogen, 223
Pasteur effect, 256
Pathway
  anaerobic, 222, 257
  anaplerotic, see also Anaplerosis, 173, 217–218
  coherence, 9, 20
  gluconeogenic, 201–202, 207, 245–246
  hydroxymethylglutaryl, 212
  pentose phosphate, 171, 187, 190, 196, 199, 204–205, 243, 245, 250

polyol, 187, 241, 243, 244
propionate, 218
sorbitol, 241, 243
Pathways, for glycogen repletion
  direct, 196, 198–199, 306–307
  indirect, 196–200, 306–307
  recycling at the triose-phosphate level, 198–199
Perchloric acid extract, see Extract
Peroxide, 243
PET, 290, 294, 317
Phase cycling, xy-4, xy-8, 74, 78, 100
Phenazine methosulfate, 205
Phenylalanine, 9, 42, 250, 310
Phosphatidylcholine, 135, 139–141, 145, 148–149, 154, 240, 274
Phosphatidylethanolamine, 274, 310
Phosphatidylserine, 137, 140
Phosphocreatine, 221, 250, 258, 291, 303, 310
Phospho-*enol*-pyruvate, 87, 171, 202–204, 246, 249, 254–257, 306
Phospho-*enol*-pyruvate carboxykinase, 202–203, 234, 245–246, 248–249, 254
Phosphoethanolamine, 291, 303
Phosphofructokinase, 217, 255
Phospholipase A$_2$, 139
Phospholipid, 134–135, 138–141, 143, 151, 153–155, 188, 240–241, 252, 313–314
Phospholipid
  bilayer, see also Model membrane, 117–118, 139–144, 146, 148
  matrix, 141
  multilayer, 142
Phosphorylcholine, 258, 291
Phosphorylethanolamine, 258
Plasmodium berghei, 203
Polarization transfer, see
  Cyclpot
  Cyclrop
  DEPT
  INEPT
  SINEPT
Polarization transfer, 180–181, 270, 275, 281–282, 303
Polarization transfer
  cyclic, see Cyclpot
  rotating frame, see Cyclrop-Losy
Polyene, 189
Polymer, glassy, 65, 68

Polyol, 241
Polyol
  dehydrogenase, 241
  pathway, see Pathway
Positional enrichment, see Enrichment
Positron emission tomography, see PET
Potassium, 225
Potassium oleate, 123
PRESS, 276
Proline, 37
Product operator, 14, 42
Propionate, 171, 187, 215–218, 255, 313
Propionate pathway, see Pathway
Proton-observe carbon-edited spectroscopy, 270, 275, 283, 291, 294
Protozoa, 251
Proximate convoluted tubule, 244–249
Pulse, radiofrequency
  adiabatic, 180, 275
  adiabatic rapid half passage, see also Pulse, sin/cos, 275
  B_1-gradient, 13
  BISEP, 275
  BIR-4, 275
  BIRD, see also Filter, 23–24
  BURP, 39
  contact, 286–287
  DANTE, 95, 104
  DEPTH, 180, 277
  Gaussian, 39, 274
  270° Gaussian, 28–29
  Gaussian cascade G3, G4, 39
  homospoil, see also Gradient, Homospoil, 16–17, 31, 33, 287
  hyperbolic secant, inversion, 277, 282, 288
  semiselective, 22, 283
  sin/cos, numerically optimized, 275, 288
  sinc, 283
  spin-lock, see Spin-lock pulse
Pyrimidine, 210
Pyruvate, 170–172, 189–190, 202–204, 206–208, 210, 215–218, 221–222, 224–226, 230, 243, 245–249, 254–255, 257, 259, 294
Pyruvate
  carboxylase, 173–174, 202–203, 207–208, 217, 227–228, 238, 246–247
  dehydrogenase, 173–174, 207–208, 230, 237–238, 246–248, 259, 294
  kinase, 202–204, 234, 245, 255

recycling, 234
recycling system, in neurons, 227–228, 237
Pyruvate dehydrogenase complex, 217, 225

Quadruple-quantum coherence, see Coherence

Radioactivity, 163
Radiochemistry, 159
Radiolabeling, 227
Radiotracer, 160, 163, 168, 170, 227, 230, 233
REDOR, 68–69, 72, 74–93, 95–97, 100, 111, 113
REDOR
  2D-, 69, 77–78
  ^{13}C-^{15}N, 75, 79, 84–85
  ^{13}C-observe, 78–79, 81
  ^{15}N-observe, 84
  double-difference, 81–82, 87
  ^{31}P-^{13}C, 87–88
Redox state, 212
Relaxation time
  T_1, see also Freeman-Hill approach, 9–11, 24, 47, 52–53, 120, 130, 162, 168, 181, 183, 185, 191, 223, 274
  T_2, 10–11, 53, 162, 181, 191, 274–276, 279, 286, 288
  T_2^*, 105, 288
  zero-quantum (T_2^{ZQ}), 105, 108, 110–112
Reperfusion, 224–226
Resolution enhancement, 192–193
Resonator, radiofrequency, 182, 286
Retinal, 108–112
Retinoid binding protein, see also Fatty acid binding protein, 69, 154
RETRO, 86
Retrovirus, 259
Reverse-INEPT, see INEPT
RF Driven Dipolar Recovery, 99
Ribose, 171
Ribulose, 205
ROESY, 36, 38, 55
Rotational-Echo, Double-Resonance NMR, see REDOR
Rotational-Echo, Triple-Resonance NMR, see RETRO
Rotational Resonance NMR, 68–69, 95, 101–104, 108–113

Rotational Resonance, Spin-Exchange Experiment, 103–105, 110–112
Rotational Resonance Recoupling, 70, 93–94

Saccharide, 315
Salt bridge, 124
Satellite technique, 184
Saturation-recovery, 52
Scatchard binding data, 123
Schiff base linkage, 108, 111
Schistosoma japonicum, 254–255
Schistosomiasis, 251
SEDRA, 95–96, 99–101
Seizure, 290
Selective decoupling, *see* Decoupling
Series, infinite convergent, 169, 172–173, 245–247
Serine, 247
Shift reagent, 177
Shikimate-3-phosphate, 87–89
Signal-to-artifact ratio, 35
Signal-to-noise ratio, 2, 10–11, 26, 35, 41, 52, 92, 167, 172, 178, 180–182, 185, 274–276, 281, 304
SINEPT, 275
Sodium
  acide, 259
  chloride, 249
  propionate, 92
Sorbinil, 249
Sorbitol, 171, 187, 241–243, 249–250
Sorbitol pathway, *see* Pathway
Specific Absorption Rate, 272–273
Spectrophotometry, 159
Spectroscopic image, *see* CSI
Spheroid, 178
Sphingomyelin, 137, 140, 240
Spin
  coupling pattern, homonuclear and heteronuclear, 165–167, 173–174, 198, 201, 208
  exchange, *see* Rotational resonance
Spin-lock pulse
  for suppressing signals from protons not bound to ^{13}C nuclei in inverse experiments, 16–19, 37, 46
  in cross-polarization experiments, 286–287
Spinning sideband, *see also* Herzfeld-Berger analysis, 67

Steady-state
  isotopic, 173, 198, 215, 218, 247
  metabolic, 173, 198, 215
STEAM, 276
Stearyl amine, 140
Steatosis, 315
Steele's equations, *see* Equation
Streptozotocin, 245
Succinate, 170–171, 173, 206, 216, 221–222, 245–249, 252, 254–257
Succinyl-CoA, 218, 222
Succinyl-CoA synthase, 222
Sugar cataract, 241, 243
Supramolecular complex, *see* Complex
Surface
  coil, *see* Coil
  gradient, *see* Gradient
Suspension, cell, 178
Synaptosome, 231
Synovial fluid, *see* Fluid

t_1 noise, 24
Taurine, 234, 291
TCA cycle
  isotopomer analysis and metabolic modeling, 169–174
  in the animal liver, 187, 198–203, 206–208, 210–212
  in the animal heart, 215–221, 223–226
  in the animal brain, 227, 228, 230–234, 237, 240, 317
  in the animal kidney, 244–249
  in parasites, 254, 256
  in tumors, 258, 259
  in the human brain, 294
Techniques for spatial localizaton of NMR signals, *see*
  CSI
  Cyclpot-Vosing
  Cyclrop-Losy
  DRESS
  gradient, electrically generated surface
  ISIS
  nulling, surface
  polarization transfer
  PRESS
  STEAM
  Vosing
TEDOR, 70, 86, 90–93, 95

TEDOR
  2D-, 91
  ^{13}C-detected, 90
TEDOR/REDOR four-channel experiment, 86, 93
Tensor, CSA, 94, 103, 109
Thiocyclosporin A, 40
Tissue
  adipose, 187, 189, 270–271, 279, 304–305, 313
  calf, 276
  connective, 258
  myocardial, 218, 223
  papillary, 249
  prostate, 315
  smooth muscular, 250
TOCSY, see also HCCH, 10, 18, 36–38, 40, 46–49, 55, 57–58
TOCSY, ^{13}C-, 10, 36, 41, 46–48
TOCSY-HMQC, 9, 19
Transaldolase, 205
Transaminase inhibitor, 222
Transketolase, 205
Transplantation, 203
α,α-Trehalose, 256
Trematode, 254–256
Triacylglycerol, 134–135, 149, 153–155, 188, 190, 314–315
Tricarboxylic acid cycle, see TCA
Triglyceride, 152, 187, 189, 208, 215, 247, 313, 315–316
Triolein, 153–154
Triose phosphate, 198, 255–256
Triose phosphate
  isomerase, 256
  recycling, 199
Trypanosomia, 251
Tryptophan, 9, 250, 310
Tumor, 203, 257–260, 278, 314–315
Tyrosine, 124–125, 250, 310
Trypanosoma brucei rhodensiense, 203

UDP-galactose, 302
UDP-gal-4-epimerase, 302

UDP-glucose, 171, 195, 302, 306
Uridine, 210
Uridinephosphate, 171
Uridyltransferase, 302
Urine, see also Body fluid, 178, 244, 248–249, 305, 307, 313
Urea, 249

Valine, 313
Vesicle
  bilayer, 139
  phosphatidylcholine, 136–137, 139–140, 142–148
  phospholipid, 137, 144, 146, 150–151
VLDL, see also Lipoprotein, 134–136
Von Gierke's disease, 309
Vosing, see also ge-HMQC, 271, 283–286

WALTZ, see also Decoupling, Spin-lock, 18, 182, 273–274, 280, 292
Water suppression with
  B_0-field gradient, 13, 47, 283
  B_1-field gradient, 13
  binomial excitation, 13
  jump-return, 13
  solvent presaturation, 13
Worm
  filarial, 256
  tape, 251

X-ray, 77, 83–84, 87–88, 110
  crystallography, 65, 124, 134
  scattering, 123

z filter, see Filter
Zellweger syndrome, 118
zeolite, 68
zinc acetate, 103–105